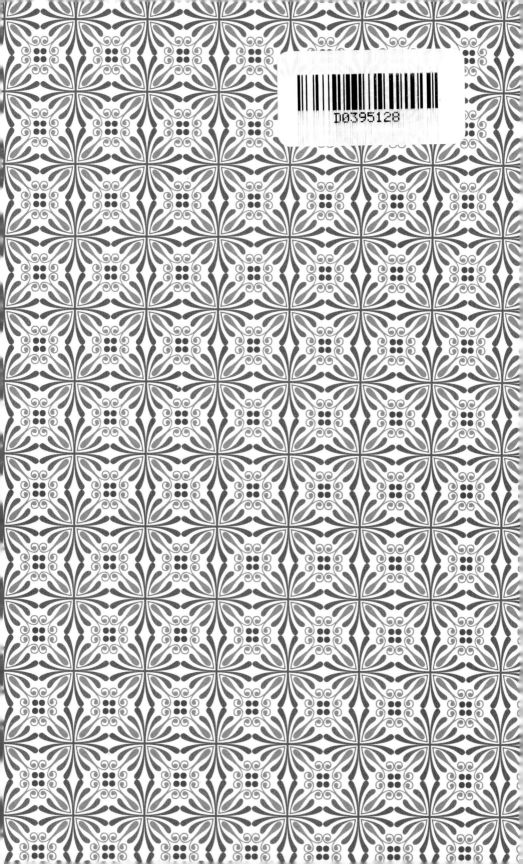

The Book of

This Day
in History

Publications International, Ltd.

Written by Jim Daley

Images from ClipArt, ThinkStock, Shutterstock.com and Wikimedia Commons

ISBN: 978-1-64030-192-4

Manufactured in U.S.A.

8 7 6 5 4 3 2 1

Contents

✳ ✳ ✳ ✳

Chapter 1

January

January 1

1772: The first traveler's checks go on sale in London. The London Credit Exchange Company became the first business to offer traveler's checks, which were secured against loss or theft. The checks could be used in 90 European cities.

1801: Giuseppe Piazzi discovers Ceres. It is the largest known object in the Asteroid belt. Piazzi initially mistook Ceres for a new star, but when he observed it moving through the night sky, Piazzi realized it was something else. He announced that it was a comet, although he privately suspected it "might be something better." He named it Ceres after the Roman goddess of grain. In 1817 Piazzi, who compiled a catalogue of over 7,500 stars, was made the director of the Naples and Sicily observatories by King Ferdinand.

1863: The Emancipation Proclamation becomes law in Confederate-held states. The law freed more than 3 million slaves in the rebelling Southern states, but excluded nearly a million slaves held in states loyal to the Union. About 50,000 slaves in areas where the Union Army was active were freed immediately.

1892: Ellis Island opens in Upper New York Bay. Three ships landed on its first day of operation and delivered seven hundred immigrants. During its first year, nearly half a million more immigrants were processed at Ellis Island, with officials reviewing up to 5,000 per day during peak times. More than 12 million people passed through Ellis Island's inspection station over the next sixty years.

1898: Greater New York City is formed. New York City consolidated the boroughs of Manhattan, Brooklyn, Queens, and the Bronx to form the Big Apple. Staten Island would join on January 25.

1908: Revelers watch the ball drop at midnight in New York City's Times Square for the first time. The ball was dropped above the roof of the New York Times building at 1 Times Square. It had one hundred incandescent light bulbs, weighed seven hundred pounds, and was five feet in diameter.

1959: Fidel Castro's forces overthrow the government of Cuban dictator Fulgencio Batista. Castro established a Communist government on the island nation and was an adversary to the United States' ambitions for regional control until his death in 2014. He was succeeded in office by his brother Raul.

1983: The Internet is created. The Advanced Research Projects Agency Network, ARPANET, officially changed to using the Internet Protocol (IP) suite to transmit information, making ARPANET an early component of the new Internet.

1993: Czechoslovakia is broken up into the Czech Republic and Slovakia. The breakup followed the peaceful Velvet Revolution of 1989 and is sometimes referred to as the Velvet Divorce.

2002: The Euro is officially adopted as the legal currency in eleven of the European Union's member states. Although the United Kingdom is a notable holdout, the Euro remains a stable currency, weathering the Eurozone crisis of 2008 and the Brexit withdrawal of 2016.

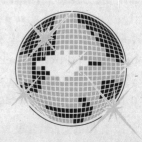

January 2

1860: The planet Vulcan is proposed to explain the peculiar orbit of Mercury. Closest to the sun, Mercury was completing its orbits faster than Newton's laws of motion predicted it should. At a meeting of the French Academy of Sciences, the mathematician Urbain Le Verrier suggested that an undiscovered planet that he named "Vulcan" was the reason why. Le Verrier, who had already discovered Neptune by studying the motion of Uranus, thought that gravitational effects of another planet must be affecting Mercury's orbit. Although no one could find the supposed Vulcan, Le Verrier died in 1877 convinced he had discovered it. It wasn't until 1915 that Einstein's theory of relativity solved the problem by modifying the predicted orbits of all the planets, including Mercury.

1942: Japanese forces capture Manila, the capital of the Philippines. The Imperial Army occupied it for the next three years until a combined assault by American and Filipino forces cleared the last Japanese troops from the city on March 3, 1945.

1958: Opera star Maria Callas walks out of a performance in Rome attended by the Italian president. Callas was forced to stop the concert because she was suffering from bronchitis and lost her voice. But because of her reputation as a temperamental diva, the press criticized her harshly, resulting in a scandal.

1963: The Viet Cong wins its first major victory against United States military forces in the Battle of Ấp Bắc. Repeated attacks by a combined force of more than 1,500 American and South Vietnamese soldiers supported by armored personnel carriers and air cover were held off by 350 Viet Cong for more than twelve hours. The Viet Cong inflicted numerous casualties before they withdrew.

1974: Richard Nixon signs the National Maximum Speed Law. The law set the U.S. speed limit at 55 miles per hour. The law was the result of gasoline shortages due to an embargo by the member states of the Organization of Arab Petroleum Exporting Countries. In 1995 Congress repealed the law, returning the authority to set the speed limit to state governments.

1980: In response to the Soviet invasion of Afghanistan in December 1979, Jimmy Carter recalls the U.S. ambassador to Moscow and asks Congress to delay passing the SALT II nuclear weapons treaty, which was ultimately abandoned. This signaled the end of détente and led to the United States boycotting the 1980 Summer Olympics held in Moscow.

2004: *Stardust*, a space probe launched by NASA in 1999, encounters the comet Wild 2. In doing so it successfully completed the first sample return mission of its kind. *Stardust* flew within 167 miles of the comet, collecting samples from its dust trail and photographing its icy core.

January 3

1521: Pope Leo X excommunicates Martin Luther from the Roman Catholic Church. Luther had famously nailed his 95 *Theses* to the cathedral door of Wittenberg five years earlier and sent copies around Germany. In them he challenged the authority of the Church to sell indulgences for sins. While he only intended to spark an academic debate on the subject, the Roman Curia and Pope Leo perceived the *Theses* to be a larger threat to their power and ordered Luther to be tried for heresy. Upon being found guilty, he burned the papal bull ordering him to recant the 95 *Theses*. Luther's excommunication and the ensuing controversy set off a religious revolution in the Catholic Church, now known as the Reformation, that gave rise to Protestantism, led to the Thirty Years' War, and had a profound impact on Christianity and European history.

1777: The Continental Army under the command of George Washington defeats the British garrison at Princeton. The defeat prompted British General Cornwallis to retreat from New Jersey. The American populace saw the battle of Princeton and the previous week's victory at Trenton as evidence the war was winnable, and the resulting morale boost led to the Continental Army recruiting many volunteers the following spring.

1952: The iconic crime drama *Dragnet* premieres on NBC. The show, written by and starring Jack Webb as the matter-of-fact Sergeant Friday, was based on a radio program of the same name and aired for eight seasons.

1959: Alaska becomes the 49th state in the Union. It was inducted nearly ninety years after the United States purchased it from Russia for 7.2 million dollars. Edward Bartlett, Alaska's territorial delegate, successfully led the push for statehood after two decades of effort in Congress. Bartlett would go on to be elected as the state's first Senator, serving until his death in 1968.

1987: Aretha Franklin becomes the first woman inducted into The Rock and Roll Hall of Fame. Franklin's career started when she was signed by Columbia Records in 1961, and skyrocketed in the 1960s with the hit songs "Respect" and "A Natural Woman." Her album *Aretha* went gold in 1986.

1990: Manuel Noriega surrenders to United States forces after a week-long siege at the Vatican's embassy in Panama City. Noriega, the military dictator of Panama with significant ties to U.S. intelligence agencies, had been indicted on charges of drug trafficking in a federal court in Miami in 1988. During the siege, the U.S. engaged in psychological warfare to convince Noriega to come out, and blared rock music at earsplitting volume for hours on end. Following his capture, Noriega was tried for drug trafficking and money laundering, and sentenced to 40 years in prison.

1993: The Buffalo Bills pull off the greatest comeback in NFL history. The Bills recovered from a 32-point deficit to beat the Houston Oilers 41–38 in overtime and advance in the playoffs. The Bills won the next two playoff games, but were defeated by the Dallas Cowboys in Super Bowl XXVIII.

2000: The final *Peanuts* comic strip is published in newspapers worldwide. Following a fifty-year syndicated run, the last strip featured its iconic character Snoopy sitting at a typewriter atop his doghouse and a thank-you note to editors and fans penned by its creator, Charles M. Schultz.

January 4

1847: Samuel Colt makes his first sale of revolver pistols. The buyer was Captain Samuel Walker of the Texas Rangers. The revolver, named the Colt Walker, helped establish Colt as the leading firearms inventor in the United States. Samuel Walker was killed later that year in action during the Mexican-American War.

1863: James Plimpton patents the first four-wheeled roller skates. Plimpton, a Massachusetts inventor, was the first to design the skates with independent axles that allowed the user to turn by pressing on one side of the skate, a huge improvement on the existing design. Plimpton's skates led to a burst of popularity in roller skating that lasted for more than fifty years. His design is still in use today.

1903: Topsy the elephant is electrocuted. The Edison film company films the electrocution of Topsy in Luna Park, Coney Island. Topsy was considered a "bad" elephant because she had killed a drunken circus spectator who had burnt her trunk with a lit cigar a few years earlier. She was strangled, poisoned, and finally electrocuted via copper bracelets attached to her ankles. The 74-second film was shown on Edison coin-operated kinetoscopes around the United States.

1954: Elvis Presley records a 10-minute demo at Sun Records in Nashville, Tennessee. The record included the songs "It Wouldn't Be the Same Without You" and "I'll Never Stand in Your Way."

1958: Sir Edmund Hillary and his party reach the South Pole. Hillary's group was the first to reach the South Pole on an overland expedition since Roald Amundsen had reached it in 1911, and were also the first explorers to use motor vehicles to do so.

1974: Citing executive privilege, President Richard Nixon refuses to hand over materials subpoenaed by the Senate Watergate Committee. Nixon would hold out on releasing the transcripts of Oval Office recordings for four months. When they were eventually made public, the growing scandal would culminate in his resignation later that year.

1998: A massive ice storm hits the northeastern United States and Canada. The storm damaged millions of trees, made roads impassable, and destroyed electrical infrastructure thorough the region, resulting in substantial power outages. Millions of residents were without power for weeks, and 35 people died.

2004: The NASA Mars Rover *Spirit* successfully lands at 04:35 UTC (Coordinated Universal Time). *Spirit* traveled 4.8 miles (7.73 kilometers)across Mars, collecting information about its planetary features and geology, and making observations of the solar system from its vantage point on the Red Planet. It became caught in soft soil during its fifth year. *Spirit* continued to perform experiments from its stationary location until March 22, 2010, when it sent its last communication back to Earth.

January 5

1781: Benedict Arnold leads British naval forces in an attack on Richmond, Virginia, during the American Revolutionary War. Arnold offered to leave the city unharmed if Thomas Jefferson would agree to surrender its tobacco stores and military arms, but Jefferson refused. In response, Arnold ordered the city to be burned. The capture and subsequent burning of Richmond was one of Arnold's most notorious actions. His actions angered

George Washington so much that Washington put an enormous bounty on Arnold's head and ordered his generals to summarily execute him if he was captured.

1875: President Ulysses S. Grant sends a company of Federal troops to Vicksburg, Mississippi. The troops were there to quell violence by anti-Reconstruction white mobs that had killed approximately 300 black citizens. Armed whites in the city had prevented black citizens from voting the year before and forced the elected black sheriff to flee the city.

1895: Alfred Dreyfus is publicly stripped of his rank and sentenced to life imprisonment. He was a French artillery captain who was convicted of treason in an incident that would come to be known as the Dreyfus affair. Dreyfus was convicted for allegedly sharing military secrets with the German Embassy in Paris and spent the next five years in prison. It soon became evident that Dreyfus had been targeted because of anti-Semitism and was not guilty. He was eventually pardoned and fully reinstated in the French Army, rising to the rank of lieutenant colonel.

1930: Bonnie Parker meets Clyde Barrow for the first time. Meeting at a mutual friend's house in West Dallas, the two fell in love at first sight. They went on to engage in a four-year crime spree, robbing banks, gas stations, and five-and-dime stores through the American Midwest. The pair and their gang killed nine police officers in getaway shootouts and several civilians during robberies. Their exploits grabbed headlines nationwide, and they were initially popular among the deprivation of the Great Depression. Their increasingly cold-blooded murders eventually turned public opinion against them and sparked a large manhunt that ended in a hail of gunfire on a rural roadside in Louisiana.

1933: Construction begins on the Golden Gate Bridge in San Francisco. The one-mile-wide bridge holds 1.2 million steel rivets and 80,000 miles of steel wire wrapped into cables that are three feet wide. It cost $35 million to build, and took over four years to complete, surviving an earthquake in the process. Eleven workers died while building it. Nineteen more who fell from the Bridge during its construction, but were saved by safety nets, founded the "Halfway to Hell" club.

1968: The Prague Spring begins. It was a short-lived period of democratic reform in Czechoslovakia during the country's domination by the Soviet Union. Reforms included granting more freedom of the press, making travel by Czech citizens easier, and decentralizing the economy. Later that year the Soviets and allies in the Warsaw Pact invaded Czechoslovakia and crushed the reform movement.

1970: *All My Children* debuts on ABC. The soap opera, created by Agnes Nixon and originally starring Susan Lucci as Erica Kane, aired for the next 41 years.

January 6

1066: Harold Godwinson is crowned King of England. Harold was the last Anglo-Saxon king of England, and was killed at the Battle of Hastings on October 14th of the same year, during the Norman Conquest. He is depicted in the Bayeux Tapestry riding a golden horse and carrying a falcon.

1832: William Lloyd Garrison founds the New England Anti-Slavery Society in Boston. It was the first abolitionist society in the United States that called for immediate and total emancipation. It would become the American Anti-Slavery Society the following year, and include Susan B. Anthony, Elizabeth Cady Stanton, Frederick Douglass and Wendell Phillips among its members. The Society published a weekly newspaper, the *National Anti-Slavery Standard*.

1893: The Great Northern Railway connects Seattle, Washington, to Saint Paul, Minnesota, and by extension the East Coast. The Railway's route was the northernmost transcontinental railroad in the United States.

1898: Samuel Lake sends the first telephone message from a submerged submarine. Lake then sailed the submarine, called the *Argonaut*, from Norfolk, Virginia, to Sandy Hook, New Jersey, a journey of more than 1,000 miles, to demonstrate its viability to the U.S. Navy.

1907: Dr. Maria Montessori opens her first school. The school, named the *Casa dei Bambini* (Children's House), was in Rome, Italy. Her experimental method of education, based on creating a nurturing environment that supports children's natural tendency toward curiosity, was radically different than most educational systems in practice at the time. In 1912 her methods spread to the United States and eventually spread to hundreds of schools across the country and around the world.

1941: President Franklin D. Roosevelt delivers his "Four Freedoms" speech. The State of the Union address outlined four basic freedoms that he argued should be enjoyed by people "everywhere in the world." They were: freedom of speech, freedom of worship, freedom from want, and freedom from fear. Roosevelt argued that the United States should help its allies in their war against Nazi Germany, breaking with the long-held tradition of nonintervention by the United States. Eleven months later, the country declared war on the Empire of Japan and its allies, formally entering World War II.

1978: The United States Post Office issues its first copyrighted stamp. The stamp commemorated the 100th anniversary of the birth of the American poet and writer Carl Sandburg. It featured Sandburg's signature and a drawing of the poet in profile, and originally cost thirteen cents.

1987: Astronomers see the birth of a galaxy for the first time. The astronomers observed a galaxy named 3C326.1, located 12 billion light years from Earth, "turning on" in a brilliant cloud of light that was a hundred times brighter than the average star.

January 7

1610: Galileo discovers the moons of Jupiter. Galileo had made improvements to his telescope the month before, increasing its magnification 20 times and allowing him to see heavenly bodies more clearly than before. He initially believed the moons to be fixed stars near Jupiter. As he continued to observe them over the next several nights, Galileo realized they were orbiting Jupiter and must be moons. The discovery challenged the existing Ptolemaic worldview, which held that the Earth was at the center of the universe and all other heavenly bodies orbited it.

1714: Henry Mill patents the first typewriter. Mill, an English inventor and waterworks engineer, submitted two patents in his life, one of them being a "Machine for Impressing or Transcribing Letters." The typewriter would not actually be built until the nineteenth century.

1782: The first American commercial bank opens. The Bank of North America, chartered by the Congress of the Confederation the previous May, opened in Philadelphia. When it sold shares to the public that year, it became the nation's first initial public offering. The shares were among the first traded on the New York Stock Exchange.

1887: Thomas Stevens finishes circumnavigating the globe by bicycle. Stevens had begun his trip three years earlier in San Francisco, and rode through Europe, Turkey, Iran, India, China, and Japan. He rode a high-wheel bicycle (also called a "penny-farthing") more than 13,500 miles around the globe.

1927: The Harlem Globetrotters play their first game in Hinckley, Illinois. The team's exhibition games featured comedy routines and amazing displays of athleticism. Former Globetrotters, including Wilt Chamberlain and Nat Clifton, would go on to play in the NBA. The team did not actually play a game in Harlem until 1968.

1929: "Buck Rogers in the 25th Century A.D." debuts as the first science fiction comic strip in a newspaper. The strip's titular hero would go on to feature in radio plays, television programs, and feature length movies for more than 60 years. Buck Rogers helped found the space opera genre and influenced later science fiction works including Isaac Asimov's *Foundation* series, the *Ender's Game* novels, and the *Star Wars* franchise.

1942: The siege of Bataan begins. Facing an attack by 75,000 Japanese troops, General Douglas MacArthur withdrew all American and Filipino forces in Luzon, the Philippines, to the Bataan Peninsula. Once there, the allies held out against constant artillery barrages and desperate hand-to-hand combat for nearly four months. Upon surrendering, they would be forced to endure the Bataan Death March, during which thousands would die.

1999: The Senate begins the impeachment trial of President Bill Clinton. Chief Justice of the U.S. Supreme Court William Rehnquist presided.

2015: Al-Qaeda gunmen attack the offices of *Charlie Hebdo*. They wounded four people and killed twelve. The attackers, who were later killed by police, claimed they were motivated by the satirical newspaper's depictions of Muhammad. The newspaper continued its normal weekly publication. Nearly eight million copies of the issue of *Hebdo* the week following the attack were printed, compared to its normal print run of 60,000.

January 8

1790: George Washington delivers the first State of the Union address. The President delivered it in New York City before a joint session of Congress. In the rather short speech, he called for the creation of a standing army, the development of agriculture and commerce, and the encouragement of science and literature by the government.

1867: African American men are granted the right to vote in Washington, D.C. Congress overrode President Johnson's veto of a bill that extended voting rights to all men in the District of Columbia age 21 or over. Welfare recipients, convicted felons, and men who voluntarily aided the Confederacy during the Civil War were excluded. Three years later the 15th Amendment to the Constitution gave the vote to all men in the country.

1901: The first National Bowling Championship is held in Chicago, Illinois. The American Bowling Congress held the event, which included team and individual competitions. Frank Brill won the individual bowling championship with a score of 648.

1918: Woodrow Wilson announces his Fourteen Points plan. This was a declaration of principles that would be the basis for negotiating peace following World War I. The Fourteen Points included the freedom of naval traffic, adjustment of colonial claims, arms reduction, territorial claims, and the establishment of the League of Nations.

1959: Charles de Gaulle is inaugurated president of France. He would remain president for the next ten years. De Gaulle had led the Free French Forces in resisting the German occupation during World War II and served as Minister of Defense and Prime Minister in 1958. While president, de Gaulle held the country together during the political turmoil of the Algerian War for Independence. He was a major figure of the Cold War, opposing U.S. control of NATO.

1968: *The Undersea World of Jacques Cousteau* premieres on ABC. In the first episode, Cousteau sails his ship the Calypso to the Red Sea and Indian Ocean to study the behavior of sharks. The series, which aired for ten years, made Jacques Cousteau a household name and brought the world of marine biology to living rooms across the United States.

1978: Harvey Milk is the first openly gay person elected to public office. Milk had run unsuccessfully three times before being elected to the San Francisco Board of Supervisors. While in office, he was instrumental in passing a city ordinance that expanded gay rights. Milk was assassinated in November of 1978 by former City Supervisor Dan White. Tens of thousands of people attended a candlelight vigil for Milk, with performances by Joan Baez and the San Francisco Gay Men's Chorus. Milk was posthumously awarded the Presidential Medal of Freedom by President Barack Obama in 2009.

January 9

1768: The first modern circus opens in London. Philip Astley, considered the father of the modern circus, pioneered the idea of performing equestrian stunts in a circus ring, rather than in a straight line. The circus ring had the dual advantage of allowing the audience to keep the rider in sight, and allowing the rider to take advantage of centrifugal force to maintain their balance while performing tricks on horseback. Astley also was the first to have acrobats, jugglers, and clowns entertain the crowd during breaks between trick riding acts. His circus was so successful that he toured Ireland and France, and was mentioned in a number of famous works of fiction, including Jane Austen's novel *Emma* and Charles Dickens' *Sketches by Boz*.

1793: The first hot-air balloon flight in the United States. Jean Pierre Blanchard took off in Philadelphia and landed across the Delaware River in Deptford, New Jersey. The flight was witnessed by President George Washington, as well as John Adams, Thomas Jefferson, and James Madison.

1839: The daguerreotype is introduced. Its creator, Louis-Jacques-Mande Daguerre, used silver halide plates that were treated with chemicals to make them sensitive to light and then exposing them with a camera.

1908: Muir Woods National Monument is established. Muir Woods is an old-growth redwood forest on the Pacific coast in the San Francisco Bay Area. President Theodore Roosevelt created the National Monument and named it after the naturalist John Muir. Muir's tireless advocacy of wilderness preservation was instrumental in the creation of the National Park system. He wrote several books about the beauty of the Sierra Mountains, and had convinced Roosevelt of the necessity of preserving Yosemite during a trip the pair took there in 1903.

1945: United States forces invade Luzon, the largest island in the Philippines. More than 70 U.S. warships supported a force of 280,000 American and Filipino soldiers in the invasion. Fifty thousand Allied soldiers and over 200,000 Japanese soldiers would be killed during the battles that raged over the island during the coming months. By March of 1945, the Allies were in control of all the strategic and economically vital areas of the island. The capture of Luzon was a major victory in the counteroffensive against the Empire of Japan and eventual victory in World War II.

1996: Chechen separatists take an entire town hostage. During the First Chechen War for Independence, guerillas launched coordinated attacks on an airfield and hospital in Dagestan, Russia. When Russian special forces counterattacked, the rebels retreated to a town called Kizlyar and took the town's 3,400 inhabitants hostage. Most of the hostages were released the next day.

2007: The iPhone is unveiled. Apple CEO Steve Jobs announced the iPhone to the Macworld Convention in San Francisco, prompting an enthusiastic reaction from customers. The iPhone reinvented the idea of the smartphone. Its use of a touchscreen, combination of a cellular phone, iPod, and wireless computer that could connect to the internet was revolutionary.

January 10

1776: Thomas Paine publishes *Common Sense*. The pamphlet called for independence for the Thirteen Colonies from Great Britain. The incendiary pamphlet immediately became a phenomenon and was widely distributed and read aloud at meeting halls and taverns. Paine linked the cause of independence to the emerging American political identity, arguing that rule by the British monarchy was tyrannical and that the idea of a King was obsolete. He went on to assert that the American colonies should draft a Continental Charter and defend their independence with arms. *Common Sense* was well-received by reviewers and hugely impacted public opinion, swaying it in favor of independence.

1863: The first subway opens in London. The Metropolitan Railway took passengers in gas-lit wooden carriages pulled by steam-powered locomotives underground from the city's financial district to the suburbs in Middlesex. The Railway's main terminal at King's Cross is still a Central London railway stop today.

1870: Standard Oil is incorporated. John D. Rockefeller established Standard Oil in Ohio, making it the largest oil refinery of its day. Standard Oil dominated its market by pioneering the concept of the corporate trust, controlling all aspects of oil refining and distribution. It operated until 1911, when the Supreme Court ruled that it violated laws against monopolies.

1927: The film *Metropolis* premiers in Berlin. The silent expressionist drama directed by Fritz Lang is widely considered a groundbreaking work of science-fiction in film. It was also one of the first feature-length movies in the genre, though it was criticized for its two hour run-time. *Metropolis* pioneered a number of special effects, including the use of miniature sets and mirrors to present actors interacting with the sets. Panned by critics and booed by some audiences, it would later be considered one of the greatest films of all time.

1964: *Introducing the Beatles* is released in the United States. The LP included the hit singles "Please Please Me," "Twist and Shout," "Do You Want to Know a Secret," and "Love Me Do." It reached Number 2 on the U.S. *Billboard* Top LP list, where it stayed for nine weeks with the Number 1 LP spot occupied by *Meet the Beatles!* The album eventually went platinum, selling over a million copies.

1967: The first popularly elected African American Senator is sworn in. Edward William Brooke III represented Massachusetts in the Senate until 1979. He helped write the Civil Rights Act of 1968 and was the first Senator to call for Nixon's resignation during the Watergate Scandal. He lost his seat to Paul Tsongas.

1999: *The Sopranos* debuts on HBO. Starring James Gandolfini as New Jersey mob boss Tony Soprano and Lorraine Bracco as his psychiatrist Jennifer Melfi, it ran for six seasons. It won 21 Primetime Emmy Awards, five Golden Globes, and two Peabody Awards.

January 11

1838: Samuel Morse publicly demonstrates his electric telegraph for the first time. With his partner Alfred Vail, Morse demonstrated the electric telegraph at the Speedwell Ironworks in Morristown, New Jersey. The early telegraph's range was only two miles, and the first transmission was witnessed by a crowd of local civilians. The first message publicly transmitted said "A patient waiter is no loser."

1878: Home delivery of milk in glass bottles begins. The Alexander C Campbell milk company pioneered the home delivery of milk in Brooklyn, New York. With the invention of the automobile, they were able to extend their business throughout the City and onto Long Island. With the invention of pasteurization and widespread use of refrigeration, fresh milk delivery eventually became obsolete.

1897: Mattie Hughes Cannon becomes the first woman state senator in the United States. The previous year, when Utah was granted statehood, she led the movement to put women's suffrage in the state constitution. In the election, Cannon defeated her own husband to become state senator. While in office she wrote a number of health and safety laws and helped found the Utah State Board of Health, which she served on after leaving the legislature.

1902: The first issue of *Popular Mechanics* is published. It was the brainchild of Henry Windsor, who published a number of other trade magazines before hitting upon *Popular Mechanics*. The magazine has published news of developments in science and technology over more than a century and achieving a circulation of more than one million readers.

1922: Insulin is used to treat diabetes for the first time. A fourteen-year-old diabetic named Leonard Thompson who was dying in Toronto was injected with insulin by the biochemist James Collip. While the first dose caused a severe allergic reaction, the second dose resulted in elimination of his symptoms. Insulin would soon be widely used to treat diabetic patients.

1927: The Academy of Motion Picture Arts and Sciences is created. Louis Mayer, head of MGM studios, announces the formation of the Academy at a banquet in Los Angeles, California. The organization was founded to improve the public image of the motion picture industry and negotiate labor disputes. Membership was limited to actors, directors, producers, technicians, and writers. An awards presentation was not part of the original concept of the Academy.

1964: The Surgeon General reports on the negative effects of smoking. The landmark report issued by the Surgeon General, which detailed the increased mortality of smokers and linked smoking to lung cancer, bronchitis, emphysema, and heart disease, had a major impact on public perception. Millions of smokers quit in the years following the report's publication.

January 12

1866: The Royal Aeronautical Society is founded. The Society, which still exists today, is dedicated to the study and promotion of the science and industry of aerospace engineering. It publishes several scientific journals, including *The Aeronautical Journal*, and has branches around the world. It awards the Notable Gold Medal for achievement in Aeronautics: its first ever was awarded to the Orville and Wilbur Wright in 1909.

1921: Kennesaw Mountain Landis becomes the first commissioner of baseball. He was given absolute authority to oversee the American and National Leagues. This was an attempt to reform baseball following a series of scandals that undermined the game's integrity, culminating in the Chicago White Sox throwing the 1919 World Series for gamblers. Landis insisted on a lifetime contract. Baseball owners, still reeling from the aftermath of the scandal, agreed. His first act was to give a lifetime ban to the eight White Sox players who had fixed the Series. Landis went on to establish an independent and powerful Commissioner's Office that, among other duties, personally approved broadcasters for World Series games. Infamously, Landis also upheld the "color line," preventing African Americans from playing in professional baseball. Jackie Robinson was signed less than a year after Landis' death in 1944.

1932: Hattie Caraway is the first woman elected to the United States Senate. Caraway represented Arkansas from 1931 to 1945 and was also the first woman to preside over the Senate as well as the first to chair a committee. While in office she joined other Southern senators in a filibuster against the Roosevelt administration's anti-lynching bill.

1959: Motown is founded by Berry Gordy Jr. in Detroit, Michigan. The name, a portmanteau of *motor* and *town*, has come to be synonymous with Detroit. The record company produced numerous artists including Jackie Wilson, Smokey

Robinson, the Supremes, the Four Tops, the Jackson 5, Gladys Knight, and Stevie Wonder. The label developed a unique musical genre that combined soul and pop elements and became known as the "Motown Sound."

1962: The first American combat operation in Vietnam begins. Eighty-two U.S. Army helicopters transported more than 1,000 South Vietnamese paratroopers to attack a Viet Cong stronghold ten miles west of Saigon. The Viet Cong were caught unawares by the swift attack and routed. The operation demonstrated the viability of helicopter troop transport and greatly changed the U.S. approach to the war.

1966: The T.V. show *Batman* debuts. The show starred Adam West as the titular character and Burt Ward as his sidekick Robin, with Cesar Romero portraying his nemesis the Joker. The show's campy themes, comedy, and formulaic plots quickly made it a hit. Although the show only aired for three seasons it was wildly popular and hugely influential.

1991: The United States Congress passes an act authorizing the use of military force to expel the Iraqi military from Kuwait. A week later, aerial and naval bombardment of Iraqi positions would commence Operation Desert Storm, after which a swift ground campaign routed Iraqi troops.

January 13

1888: The National Geographic Society is founded. Thirty-three scientists and explorers met at the Cosmos Club in Washington, D.C., to organize the Society "to increase and diffuse geographic knowledge." The Society published the first issue of *National Geographic Magazine* in October of that year. It funded many research expeditions, including the excavation of Machu Picchu, explorations of the Arctic, and Jane Goodall's studies of chimpanzees. The Society began producing television series in 1964, launched a TV channel in 2001, and has produced feature films including 2005's *March of the Penguins*.

1910: The first public radio show is broadcast. A live opera performance starring Enrico Caruso and Emmy Destinn was broadcast from the New York City Metropolitan Opera House. Ships in New York Harbor, hotels in Times Square, and other locations received the broadcast. The microphones used in the broadcast were not of good quality, and were unable to pick up much of the singing, resulting in a broadcast that was mostly static. The event was nevertheless regarded as the birth of modern radio.

1927: Ernst Alexanderson displays the first television. Alexanderson, an inventor and pioneer of radio technology, realized that video images could be similarly transmitted and played a key role in television's development.

1938: The Church of England formally accepts the Theory of Evolution. After opposing the Theory as antithetical to Scripture for decades, the Church finally admitted that the Creation narratives were "mythical in origin" and therefore were not at odds with Darwin's theory.

1960: The USSR abolishes the Gulag system. Gulag labor camps had been established in 1929, and millions of petty criminals and political prisoners were sent to them during their tenure. As part of Premier Nikita Khrushchev's campaign of de-Stalinization and partial reform, the Gulag system was largely dismantled.

1968: Johnny Cash performs at Folsom Prison. The performance included the iconic songs "Folsom Prison Blues," "The Long Black Veil," and "25 Minutes to Go." The show was recorded, and the album *At Folsom Prison* was released later that year and reached number 9 on the Pop Albums chart. Cash's gritty songs about the hardships of prison life and the audience's wildly enthusiastic reception heard in the background earned the album rave reviews. The album was certified as a gold record a few months after its release. The success of the album revitalized Cash's singing career. He followed it with a performance at San

Quentin State Prison in 1969, which reached number one on the Pop Albums chart. *At Folsom Prison* is widely considered one of Cash's best albums, as well as one of the greatest country music records of all time.

January 14

1539: Spain annexes Cuba. The Conquistador Hernando de Soto claimed Cuba for the Kingdom of Spain, which would hold the island until the Spanish-American War forced it to relinquish all claims in 1898.

1784: The Treaty of Paris is ratified by Congress. The treaty, signed by representatives of Great Britain and the United States the previous September, ended the American Revolutionary War. The American delegation to Paris included John Jay, John Adams, and Benjamin Franklin.

1900: *Tosca* premiers in Rome. The opera by Giacomo Puccini has some of Puccini's most famous arias, including "E Lucevan le Stelle" ("And the Stars Were Shining") and "Vissi d'arte" ("I Lived for my Art"). The opera was panned by critics, but was a huge success with the public.

1943: The Casablanca Conference begins. Franklin D. Roosevelt and Winston Churchill met in Casablanca, Morocco, for ten days to discuss the Allied strategy for World War II, including the invasion of mainland Europe. In attending the Conference, FDR was the first U.S. president to travel by airplane while in office, as well as the first president to travel during wartime. Josef Stalin was unable to attend due to the Battle of Stalingrad. At the Conference, Roosevelt argued for an invasion across the English Channel, while Churchill favored an invasion of Sicily followed by mainland Italy. Roosevelt eventually agreed to Churchill's plan, provided that the British would commit troops and resources to the Allied troops in fighting in the Pacific.

1953: Tito is inaugurated president of Yugoslavia. As dictator, he would remain in office until his death in 1980. Tito's communist government split with Stalin and the USSR in 1955. Along with India and Egypt, Yugoslavia became one of the leading states of the Non-Aligned Movement during the Cold War. Tito also suppressed nationalist sentiment in the areas of Yugoslavia that would become independent nations following the country's breakup in 1990.

1967: The Human Be-In is held in San Francisco. The event was the first major counterculture gathering and foreshadowed the hippie's Summer of Love that would take place later that year. The Be-In was organized by the artist Michael Bowen and the poet Michael Cohen, and promoted in their underground newspaper, the *San Francisco Oracle*. Speakers included the counterculture guru Timothy Leary, who coined his famous phrase "tune in, turn on, drop out" at the gathering. Also in attendance onstage were the poet Allen Ginsberg, comedian Dick Gregory, and author Lawrence Ferlinghetti. Musical guest Jefferson Airplane, the Grateful Dead, Big Brother and the Holding Company, and Blue Cheer performed. Owsley Stanley, an audio engineer for the Grateful Dead and underground chemist, provided huge amounts of LSD to the assembled crowd of twenty to thirty thousand hippies. That summer over 100,000 young hippies gathered in San Francisco's Haight-Ashbury neighborhood. The Summer of Love popularized hippie subculture and the exploration of new modes of expression.

1973: Elvis Presley broadcasts a live concert via satellite from Hawaii. The concert, *Aloha from Hawaii*, set the record for the most watched broadcast by a live performer, with estimates of more than 1 billion viewers worldwide. It was also the most expensive television special at the time, costing more than $2.5 million to produce.

January 15

1559: Elizabeth I is crowned Queen of England. Following her coronation, she established the English Protestant Church, the forerunner of the Church of England, and made herself Supreme Governor of it. She ruled England and Ireland for the next forty years. During her reign she oversaw the defeat of the Spanish Armada and rise of English naval dominance. She imprisoned and executed her chief political rival, Mary Queen of Scots. Her long rule helped establish a British national identity and is often considered a one of Britain's golden ages.

1797: John Etherington creates the first top hat. According to popular lore, the English haberdasher wore his silk top hat in public, causing a riot. He was arraigned for breach of the peace, and the witnesses for the prosecution claimed that several women fainted at the sight of the hat, children cried, and dogs barked at him. Etherington posted a bond of 500 pounds and was released.

1870: The donkey becomes a popular symbol of the Democratic Party. Prior to 1870, the political enemies of President Andrew Jackson mocked him by calling him "Andrew Jackass." The insult backfired, however, as Democrats embraced the populist association with the hardworking, stubborn donkey, and the image stuck. The most enduring association of the donkey with the Democratic Party appeared in the magazine *Harper's Weekly*, in a political cartoon by Thomas Nast. The cartoon depicts the donkey kicking a dead lion. The donkey represented Southern Democrat "copperhead" press, and the lion represented President Lincoln's Secretary of War during the Civil War. Other political cartoonists followed Nast's lead, and the donkey was cemented as the symbol of the Party.

1908: Alpha Kappa Alpha is founded. The sorority was the first established for African-American college women. Founded by Ethel Hedgeman Lyle and sixteen other undergraduate students at Howard University, it now has over nine hundred chapters and more than a quarter of a million members. The sorority's motto is "By Culture and By Merit."

1919: The Great Molasses Flood hits Boston, Massachusetts. A fifty-foot-tall storage tank containing more than two million gallons of molasses suffered a catastrophic failure that sent a wave of molasses 25 feet high rushing through the streets of Boston's North End neighborhood at 35 miles per hour. Twenty-one people were killed and 150 were injured. Cars and trucks were washed into Boston Harbor by the wave of molasses, and a firehouse, office building, and the elevated railway were all severely damaged. The cleanup effort took several weeks, and the harbor was stained brown by the molasses through the rest of the summer. Rescue workers and cleanup crews leaving the scene tracked molasses throughout the whole city, and residents were still finding sticky residue months later.

1967: Super Bowl I is played. The NFL's Green Bay Packers defeated the AFL's Kansas City Chiefs 35–10 at the Los Angeles Memorial Coliseum.

2001: Wikipedia goes online. The free, online encyclopedia was unique in that anyone could edit and enter new articles. Originally an English-language edition only, it later grew to include many other languages and millions of articles.

January 16

1547: Ivan the Terrible becomes Tsar of Russia. During his reign, Ivan greatly expanded Russia's territory by conquest, capturing Siberia, Kazan, and Crimea, and developed an extensive bureaucracy to administer them. He brought Russia out of the Medieval Ages and transformed it into an Imperial world power. He established close ties with England, corresponding with Queen Elizabeth, and sent delegations to Egypt and Germany. Ivan was prone to outbursts of rage: during one he killed his own son and heir to the throne, bludgeoning him to death with his royal scepter. Ivan would remain on the throne until his death in 1584, when he was succeeded by his younger son Feodor.

1605: The first edition of *Don Quixote de la Mancha* by Miquel de Cervantes is published in Madrid. The novel is considered one of the works that ushered in modern Western literature and remains one of the most influential pieces of literature written during the Spanish Golden Age.

1864: General William Sherman issues Special Field Order Number 15. The order provided for the confiscation of Confederate farmland along the coast of South Carolina, Georgia, and Florida and its distribution to freed slaves and their families in 40-acre parcels. Special Field Order Number 15 is the primary source of the expression "forty acres and a mule."

1893: U.S. Marines invade Hawaii. The Marines supported the overthrow of the Hawaiian Queen Lili'uokalani by white Americans who wished to annex the island archipelago to the United States. Wanting to avoid the loss of life of her subjects, the Queen surrendered. The rebels, backed by U.S. military might and led by Sanford Dole, declared a provisional government. Dole would become the first territorial governor of Hawaii after annexation by the United States in 1898.

1939: The comic strip *Superman* debuts. The strip ran until 1966 and was carried in over 300 daily newspapers at its peak, with a readership of more than 20 million. The comic strip included the introduction of the villain Lex Luthor and Superman's first costume change in a telephone booth.

1964: Whiskey a Go-Go opens on the Sunset Strip in West Hollywood, California. The club pioneered go-go dancing as well as the "go-go-girl" costume consisting of knee-high boots and miniskirts. It is widely credited with having given birth to the rock and roll scene in California. Whiskey a Go-Go gave many famous musical acts their start, including The Byrds, the Doors, Frank Zappa, Van Halen, and Guns 'n' Roses.

January 17

1377: Pope Gregory XI moves the Vatican back to Rome. The Papal court had resided in Avignon, France, for nearly 70 years. After Gregory's death the following year, the Papal Schism occurred, during which three different men all claimed to be the true pope. The Schism was ended by the Council of Constance when they elected Martin V to the papacy.

1562: The Edict of St. Germain grants limited recognition to the Huguenots, an ethno-religious minority group of French Protestants. The Huguenots followed the teachings of the Protestant Reformer John Calvin, and were heavily persecuted for their beliefs in the largely Roman Catholic realm. The Edict, which was decreed by Catherine de'Medici, the regent of France, was intended to appease the growing Huguenot minority without upsetting the Catholic Church. The Edict was ineffective, however, and a massacre of Huguenots by the Duke of Guise later that year provoked the French Wars of Religion. Huguenots continued to be persecuted for centuries and launched numerous unsuccessful rebellions. In 1789 the Declaration of the Rights of Man and of the Citizen finally granted equal rights to the group as full citizens.

1912: Captain Robert Falcon Scott reaches the South Pole. Scott and his companions had set out on their journey from Cardiff, Wales, on the ship *Terra Nova* two years before. While making their way to Antarctica, Scott's party received a telegram from Amundsen in which the Norwegian explorer informed them he was "proceeding South," alerting them to the fact that they were in a race to be the first to reach the South Pole. Scott's party traveled overland, and waited out the winter of 1911 in a permanent settlement of huts at Cape Adare. When they finally arrived at the South Pole, they found a flag planted by Amundsen's party and a note stating Amundsen had already arrived there more than a month earlier. Turning home, they traveled for more than two months to a rendezvous point where teams on dogsleds were scheduled to pick them up. They never made it back. Trapped by a fierce blizzard, the party was unable to advance. Scott's final diary entry was dated March 29, 1912. His body was found on November 12.

1917: The United States buys the Virgin Islands from Denmark. The U.S. was concerned that Germany might be able to use the islands to launch U-boat attacks during World War I if it captured Denmark. The U.S. paid $25 million in gold for the Islands, then called the Dutch West Indies, which had been owned by Denmark since 1733.

1929: Popeye the Sailor Man debuts in the *Thimble Theatre* comic strip. The strip had been in syndication for ten years, but Popeye quickly became its main character. Popeye would go on to be the subject of numerous theatrical cartoons, comic books, arcade games and a live-action film.

1946: The U.N. Security Council holds its first meeting. The World War II Allies—the United States, France, Britain, the Soviet Union and the Republic of China—became permanent members of the Security Council, with veto powers. Ten other seats are rotated through non-permanent member states.

1995: An earthquake hits Kobe, Japan. The 6.9 magnitude earthquake was Japan's worst in the previous sixty years: more than 6,000 people were killed and over $200 billion in damage was caused to the city of Kobe and area around it.

January 18

1778: Captain James Cook arrives in the Hawaiian Islands. Cook was the first European to reach the islands, and he names the archipelago the "Sandwich Islands" in honor of the Fourth Earl of Sandwich.

1788: The first ships carrying convicts arrive in Australia. The first shipment of 736 convicts arrived in Botany Bay to found a penal colony that would become Sydney, Australia. Over the next ninety years, about 160,000 convicts were transported to Botany Bay and other penal colonies in Australia. Convicts were sent to Australia both for petty crimes such as theft and as political prisoners who had participated in rebellions or riots. Some managed to escape the penal colonies. Notable escapees included Mary Bryant, who with a group of others made her way to West Timor 5,000 miles away by rowboat, and William Buckley, who escaped and lived among Indigenous communities. Once they completed their sentences, most convicts joined the colonies of free settlers rather than return to Britain. Although there was stigma associated with being descended from a convict through the nineteenth century, these attitudes evolved over time, and eventually became a source of pride for many Australians.

1815: The Battle of New Orleans ends nearly three weeks after the War of 1812. Although the Treaty of Ghent had been signed on December 24th, 1814, news of the armistice had not yet reached the combatants. During the battle, Major General Andrew Jackson commanded a force of fewer than five thousand combatants who held out against an army three times as large and supported by sixty British ships. The Battle made Jackson famous in the United States and helped launch his political career, which would culminate in his election as President.

1896: The X-ray machine is unveiled. H.L. Smith displayed an X-ray machine to the public, along with an image made by Wilhelm Rontgen of his wife's hand, which showed her wedding ring and bones.

1911: The first naval landing of an aircraft. Eugene Ely flew a biplane from an airfield in San Bruno, California, to the cruiser USS *Pennsylvania* stationed in San Francisco Bay, landing on a platform attached to the ship's afterdeck. The landing ushered in the era of naval aviation and the use of aircraft carriers in warfare.

1943: The Warsaw Ghetto Uprising begins. When Nazi soldiers began deporting Jewish residents of the Warsaw Ghetto, hundreds of residents fought back using handguns, Molotov cocktails, and in hand-to-hand combat. The resistance fighters suffered heavy casualties but managed to temporarily halt the deportation from proceeding.

1974: The Yom Kippur War ends. Although Egypt and Israel had signed a cease-fire agreement in October of 1973, skirmishes continued along the Sinai peninsula until the two sides agreed to withdraw their forces from the battle lines.

2005: The Airbus A380 is unveiled in Paris, France. The commercial jet is the world's largest passenger airliner. It is a wide-body, double-decker plane that can carry up to 853 passengers.

January 19

1915: Neon advertising signs are invented. The French inventor and engineer Georges Claude filed a U.S. patent for the use of electrodes in neon lights. Claude's design included a completely sealed tube, which was an improvement on earlier designs. The electrodes in his patent transferred power to the glowing neon gas in the signs' closed tubes and minimized the sputtering effect that other designs had suffered from. Based on this patent his company, Claude Neon Lights, held a near-monopoly on the design and production of neon gas signs. The company's first customer in the United States was a Packard car dealership in Los Angeles, California. The dealership purchased two neon signs that read "Packard" for $2,500. Neon signs were an overnight sensation. Brightly lit and visible even in daylight, new installations would draw crowds that would marvel at the signs for hours.

1920: The United States Senate rejects the League of Nations. Following World War I, Woodrow Wilson and leaders of the other Allied nations drew up plans for an international body that would settle disputes between countries. The League of Nations was formally proposed at the Paris Peace Conference in 1919. Wilson campaigned tirelessly for the League in the United States, going on a nationwide tour to promote it, but was unable to convince enough Senators. During his tour Wilson suffered a stroke, and died four years later.

1940: *You Natzy Spy!* debuts. The slapstick film, starring The Three Stooges, was the first Hollywood film to satirize Adolf Hitler and the Third Reich. It helped expose the threat of the Nazi regime in a time when public opinion was still neutral regarding World War II.

1953: Lucy goes to the hospital. In one of the most-watched episodes of *I Love Lucy*, nearly 44 million American households (72%) tuned in to watch Lucy give birth. The episode was the conclusion of the character's fictional pregnancy, which was paired with the actress Lucille Ball's real-life pregnancy. It was the first time a pregnancy was shown in a fictional TV show.

1983: The Apple Lisa is announced. The Lisa was one of the first desktop computers that had a graphical user interface (GUI) and a mouse controller. The computer's high price of $9,995 and poor performance made it a commercial failure, with fewer than 100,000 units sold.

1986: The first computer virus targeting MS-DOS is released. The virus, called ©Brain, attacked the boot sector of the DOS file system and made the computer's memory unavailable to the operating system, rendering it unusable. The virus was created by Basit and Amjad Alvi, two brothers living in Pakistan. They were identified as the creators because virus also made the user's computer display the brothers' home address and phone numbers.

January 20

1841: Hong Kong Island is occupied by the British following the First Opium War. The island would remain under British control for a century until Japanese forces occupied it during World War II.

1887: Pearl Harbor becomes a naval base. The United States Navy was granted exclusive rights by the Senate to begin leasing the Harbor as a fueling and repair station for its ships. The United States government wanted to build a military presence in the Pacific, and the base was necessary for refueling its coal-powered ships at the time.

1920: The American Civil Liberties Union (ACLU) is founded. The ACLU was founded by Helen Keller, Jane Addams, Morris Ernst and other progressives of the day. Its initial focus was on defending anti-war protestors and advocating for freedom of speech. It eventually widened its scope to defend the rights of striking laborers, artists, and Native Americans during the 1920s. In 1925 the ACLU, with Clarence Darrow as lead attorney, defended John Scopes, who was on trial for teaching evolution in Tennessee. In the 1930s the ACLU began working with the National Association for the Advancement of Colored People (NAACP) to fight Jim Crow segregation. In 1954 the two organizations won *Brown vs Board of Education*, which declared that racially segregated schools were unconstitutional. The ACLU has taken controversial cases as well, defending the free speech rights of a Nazi group in Skokie, Illinois in the 1970s. The organization has more than half a million dues-paying members.

1937: Franklin Delano Roosevelt is sworn in to his second term. The 20th Amendment to the Constitution made the president's term begin on January 20th. Roosevelt was the first president to begin a term on this day. Before the passage of the 20th Amendment the president's term began on March 4. The change reduced the duration of the lame-duck session of Congress.

1981: Iran releases 52 American hostages. The hostages were released twenty minutes after president Ronald Reagan was inaugurated and 444 days after they were taken captive. The hostage crisis began when radical students supporting the Iranian revolution attacked the United States embassy in Tehran. The students were angry that the Carter administration had allowed the deposed Iranian Shah to travel to New York for medical treatment, and saw it as evidence that the U.S. intended to depose the revolutionary government. They decided to seize the embassy in an effort to gain leverage to demand the Shah be sent back to Iran to be tried for crimes against his people. On November 4, 1970, hundreds of students broke through the embassy's main gates and stormed the building. They bound and blindfolded Marines and embassy staff and paraded them before the press. Six Americans were able to escape and took refuge in the Canadian Embassy, later escaping the country disguised as a Canadian film crew. The Iranian government refused all appeals to release the hostages, and the Carter administration began making preparations to rescue them in a military operation. During the attempt, called Operation Eagle Claw, two helicopters crashed in the Iranian desert and eight U.S. soldiers died. Following negotiations mediated by the Algerian government, an agreement was reached to release the hostages in exchange for unfreezing Iranian assets in overseas banks.

2007: Skiing to the South Pole. A four-man team led by Henry Cookson and Paul Landry arrived at a research station near the South Pole using only skis and kites, becoming the first explorers to do so without mechanical vehicles.

January 21

1789: The first American novel is printed. *The Power of Sympathy of the Triumph of Nature Founded in Truth*, an epistolary novel written by William Hill Brown, was published in Boston. The novel advocated for moral education and warned against the dangers of giving into one's passionate urges. The novel was a thinly veiled retelling of a New England scandal that involved the author's neighbors, Perez Morton and Fanny Apthorp. Morton, Apthorp's brother-in-law, had an affair with her. She became pregnant and committed suicide.

1793: Louis XVI is guillotined. The deposed monarch was found guilty of treason by the French National Convention following the French Revolution. The Convention found over-whelming evidence that Louis had conspired with a force of Prussian invaders who had attempted to stop the Revolution. Of the 721 deputies to the Convention, 693 voted to find him guilty. In a tense vote to determine his fate, 361 voted in favor of execution. Louis was condemned by a majority of a single vote. At the *Place de la Revolution*, where he was executed, Louis declared himself innocent of the charges, and announced that he pardoned those who had voted for his execution.

1911: The Monte Carlo Rally is held for the first time. The Rally was organized by the Automobile Club of Monaco and was the creation of Prince Albert I of Monaco. Twenty-three cars set out from eleven locations around Europe, speeding towards Monaco. Henri Rougier made it there first, driving from Paris, France, 634 miles away. The rally continues to be held each year, with competitors still starting from one of several different points around Europe.

1919: The Irish War for Independence begins.
Following a landslide electoral victory, Sinn Fein, the politi-cal wing of the Irish Republican Army, declared independence from the United Kingdom. The party refused to fill its seats in the British Parliament, establishing a separate provisional Irish government. The same day, a "flying column" of Irish Volunteers ambushed a British arms transport, killing two policemen and seizing explosives.

1954: The nuclear submarine is born. The USS *Nautilus* was launched from Groton, Connecticut after being christened by First Lady Mamie Eisenhower, ushering in the era of nuclear subs. The Nautilus' nuclear-powered propulsion produced no emissions and consumed no air, allowing it to remain underwater far longer than other submarines of the day. It broke numerous speed and distance records for underwater travel.

1968: The Battle of Khe Sanh begins. The seven-month battle between United States forces and the North Vietnamese Army was one of the most highly controversial battles of the Vietnam War. More than 1,400 U.S. soldiers died during the battle, fueling public opposition to the war.

1976: The Concorde takes off. The first scheduled flights of the supersonic passenger jet began on routes from London to Bahrain and from Paris to Rio de Janeiro. The Concorde continued to fly until 2003, when Air France and British Airways simultaneously announced they would retire the plane.

1981: The DeLorean DMC-12 is created. The car's gull-wing doors and fiberglass body made the car iconic. Fewer than 9,000 were built, and production was halted in 1983. The DeLorean was forever immortalized as the time machine in the 1985 film *Back to the Future*.

January 22

1506: Swiss Guards arrive at the Vatican. At the time, the Pope was a political force on par with kings, and the Vatican was often embroiled in conflicts with other powers of Europe. Pope Julius II requested the contingent of mercenaries from the Swiss government to defend the Papal city from invasions by rival factions during the Italian Wars of the sixteenth century. After the end of the Italian Wars, the Swiss Guard ceased to be a military unit and shifted its purpose to serving as the Pope's personal bodyguards. By the nineteenth century the Guard took on a purely ceremonial role, but after the attempted assassination of Pope John Paul II in 1983, they established a plainclothes detachment of armed bodyguards who accompany the Pope.

1887: Columbia Records is formed. Originally called Columbia Phonograph, the company was founded by Edward Easton and a group of investors. The company started by selling and servicing Edison phonographs as well as producing cylinder recordings. By 1891 it was distributing a ten-page catalogue of musical records. In 1902, Columbia introduced disc-shaped wax records and began working with the New York Metropolitan Opera to make recordings. By 1912 the company had stopped selling phonographs altogether and was concentrating solely on producing records. In the 1930s suffering from low sales, Columbia recruited the talent scout John Hammond. Hammond made numerous artists famous and popularized genres from swing jazz to rock music. Under Hammond's direction, Columbia signed artists including Benny Goodman, Count Basie, Pete Seeger, Frank Sinatra, Aretha Franklin, and Bruce Springsteen to Columbia.

1905: Bloody Sunday in Saint Petersburg, Russia. A crowd of unarmed demonstrators marched to the Winter Palace to present a petition to Tsar Nicholas II, and were fired upon by the Imperial Guard. The killings caused widespread public outrage and provoked a series of large strikes throughout the country. The event was a major turning point in the Russian Revolution of 1905, which severely weakened the Tsar's power and would eventually lead to his downfall.

1927: The first radio broadcast of a soccer game. Teddy Wakelam provided play-by-play and commentary of a match between Arsenal F.C. and Sheffield United on BBC radio. The match was a 1–1 tie.

1957: The "Mad Bomber" is captured. George Metesky, angry about a workplace injury, planted dozens of bombs around New York City for sixteen years until he was caught. He was found legally insane and committed to an asylum until 1973.

1968: Apollo 5 lifts off. The mission carried the Apollo Lunar Module in its first unmanned flight on a Saturn IB rocket. The

Lunar Module orbited Earth and was tested to determine its descent and ascent capabilities, and was a vital step in the Apollo Program.

1973: With *Roe v. Wade*, the U.S. Supreme Court legalizes elective abortion in all fifty states. The court ruled 7–2 that the right to privacy under the 14th Amendment extended to a woman's decision to have an abortion.

1984: The Apple Macintosh is unveiled. The Macintosh was the first computer to be unveiled in a Super Bowl commercial. The commercial was considered a watershed event of advertising. It featured the Macintosh as a means of saving humanity from the conventionality of IBM computers, backed by a "Big Brother" theme that alluded to George Orwell's novel *1984*.

January 23

1849: Elizabeth Blackwell receives a medical degree. Blackwell, a British-born physician, was the first woman to be granted a medical degree in the United States as well as the first female doctor registered in the United Kingdom. She tirelessly promoted the medical education of women in the U.S. and U.K. During the Civil War, Blackwell organized the Women's Central Relief Association and worked with Dorothea Dix to train nurses. In 1874 she opened the London School of Medicine for Women.

1943: The Allies capture Tripoli. General Bernard Montgomery led the British Eighth Army in an attack on the city, which was defended by Rommel's German Fifth Panzer Army. The battle lasted for six weeks, until the British were reinforced by American and French troops. The Allies swept into Tripoli after fierce fighting, and Rommel retreated rapidly across North Africa. The Allies continued their advance, driving Rommel out of Africa in May of 1943.

1950: Jerusalem is made the capital of Israel. The Israeli Knesset declared Jerusalem its capital despite the fact that Jordan had annexed the Eastern part of the city. The stalemate would persist until the Six-Day War in 1967, when Israel captured the remainder of the city.

1957: The Frisbee is born. Although variations on the frisbee date as far back as 1871, the modern version was designed by Fred Morrison, who sold the rights to the Wham-O company. Wham-O initially called it the Pluto Platter, but renamed it the Frisbee after discovering that students at Yale tossed pie tins from the Frisbie Baking Company to one another. The Frisbee did not catch on until "Steady" Ed Headrick was hired by Wham-O. Headrick redesigned the disc, increasing its mass and making the rim thicker. The new design made the Frisbee easier to catch and throw accurately. Sales skyrocketed.

1964: The 24th Amendment is ratified. The Amendment outlawed the use of poll taxes in federal elections. Poll taxes were a common tactic in Southern states used to prevent African Americans and poor whites from voting. When the Amendment was ratified, Alabama, Arkansas, Mississippi, Texas, and Virginia still had poll tax laws on the books.

1997: Madeleine Albright is the first female U.S. Secretary of State. Prior to being appointed to the position, Albright served as the U.S. Ambassador to the United Nations. She served on the National Security Council during the Carter Administration.

January 24

41: Caligula is assassinated. The Roman Emperor had grown increasingly irrational, claiming to be a god and referring to himself as "Jupiter." His reign was marked by a mix of sadistic cruelty, lavish extravagance, and sexual perversion. He attempted to increase his own power and feuded with the Senate and nobility, going so far as to have several senators put to death. His tyranny resulted in numerous plots to overthrow him. Cassius Chaerea, a centurion in Caligula's Praetorian Guard, had grown increasingly angry at the emperor's humiliating treatment, and led several of his fellow guardsmen in stabbing Caligula to death.

1848: The Gold Rush begins. James W. Marshall found gold at Sutter's Mill in Coloma, California. While Marshall and Sutter attempted to keep the discovery quiet, news of it quickly spread. A newspaper publisher in San Francisco named Samuel Brannan opened a store selling prospecting supplies and then publicized the discovery. Over 300,000 people flocked to San Francisco in hopes of striking it rich. The population grew from just over 1,000 residents in 1848 to more than 25,000 by the end of 1849. An estimated 90,000 people arrived in California in 1849 and were called "forty-niners."

1908: The Boy Scouts are organized. British Army Lieutenant General Robert Baden-Powell started the Boy Scouts after his military manual *Aids to Scouting* was widely popular. He opened a scouting camp the year before to test his ideas about scouting, and published *Scouting for Boys*, which became one of the best-selling books of the 20th century. The Scouts organization grew rapidly.

1946: First United Nations resolution is passed. The resolution established the U.N. Atomic Energy Commission (UNAEC) to oversee the international development of atomic energy and attempt to ensure it was used only for peaceful purposes.

1972: Shoichi Yokoi surrenders. Yokoi was a Japanese sergeant in the Japanese Army during World War II in Guam. American forces liberated the island in 1944 after three weeks of battle, and most of the 7,500 Japanese soldiers on Guam were captured. Yokoi and nine other soldiers went into hiding, intending to continue fighting. By 1952 it had become clear to the men that the war was over, and seven of them left, but Yokoi and two others remained. After the two other men were killed in a flood in 1964, Yokoi was alone. He survived by hunting, living in a cave near a waterfall, and only coming out at night. He was discovered by two fishermen while they were checking their traps. Yokoi attacked the men, but they were able to subdue him and take him out of the jungle. Upon returning to Japan, Yokoi famously said "it is with much embarrassment that I return," and his comment became a popular Japanese saying. He wrote an autobiography and was featured in a documentary about his life that aired in 1977. Yokoi died in 1997 at the age of 82.

2003: The Department of Homeland Security opens. It was created in response to the terrorist attacks of September 11th and incorporated 22 agencies, including U.S. Customs, Immigration and Naturalization, the Transportation Security Administration, and FEMA.

January 25

1533: Henry VIII marries Anne Boleyn. Henry had his marriage to Catherine of Aragon annulled by Thomas Cranmer, which prompted the Pope to excommunicate both Henry and Cranmer. In response, Henry simply declared himself the head of the Church of England. Anne gave birth to a daughter, the future Queen Elizabeth I. But within a few years, Henry had met and begun courting Jane Seymour. He had Anne arrested for treason and sent to the Tower of London, where she was executed in 1536. When Elizabeth ascended the throne, Anne was made a martyr of the English Reformation.

1858: "The Wedding March" is made famous by Princess Victoria. Written by Felix Mendelssohn as a suite of musical pieces to accompany Shakespeare's play *A Midsummer Night's Dream*, it was played at the wedding of Princess Victoria to Emperor Frederick III, and has been popular at weddings since.

1890: Nellie Bly circumnavigates the globe. Nelly Bly was an American investigative journalist and inventor. In 1888 she pitched the idea of attempting to recreate the trip around the world described in Jules Verne's novel *Around the World in Eighty Days*. She traveled first to England, taking only a change of clothes, a sturdy overcoat, and cash. While traveling through France she met and interviewed Verne in Paris. She traveled by steamship and railroad, sending short reports of her progress by telegraph and longer articles by mail. She arrived in San Francisco two days behind schedule because of rough seas during her crossing of the Pacific. Pulitzer chartered a private passenger train, dubbed the *Miss Nelly Bly Special*, to help her make the final, transcontinental leg of her journey in record time. The train made the trip in less than three days. Bly arrived in New Jersey at 3:51 P.M., having traveled around the world in just 72 days.

1924: First Winter Olympic Games. The inaugural games were held at the foot of Mont Blanc in Chamonix, France, and included curling, bobsledding, ice hockey, skiing, and ice skating events. Athletes from sixteen nations participated in the Games. The Nordic countries dominated the games, winning more than half the medals between them.

1949: First Emmy Awards ceremony. The Emmy Awards initially only honored television shows produced and aired locally in Los Angeles, California. The Awards were presented at the Hollywood Athletic Club and included winners for the Best Film Made for Television, the Most Outstanding Television Personality, and Most Popular Program.

1964: The forerunner of Nike is founded. The company, called Blue Ribbon Sports, was started by athletes from the University of Oregon track and field team. It originally was only a shoe distributor for the company that became ASICS footwear, but later moved into designing its own shoes. In 1971 the company changed its name to Nike and adopted the familiar Swoosh logo, designed by Carolyn Davidson.

1971: Idi Amin leads a coup in Uganda. While he was Commander of the Army, the President of Uganda, Milton Obote, discovered that he was embezzling army funds and planned to have him arrested. Amin seized power while Obote was out of the country. The United Kingdom supported his coup because Obote was moving to nationalize British businesses in Uganda. Amin remained in power until 1979 when he was overthrown and fled to exile in Saudi Arabia until his death.

1998: Pope John Paul II visits Cuba. It was the first time a Pope had visited the island nation, and closed one of the final chapters in the Cold War. John Paul met with Fidel Castro, Cuba's Communist dictator, and performed an open-air mass in Santiago, where he delivered a message of freedom of expression and freedom of association.

2011: The Egyptian Revolution begins. The Revolution was a major incident in the series of uprisings across the Middle East and North Africa that would come to be known as the Arab Spring. Thousands of demonstrators occupied Cairo's Tahrir Square, demanding the resignation of President Hosni Mubarak and a spate of democratic reforms. Mubarak stepped down in February.

January 26

1500: Vicente Pinzòn arrives in Brazil. The Conquistador was the first European to set foot on Brazilian territory. He had arrived in the New World as captain of the *Niña* during Columbus' 1492 expedition. Pinzòn also traveled fifty miles up the Amazon River while he was surveying the coast of Brazil.

1856: The First Battle of Seattle is fought. The battle was an incident in the Puget Sound War between white settlers and Natives. An alliance of Native tribes attacked Seattle, then only a settlement of only a few hundred. The settlers had the advantage of knowing the attack was imminent, and were supported by the warship *Decatur,* a contingent of U.S. Marines, and artillery. They suffered only two casualties, while the attackers lost at least 28.

1870: Virginia rejoins the Union. Once the seat of the Confederacy, Virginia was allowed back into the Union due to efforts by the Committee of Nine, a group of Republican political leaders who favored allowing former Confederates to vote and participate in politics. During Reconstruction, the legislature was dominated by the Readjuster Party, which was made up of free African Americans and progressive whites. The legislature adopted a constitution that guaranteed political enfranchisement to all citizens and provided for public education. The Readjusters lost power to Democrats in the election of 1883. They set out to dismantle many of the Reconstruction-era initiatives that had been gained, instituting a poll tax, segregating schools, and passing Jim Crow laws.

1905: The Cullinan diamond is found. It was the largest diamond ever unearthed weighing more than 3,100 carats (1.3 pounds). The diamond was found in a mine in South Africa and named after the mine's owner, Sir Thomas Cullinan. The diamond was transported to England by parcel post, with a huge diversion staged by sending a fake copy by a heavily guarded steamer ship. It was presented to the British King Edward VII as a birthday gift and token of the loyalty of the South African settlers, who had just fought and lost a war of colonial independence with the British. A large part of the diamond was cut and polished to make the Great Star of Africa diamond, which weighs 530 carats and adorns the Sovereign's Scepter, part of the British Royal Crown Jewels.

1911: The first seaplane takes off. While a number of seaplane prototypes that could land on water had been built, none had managed to take off from the water before then. Glenn Curtiss, an early aviator and one of the founders of the aviation industry, developed a pontoon design that allowed him to take off and land in the water.

1965: Hindi becomes the official language of India. Although India has the world's second highest number of spoken languages, at 780, the Indian constitution established Hindi as the official language, spoken by more than 260 million people. Hindi is not, however, the national language: India has no national language, reflecting its linguistic diversity.

January 27

1302: Dante Alighieri is exiled from Florence. The Italian poet and author had sided with Pope Boniface in his quarrel with the powerful Florentine Gherardini family, and he was exiled along with other supporters when the Gheradinis consolidated their power. During his exile, Dante conceived of the *Divine Comedy*. It would become his greatest work.

1606: Trial of Guy Fawkes. Fawkes and a group of English Catholics had attempted to assassinate the Protestant King James I of England by blowing up Parliament. The Gunpowder Plot, as it became known, was revealed to the authorities by an anonymous letter, and Fawkes and his co-conspirators were arrested and tried for treason. The trial was short, as Fawkes and the others had all signed confessions extracted under severe torture. They were executed only a few days later.

1880: Edison patents the incandescent lamp. He had been working on the problem of creating a long-lasting electric light for two years. In 1879 he hit upon using a carbon filament for his light bulb, and in the first successful test of it, the bulb burned continuously for 13 hours.

1944: The Siege of Leningrad ends. The city's inhabitants had held out against the German siege for two years, four months, two weeks, and five days. It was the longest siege of a city in history, as well as the costliest. The Soviets suffered more than 3.4 million casualties. The Red Army finally pushed the Germans out of Leningrad and regained control of the Leningrad railway following several weeks of heavy fighting.

1967: Apollo 1 disaster. The astronauts Gus Grissom, Edward White, and Roger Chaffee were killed when they became trapped in the spacecraft's Command Module cabin as a fire raged through it. The three were in the Command Module to run tests for the planned launch of the spacecraft. The tests were considered at the time to be non-hazardous, but while the astronauts ran through their checklists, an electrical fire was ignited. The cabin was filled with pure oxygen, which caused the fire to burn so intensely that it ruptured the Module's inner wall in an explosion. The astronauts were unable to open the hatch of the pressurized cabin, and quickly died from smoke inhalation. The fallout from the incident resulted in the Apollo Program to make significant changes to its procedures, valuing safety above all else.

1973: Paris Peace Accords end the Vietnam War. The treaty was signed by representatives of North Vietnam, South Vietnam, and the United States. It ended direct U.S. military combat in Vietnam; the U.S. began withdrawing troops that same month. Two years later, North Vietnamese forces invaded South Vietnam, captured Saigon, and established control of the entire country.

1996: Germany observes the first Holocaust Remembrance Day. The date was selected to commemorate the liberation of Auschwitz-Birkenau by the Soviet Red Army in 1945. The concentration camp was the largest during the Holocaust. Holocaust Remembrance Day is observed around the world following a United Nations resolution passed in 2005.

January 28

1754: "Serendipity" is coined. Horace Walpole, an English art historian, coined the word in a letter to his friend Horace Mann. Walpole was inspired by a Persian fairy tale, "The Three Princes of Serendip," who were regularly enjoying happy accidents. The name Serendip comes from the old name of Sri Lanka, Sarandib.

1813: *Pride and Prejudice* is published. Its author, Jane Austen, had sold the copyright to the book to its publisher for a single payment of 110 pounds sterling. The first edition quickly sold out, as did a second edition published later the same year. The book became one of the most popular novels ever written in the English language; over 20 million copies have been sold, and it is widely considered a cornerstone of modern English literature.

1820: Antarctica is discovered. Fabian Gottlieb von Bellingshauen and Mikhail Petrovich Lazarev discovered the continent while attempting to sail around the world. Their expedition was exploring the Southern Ocean and searching for land near the South Pole when they sighted the coastline of Antarctica from their ships, the *Vostok* and the *Mirny*.

1896: The first speeding ticket is given. Walter Arnold owed one shilling plus court costs for speeding through the village of East Peckham, England, in a motor vehicle. He was stopped by a policeman who caught up to him while riding a bicycle. Arnold's reckless speed was…eight miles per hour, six over the speed limit. Exactly forty-two years later, the World Land Speed Record on a public road would be broken by Rudolph Caracciola, driving his Mercedes-Benz 268.9 miles per hour.

1915: The U.S. Coast Guard is founded. The modern Coast Guard was formed by a merger of the Lifesaving Service, which saved shipwrecked mariners, and the Revenue Cutter Service, which acted as a customs enforcement service. The Coast Guard engages in rescue operations and law enforcement, as well as maintaining all the lighthouses on the American coastline.

1958: Lego bricks are patented. The Lego toy company had begun manufacturing construction sets in 1949, but it did not develop the design modern Lego brick until later. After the design was perfected, it took another five years to find the right plastic for the brick. The same design remains in use more than half a century later, and Lego bricks became one of the most popular toys in the world.

1985: USA for Africa records "We Are the World." The song was written by Michael Jackson and Lionel Richie and performed by a supergroup of more than thirty singers, including Diana Ross, Tina Turner, Stevie Wonder, Kenny Rogers and Billy Joel. Sales of the recording, which sold more than 20 million copies, were used for humanitarian aid in Africa.

1986: The Space Shuttle *Challenger* explodes. The shuttle was on its tenth mission, and its explosion on liftoff killed all seven crew members aboard. The *Challenger* disaster was broadcast live in classrooms around the country, as the shuttle was carrying a schoolteacher named Sharon Christa McAuliffe among the crew members. McAuliffe had been selected from over 11,000 applicants to participate in the NASA Teacher in Space Project and would have become the first teacher in space. The explosion resulted in the space shuttle program being grounded for nearly three years while NASA studied new safety measures and rocket booster designs.

January 29

1845: "The Raven" is published. The poem, written by Edgar Allen Poe, was first published in the *New York Evening Mirror*. While it did not earn Poe much money, it made him a household name in New England and greatly contributed to his popularity. The poem's haunting, supernatural nature and stylistic language has made it one of Poe's most enduring works. "The Raven" remains one of his most influential works, and has been referenced in novels, television, and film ever since.

1850: The Compromise of 1850 is proposed. Senators Henry Clay and Stephen Douglas drafted a series of bills designed to resolve a four-year standoff between free states and slave states. The Compromise provided for California being admitted to the Union as a free state, Texas surrendering its claims to New Mexico, the banning of the slave trade in the District of Columbia, and strengthening the Fugitive Slave Law. New territories were allowed to decide for themselves whether to permit slavery in their borders. The Compromise was politically popular, but it also led directly to the formation of the Republican Party. Ten years later, when the Republican Party's nominee Abraham Lincoln was elected to the Presidency, the Secession Crisis provoked the outbreak of the Civil War.

1856: The Victoria Cross is introduced. Queen Victoria introduced the award to honor acts of bravery by British soldiers during the Crimean War. She presented the first awards in 1857. More than 1,300 have been given to British soldiers, sailors, aircraft pilots, and civilians. The Cross continues to be the highest military decoration in the United Kingdom.

1907: The first Native American Senator is elected. Charles Curtis was an attorney from Kansas who had Kaw, Osage, Potawatomi, and French ancestry. Curtis served in the U.S. Senate from 1907 to 1929. He was President *pro tempore* of the Senate in 1911, the Minority Whip from 1919 to 1924, and the Majority Leader from 1924 to 1929. Curtis served as Herbert Hoover's Vice President from 1929 to 1933.

1916: Zeppelins bomb Paris for the first time. Until then, the German air force primarily used the giant airships to attack targets in Great Britain. The Zeppelins approached Paris under cover of darkness, and while they did minimal damage to their targets, the aerial bombing caused a panic in the city. Air defenses were hastily organized. With the French preoccupied with securing the capital, they ignored indications the Germans were preparing to attack Verdun.

1936: The first Baseball Hall of Fame inductees are announced. Honus Wagner, Walter Johnson, Christy Mathewson, Babe Ruth ,and Ty Cobb were the first baseball players inducted into the hall of fame. Twenty more ballplayers would be inducted before the Hall opened in Cooperstown, New York, three years later.

2002: The "Axis of Evil" is announced. George W. Bush used the term during his State of the Union address to describe North Korea, Iran, and Iraq as state sponsors of terrorism who were seeking to acquire weapons of mass destruction. The Axis of Evil would be repeatedly referred to in order to rally support for his administration's so-called War on Terror as well as the invasion of Iraq the following year.

January 30

1703: The 47 Ronin avenge their leader. The *ronin*, or leader-less samurai, had hatched the plot after their feudal lord, Asano Naganori, was forced to commit ritual suicide the year before by the shogun for assaulting a member of the royal court named Kira Yoshinaka. His samurai retainers became ronin. Their leader, Oishi Yoshio, convinced the other samurai to leave the castle and begin planning their vengeance. To cover their actions, they milled about in Geisha houses, giving the appearance of being distracted and incompetent. On the day of the assassination, Yoshio led 46 of his comrades to Yoshinaka's house, where they killed and decapitated him. After informing Naganori's widow that they had avenged his death, the 47 ronin went to his gravesite, where they committed ritual suicide. The story was popularized by numerous plays, literature and films.

1826: The first modern suspension bridge opens. The Menai suspension bridge in Wales connected the island of Anglesey to the mainland. It was designed by Thomas Telford. It has a span of 577 feet and is 98 feet tall. Its extreme height was intended to allow sailing ships to pass underneath.

1835: The attempted assassination of Andrew Jackson.
While Jackson was leaving the United States Capitol, an unemployed English housepainter named Richard Lawrence approached him and aimed a pistol at him. The gun misfired, as did the second one Lawrence attempted to shoot him with. Enraged, Jackson attacked Lawrence with his cane. The lawmakers at the scene, including Davy Crockett, grabbed Lawrence and restrained him. Lawrence later claimed to be King Richard III and was found to be insane.

1862: The first ironclad is launched. The USS *Monitor*, designed and built by John Ericcson in Brooklyn in less than three months, was a brand-new concept in naval warfare. The ship had a revolving gun turret and was protected from attacks by heavy iron plates. It is famous for its battle with the Confederate ironclad *Virginia* during the Civil War. During the battle, neither ship was able to damage the other, and after four hours, both withdrew. The *Monitor* engaged in several other battles during the Civil War, but sank less than a year after she launched during a heavy storm at sea.

1933: Hitler becomes Chancellor of Germany. His appointment by President Hindenburg was a major turning point in German and world history. While Chancellor, Hitler's agents burned the Reichstag, and he prompted Hindenburg to suspend basic rights. Hitler seized power as *Fuhrer* of the Reich the following year.

1948: Gandhi is assassinated. As Gandhi walked to the Birla House, where he planned to conduct an interfaith prayer meeting, the ultra-right Hindu nationalist Vinayak Godse stepped out of a crown and shot Gandhi three times at point-blank range. Godse was angry at what he perceived as Gandhi's acquiescence to Muslim demands during the Partition of India and Pakistan. The site of his assassination is now a memorial to his legacy.

1968: The Tet Offensive is launched. More than 80,000 North Vietnamese soldiers and Viet Cong irregulars attacked over one hundred targets throughout South Vietnam, including the capture of the city of Hu and an assault on the United States Embassy in Saigon. It was one of the largest military offensives of the war. The United States and South Vietnamese forces were caught completely off guard, and initially lost control of several strategic targets. The Tet Offensive had the lasting effect of turning American public sentiment against the war.

1969: The Beatles perform their last show. They performed a 42-minute set on the rooftop of the Apple Corporation in Central London. As news of the performance spread, crowds of spectators flocked to the street below and roofs of adjacent buildings. The police, concerned about traffic issues, climbed to the roof and shut the performance down. The show was later included in the documentary *Let It Be*.

January 31

1801: John Marshall is appointed Chief Justice of the Supreme Court. The fourth Chief Justice of the United States, Marshall's landmark opinions would lay the foundations of U.S. constitutional law. Marshall is widely credited with elevating the status of the Supreme Court to be equal to that of the legislative and executive branches of government. He also established the supremacy of the federal government and federal laws over those of the states and set the standard for judicial review of laws passed by the Congress. Marshall was the longest-serving Chief Justice in history, serving until his death in 1835.

1862: Sirius B is discovered. The tiny, incredibly heavy white dwarf star has a mass equal to the sun's compressed into a volume the size of the Earth. It was discovered by the astronomer Alvan Graham Clark when he saw it faintly orbiting its brighter, larger companion, the star Sirius.

1915: Poison gas is introduced as a weapon of war. During the Battle of Bolimòw in World War I, German troops fired artillery shells filled with xylyl bromide, a respiratory irritant similar to tear gas, at Russian positions. The attack turned out to be a failure because the weather was too cold for the chemicals to properly aerosolize.

1930: 3M unveils Scotch Tape. Richard Drew, an inventor who worked for Johnson & Johnson before joining 3M, came up with the idea for the clear tape. At that time, 3M was primarily in the business of manufacturing sand paper. Drew invented masking tape in 1922 for use in making straight borders in automobile paint jobs. When he first  tested the tape, it was not adhesive enough to stay on the car, and the angry auto painter told Drew to take the tape back to his "Scotch," meaning stingy, bosses. The name stuck, and was soon applied to 3M's entire line of adhesive tapes. When Drew invented the clear cellophane tape, it was an instant hit, and continues to be widely used today.

1953: The North Sea Flood kills over 2,500 people. The flood was caused by a major storm in the North Sea that pushed a storm tide ashore in the Netherlands, Belgium, England and Scotland. The storm tide was more than eighteen feet high in some locations and came ashore at night. Most of the victims were killed in the Netherlands, as most of its territory was below sea level. Major changes to Dutch sea defenses were made as a result.

1971: The Winter Soldier Investigation begins. The event was held by Vietnam Veterans Against the War and was designed to expose war crimes and atrocities committed by the United States during the war. Servicemen from every branch of the military provided testimony about war crimes they had either witnessed or committed themselves during the war. Mainstream media largely ignored the event, but the counterculture and underground press extensively covered it.

Chapter 2

February

February 1

1865: Abraham Lincoln signs the 13th Amendment. The Amendment abolished slavery and involuntary servitude except as punishment for a crime. It was one of three Reconstruction Amendments; the 14th and 15th Amendments provided for equal protection of all citizens and addressed voting rights, respectively.

1942: Voice of America goes on air. The radio station played news reports, music, and propaganda pieces and was initially broadcast to Germany and Occupied Europe. By the end of World War II, it was being broadcast in more than forty languages around the world. The station remained in use through the Cold War, targeting citizens of the Soviet Union. During the Cuban Missile Crisis, Voice of America played Spanish language shows around the clock. The station continued to broadcast in English through the end of the Cold War and into the 21st century.

1960: The lunch counter sit-ins begin in Greensboro, North Carolina. Four African-American freshmen at North Carolina A&T University, David Richmond, Franklin McCain, Joseph McNeil, and Ezell Blair, Jr., participated in the action. They sat down at a lunch counter in the Woolworth's Department Store in downtown Greensboro and asked to be served. The staff, citing the store's "whites only" policy, refused and asked them to leave. They stayed until the store closed for the night. Over the next several days, they continued returning to the lunch counter, and by the fourth day more than 300 people had joined the sit-in. While it was not the first such demonstration, the sit-in was instrumental in sparking a movement of similar actions that spread across Southern cities at swimming pools, libraries, parks, and beaches. Some sit-ins were met by hostile and violent mobs of angry whites. Many were successful in achieving their aim of desegregating public places, and the movement directly contributed to the establishment of the Student Nonviolent Coordinating Committee (SNCC). After four years of demonstrations throughout the Deep South, the Civil Rights Movement won a victory in the passage of the Civil Rights Act of 1964. The Act outlawed segregation in public places.

1968: Eddie Adams photographs an execution in Vietnam. The photograph depicts the Saigon chief of police summarily executing a suspected Vietcong prisoner during the Tet Offensive. It won the 1969 Pulitzer Prize for Photography and became one of the most infamous, enduring images of the war.

1979: Ayatollah Khomeini returns to Iran. Khomeini had been in exile for fifteen years, and following a popular revolution that overthrew the Shah of Iran, Khomeini returned as the revolution's leader. He became the Supreme Leader of the newly formed Islamic Republic until his death in 1989.

2003: The Space Shuttle *Columbia* disaster happens. The shuttle disintegrated as it reentered Earth's atmosphere, killing all seven astronauts aboard.

2013: The Shard opens in London. At 1,016 feet, it was the tallest building in the United Kingdom when it opened. Construction of the skyscraper took four years to complete. The building has 44 elevators serving 95 floors and was designed by architect Renzo Piano.

February 2

1653: New Amsterdam is incorporated. The Dutch colonial settlement was situated at the southern end of Manhattan Island. After English forces captured it in 1664, it was renamed New York.

1709: Andrew Selkirk is rescued. He had spent more than four years on a desert island in the South Pacific after he was marooned there by his captain. Selkirk had objected to putting to sea on a British privateer ship that had been badly damaged in a battle with a French warship. When he said he would prefer to be left on a desert island rather than stay on the ship, his captain obliged him. He left Selkirk on an island that was 420 miles west of the Chilean coast, with only a rifle, a hatchet, a knife, and a Bible. The ship later sank off the coast of Colombia. Selkirk lived on the island, building a hut for himself and hunting goats to survive. He was rescued by Woodes Rogers, captain of the British privateer ship *Duke*. Selkirk returned to England in 1711, eight years after he had left. His ordeal and rescue were publicized in English newspapers and in a book written by Rogers. Selkirk became the inspiration for the book *Robinson Crusoe* by Daniel Defoe. The book was one of the most popular novels written in the 18th century and sparked the beginning of the realistic fiction genre. The island Selkirk lived alone on for four years was renamed Robinson Crusoe Island in 1966.

1848: The Treaty of Guadalupe Hidalgo is signed. The treaty ended the Mexican-American War and ceded much of Mexico's territory to the United States. The border between Texas and Mexico was established at the Rio Grande, and the United States annexed California, much of modern-day New Mexico, Arizona, Nevada, and Utah, and parts of Colorado and Wyoming. The U.S. paid Mexico $15 million. Mexican citizens living in the newly annexed territory were given the choice of becoming American citizens or moving.

1876: Baseball's National League is founded. The NL replaced the National Association of Professional Baseball Players and is the oldest surviving professional baseball league in the world. Of the eight founding members teams, the Chicago Cubs (formerly the Chicago White Stockings) and Atlanta Braves (formerly Boston Red Stockings) are the only two still playing baseball today.

1887: The first Groundhog Day is observed in the town of Punxsutawney, Pennsylvania. The tradition of the groundhog predicting the start of spring weather grew out of Pennsylvania German customs in the 18th century. The celebration in Punxsutawney is the largest in the world, attracting as many as 40,000 visitors for the event. It was popularized by the 1993 film *Groundhog Day*, starring Bill Murray.

1935: The polygraph is used in a criminal investigation for the first time. Its inventor, Leonard Keeler, used the polygraph in an interrogation of two murder suspects. The pair were later convicted based partly on the evidence gleaned from the polygraph examination. Keeler went on to head the Scientific Crime Laboratory at Northwestern University.

1990: The African National Congress is recognized by the government of South Africa. After the organization operated as an illegal, clandestine organization for 78 years, the ban on the ANC was lifted by President F.W. de Klerk.

February 3

1690: The first paper money in America is issued. The Massachusetts Bay Colony issued the first banknotes to pay soldiers to fight the French in Canada. British laws prevented the Colony from minting its own coins—even going so far as to shut down a mint that issued silver coins—and the Colony issued notes of credit instead. The IOUs could be redeemed for their value in silver.

1897: The First Greco-Turkish War begins. The breakout of hostilities between the Kingdom of Greece and the Ottoman Empire was triggered by a rebellion of the ethnically Greek population of Crete. The citizens of Crete, then a part of the Ottoman Empire, wished to join the independent Greek state. Ottoman troops arrived on the island to quell the uprising, and Greek troops arrived to support it. The Greeks were soundly defeated by the numerically superior, better equipped, and better organized Ottoman Army in a few months.

1918: The Twin Peaks Tunnel opens. The 2.3-mile-long tunnel was the longest streetcar tunnel in the world when it opened. It runs under the Twin Peaks, connecting the Downtown San Francisco, California, to the city's West Portal neighborhood.

1959: The Day the Music Died. Buddy Holly, Richie Valens, and J.P. "The Big Bopper" Richardson were killed when their plane crashed during an ice storm. The three musicians were on the Winter Dance Party tour with Waylon Jennings, Tommy Allsup, and Carl Bunch. The musicians typically took a chartered bus from one venue to the next, but the freezing weather made the rides cold and uncomfortable, with some performers suffering from frostbite and pneumonia as a result. After playing a show in Clear Lake, Iowa, Buddy Holly decided to charter a plane to the next venue, in western Minnesota, rather than take the bus. Jennings gave his seat on the plane to Richardson, who had recently come down with the flu. Allsup and Valens flipped a coin to see who would get the last seat on the plane, and Valens won the toss. The small plane took off late at night in a snowstorm under poor visibility. Soon after takeoff, it crashed into a cornfield, killing everyone aboard.

1971: Frank Serpico is shot during a drug bust in Brooklyn. Serpico, a New York City Police Department (NYPD) officer, blew the whistle on widespread corruption and graft in the NYPD, cooperating with an internal investigation and a *New York Times* story on the subject. He was shot in the face during a raid on an apartment where a drug deal was taking place after his fellow officers left him alone with armed suspects. Serpico survived the shooting and later testified about corruption before the Knapp Commission. He retired and moved to Italy. In 1973 he was portrayed by Al Pacino in the film *Serpico*.

1984: The first successful embryo transfer takes place. A team at the UCLA Medical Center led by Dr. John Buster transferred an embryo from the uterus of one woman to another's. The embryo was conceived by artificial insemination using the egg of the recipient, who gave birth 38 weeks later.

1995: Eileen Collins becomes the first woman to pilot the Space Shuttle. The astronaut flew the space shuttle *Discovery* in orbit to a rendezvous with the Russian space station *Mir*. In 1999 Collins also became the first female Mission Commander of a Shuttle mission.

February 4

1789: George Washington is elected President. The Commander of the Continental Army during the American Revolution was unanimously elected by the Electoral College, and John Adams was elected his vice president. Washington would serve two terms, being unanimously reelected, before declining to run for a third term. While in office, he set a number of precedents for the office.

1797: The Riobamba earthquake happens. The 8.3 magnitude quake destroyed the city of Riobamba and killed 40,000 people. It was the most devastating natural disaster to strike Ecuador.

1859: The Codex Sinaiticus is discovered. The handwritten copy of the Greek Bible was written in the 4th century and discovered by Konstantin von Tischendorf in a monastery at Sinai, Egypt. It is one of only four known copies of the Greek Bible in existence.

1941: The United Service Organization (USO) is created.
The organization was founded by Mary Ingraham, the president of the Young Women's Christian Association (YWCA), at the request of President Franklin D. Roosevelt. The USO brought several civilian organizations together to organize recreation and entertainment for U.S. military personnel. They brought Hollywood celebrities, famous musicians, and entertainers to perform for troops stationed abroad, and continue to do so.

1974: Patty Hearst is kidnapped. Hearst, the 19-year-old heiress to the publishing magnate William Randolph Hearst, was kidnapped by members of the Symbionese Liberation Army, a radical left-wing organization in the San Francisco Bay Area. The SLA initially attempted to pressure the Hearst family into using their political influence to free two of their members who were in prison for murder. When they were unable to do so, the group demanded that they distribute food to needy Californians. Hearst's father donated $2 million worth of food to poor people in the Bay Area, but the SLA refused to release Patty. Hearst was held for nineteen months by the group, who subjected her to torture and intimidation. She eventually was coerced to join the SLA, and was seen participating in a bank robbery with other members of the group. The authorities no longer considered her a kidnapping victim after the robbery, and she was now a fugitive. She was arrested with another SLA member the following year and convicted of bank robbery. Her sentence was commuted by President Jimmy Carter, and she was later pardoned by President Bill Clinton.

2004: Facebook is founded. Founder Mark Zuckerberg initially called the social networking site for students at Harvard University "thefacebook.com" but later changed the name on the advice of entrepreneur Sean Parker. When the site first launched, Zuckerberg was accused of stealing the idea from fellow students Cameron and Tyler Winklevoss and Divya Narendra. They later settled for 1.2 million shares in the company. Facebook went on to become the most popular social networking site in the world, .

February 5

1632: Roger Williams arrives in the Massachusetts Bay Colony. The religious leader would later publicly oppose the Puritan administration's treatment of Native Americans and confiscation of their lands for use by the Colony. The Colony's leaders had him banished from Massachusetts as a result, and Williams resettled, founding the colony of Rhode Island.

1869: The Welcome Stranger gold nugget is found. It was the largest alluvial nugget ever found, weighing more than 214 pounds. The nugget was found in Victoria, Australia, by the prospectors John Deason and Richard Oates. They were paid £9,381 for their discovery.

1901: The modern roller coaster is born. Ed Prescot opened the Loop the Loop Centrifugal Railroad on Coney Island. Safety inspectors banned the use of more than one car at a time, and the low volume of riders led to the Loop the Loop being a commercial failure. It closed in 1910.

1909: Bakelite, the world's first synthetic plastic, is invented. The compound was discovered by Leo Baekeland in Yonkers, New York, while he was experimenting with formaldehyde and phenol in his home laboratory. He was attempting to synthesize a replacement for shellac, a resin used to glaze wood. By experimenting with the pressure and temperature applied to his formaldehyde-phenol compound, he was able to produce a solid, malleable material that he dubbed "Bakelite." He immediately recognized its potential as a material that could be molded into a wide array of useful items, and filed several patents for it in the United States and internationally. Within a year, the General Bakelite Company was producing the material on a large scale. The production of Bakelite led to an explosion in all kinds of molded plastic products, from cigarette holders and costume jewelry, to children's toys and even telephones. Items made with Bakelite plastic later became highly valued as collectibles.

1945: Douglas MacArthur returns to Manila. MacArthur had been forced to declare Manila an open city and flee the Philippines three years earlier as the Imperial Japanese Army overran Allied positions in Bataan. He had famously vowed to return, and did so with the Sixth U.S. Army in 1944. The Allied troops fought their way through the Philippines, recapturing Manila after one of the bloodiest battles of the war; more than 100,000 civilians were killed during the fighting.

1952: "Don't Walk" signs are invented. New York City, in response to a slew of pedestrian fatalities in busy Manhattan streets, began erecting the first automatic "Don't Walk" signs.

1971: Alan Shepard lands Apollo 14 on the moon. It was the eighth manned mission in the Apollo Program, and the third time that astronauts were landing on the moon. They landed in the Fra Mauro formation and made two moonwalks, during which they collected moon rocks, performed scientific experiments, and hit golf balls.

February 6

1815: The first railroad charter is granted. The New Jersey Railroad Company, owned by John Stevens and others, is granted a charter to build a rail line running between New Brunswick, and Trenton, New Jersey. Stevens was the inventor of the first U.S. steam locomotive, the first steam-powered ferry, and the first commercial U.S. ferry service. Stevens' New Jersey Railroad continued to operate until 1867.

1820: The first group of free African Americans leaves for Liberia. The American Colonization Society (ACS) sponsored a group of 88 African Americans to resettle in Africa. The group received a grant of $100,000 from Congress to establish a settlement on the continent's west coast near Sierra Leone. The ACS was not an abolitionist organization, but was opposed to free African Americans living among whites, and argued that they should be repatriated to Africa.

1862: The Union Army captures Fort Henry. The Battle of Fort Henry, Tennessee, was the Union's first victory during the war. General Ulysses S. Grant led a force of 15,000 Union soldiers and seven ironclad gunboats. Following a bombardment of just over an hour, the Fort's commander, General Lloyd Tilghman, surrendered. The victory was a significant morale boost for the Union, and ensured Union control of important waterways in the western theater of the war.

1911: The hood ornament is born. The British automobile manufacturer Rolls-Royce began decorating its cars with the silver-winged "Spirit of Ecstasy" mascot. The hood ornament later became the company's official brand; hood ornaments identifying luxury cars' brands remain popular more than a century later.

1919: The Weimar Republic is established. The German constitutional federal democracy ushered the country through post-World War I economic recovery and rebuilding. It became politically vulnerable during the worldwide depression that occurred in the 1930s, and was overthrown from within by Adolf Hitler.

1952: Elizabeth II is crowned. Elizabeth became the Queen Regnant of the United Kingdom and Commonwealth upon the death of her father, King George VI. She was 26 years old.

1959: The integrated circuit is patented. Jack Kilby, an inventor and electrical engineer working for Texas Instruments (TI), patented the integrated circuit, more commonly known as the computer chip. The integration of multiple transistors into a single circuit made the production of computers far faster, smaller, and cheaper than had previously been possible and revolutionized the world of electronics. The advance made it possible to mass produce computer components for the first time and standardized computer chip design. Integrated circuits are the main component of nearly every piece of electronic equipment, from smartphones to home appliances. He received the 2000 Nobel Prize in Physics for his invention.

1989: The Round Table Talks begin. The negotiations between Lech Walesa, leader of the banned Polish trade union Solidarity, and the Polish government were an attempt to control growing instability in Warsaw. The talks resulted in the legalization of independent trade unions, the formation of a Senate, and the replacement of the office of Communist Party general secretary with a national President. The talks led to the downfall of communism in Poland.

February 7

1497: The Bonfire of the Vanities occurs in Florence. Supporters of Girolamo Savonarola, a Catholic friar who preached a form of asceticism, collected and publicly burned thousands of objects they saw as sinful. The crowd burned cosmetics, mirrors, musical instruments, fine clothing, art objects, and books. Savonarola later fell out of popular favor and was arrested, tried and executed.

1882: Bare-knuckle boxing era ends. The last officially sanctioned world heavyweight bare-knuckle boxing championship was fought in Mississippi between John L. Sullivan and Patty Ryan. Sullivan, known as the "Boston Strong Boy," won the fight. He later went on to become the first heavyweight champion of gloved boxing, and held the title for the next ten years.

1898: Emile Zola goes on trial for his *J'Accuse* letter. Zola, an influential French writer, had published the editorial letter in the newspaper *L'Aurore* in defense of Alfred Dreyfus, the French Army captain convicted of espionage. Zola accused the Army of suppressing evidence that proved Dreyfus was innocent of the charges. Although he was correct in his accusations, he was still found guilty of libel, fined 3,000 francs and sentenced to a year in prison. The term "J'Accuse" has since become associated with the act of standing up to government injustice.

1914: The Little Tramp is born. British actor Charlie Chaplin performs as the Little Tramp for the first time in the silent film

Kid Races at Venice by Keystone Studios. The character, with his mustache, bowler hat, baggy clothes and cane, quickly became one of the most popular ones Keystone had ever produced. The following year Chaplin would write, direct, and star in *The Tramp*, which made the character timeless.

1942: Automakers are ordered to join the war effort. The United States government issued an order to all automobile manufacturers directing them to stop producing passenger cars and switch to manufacturing military hardware, including tanks, jeeps, and munitions. Factories all over the country change production very quickly, turning the nation's industry into a wartime footing. The government made it easier to comply with the order by guaranteeing auto manufacturers' profits regardless of the cost of production throughout the war, and helped fund the investment with nationwide sales of war bonds.

1984: First untethered spacewalk. American astronaut Bruce McCandless exited the Space Shuttle *Challenger* and used the Manned Maneuvering Unit (MMU) to perform the first completely self-contained space walk in history. McCandless designed the MMU, a large jet-powered backpack, and used it to fly outside the shuttle for over five hours on the first spacewalk. A photograph of McCandless floating untethered in space was on the front page of newspapers around the world, and has become one of NASA's best-known images of an astronaut in space.

1992: The Maastricht Treaty is signed. The treaty established the European Union and called for integrating the economies of twelve Western European nations. The treaty directly led to the creation of the Euro in 1995.

2013: Mississippi becomes the last state to officially certify the Thirteenth Amendment, which outlawed slavery. Mississippi had formally ratified the Amendment in 1995.

February 8

1692: Salem searches for witches. A doctor in Salem, Massachusetts, publicly declares that the nine-year-old daughter and eleven-year-old niece of a local pastor are under the influence of "an evil hand." The girls had been having strange episodes in which they screamed, threw objects, crawled around the room and under furniture, and contorted their bodies into bizarre positions. They complained that they felt like they were being pricked by invisible pins. The doctor, William Griggs, was unable to diagnose any physical ailment and declared supernatural causes must be to blame. Soon, Sarah Good, Tituba, and Sarah Osborne were arrested and charged with using witchcraft to torment the girls. The three are believed to have been targeted in part because of their social status: Good was a homeless woman, Titibu was a Native American slave, and Osborne rarely attended church meetings. Accusations of witchcraft became rampant, and the jails were soon filled with suspected witches. One hundred were imprisoned, and by the end of the panic, nineteen people had been executed.

1693: William and Mary College is founded. King William III and Queen Mary II of England granted a charter to the college to educate civil servants and members of the clergy in the colony of Virginia. It is the second oldest institution of higher learning in North America.

1862: Union forces capture Roanoke Island in North Carolina during the American Civil War. This relatively minor military victory would become a pivotal part of the war. By controlling the island, the Union was able to control shipping in the area and directly threaten the Confederate capital of Richmond, Virginia.

1908: The military aircraft is born. Orville and Wilbur Wright signed a contract with the United States Army to build a Model A flying machine for $25,000. The Army added a bonus of an extra $5,000 if the brothers could design the plane to fly faster than 40 miles per hour.

1922: The White House gets a radio. President Warren G. Harding installed the first radio set in the White House. At the time, the radio was cutting edge technology. The White House would not broadcast anything by radio for another two years.

1928: Transatlantic television is born. The first image was broadcast across the Atlantic Ocean, ushering in a new era of telecommunications. The image, a picture of a woman named Mia Howe, was broadcast from London, England, to Hartsdale, New York, by John Logie Baird.

1946: The Revised Standard Version of the Bible is published. The Revised version was the first serious challenge to the predominance of the King James Version of the Bible, which had been the standard for centuries. The Revised version was written in colloquial modern English to make it more accessible to the reader. It was received with some controversy regarding content choices, but ultimately became popular among Catholics and Protestants alike.

1950: The Stasi is founded. The notorious East German secret police force would gain a reputation as one of the most repressive and ruthless state intelligence agencies in history. It built an extensive network of informants, spied on and imprisoned countless East German citizens, and remained in power until the breakup of the Soviet bloc in 1990.

February 9

1861: Jefferson Davis is elected President of the Confederacy. Davis originally wanted to serve as the commander in chief of the Confederate Army, but his fellow Southern lawmakers opted to make him president instead. He remained president until the Confederacy was defeated in 1865, when he was captured and imprisoned in Fortress

Monroe, Virginia. He was released on bail in 1867 pending trial and fled to Canada. He remained a fugitive until President Andrew Johnson issued a general pardon and amnesty for all Confederates.

1895: It's a slam dunk! The first American college basketball game is played between the Minnesota State School of Agriculture and Hamline College. At the time, the game was still played using peach baskets for hoops. The rules were not quite formalized yet: there was no three point line, baskets were worth one point, and any ball that went out of bounds would be given to whichever team could grab it first. Minnesota State won the game, 9 to 3.

1950: McCarthyism begins. United States Senator Joseph McCarthy declared that he had a list of more than two hundred "known communists" who were working in the U.S. State Department during a speech to the Republican Women's Club of Wheeling, West Virginia. The speech was immediately national news, as communist infiltration and espionage were a subject of significant concern at the time. McCarthy, overwhelmed by the media response to his speech, quickly revised his figure to only 57 communists, later changing it to 81 "loyalty risks." The media storm prompted the U.S. Senate to unanimously vote to investigate McCarthy's charges, establishing the Tydings Committee to do so. The Committee eventually found that his charges were entirely baseless, officially calling them a "fraud and a hoax." McCarthy continued to level accusations that the government was not taking the threat of communist infiltrators seriously for several years. Following the Army-McCarthy hearings of 1954, and the suicide of Senator Lester Hunt after McCarthy outed his son as gay, McCarthy's popularity declined. He was censured by the Senate for his actions, and his name became synonymous with leveling baseless accusations.

1964: The Beatles appear on *The Ed Sullivan Show*. They played five songs, including "Love Me Do" and "All My Lovin'" for an in-house audience 728 screaming fans and an estimated television audience of 73 million viewers in 23 million households, or 34% of the American population.

1971: Satchel Paige is inducted into the Baseball Hall of Fame. Paige was the first player from the sport's Negro League to be selected to join the Hall of Fame, having played for such notable Negro League teams as the Kansas City Monarchs and Philadelphia Stars. At age 42 he had become the oldest rookie in Major League Baseball, pitching for the Cleveland Indians. His earned run average of 3.29 remains one of the best in history.

1986: Halley's Comet passes by. The comet passes through the inner Solar System and is visible from Earth to the naked eye once every 74–79 years.

1996: Copernicium, the 112th element on the Periodic Table, is discovered. It was created by the Center for Heavy Ion Research in Darmstadt, Germany, and was named for Nicolaus Copernicus. The element has a half-life of approximately 29 seconds.

February 10

1667: Fossils are discovered to be remains of living organisms. The Royal Society in London publishes the first paper that identifies fossils as having come from living creatures. The paper, titled "Head of a shark dissected," was written by paleontologist Nicolas Steno, and revolutionized the field.

1720: Edmund Halley, the discoverer of the comet that bears his name, is officially appointed the Royal Astronomer of Great Britain. He was the second person to hold the position.

1763: The French and Indian War ends. Known as the Seven Years' War in Britain, the global conflict between France, England, and their respective colonies ended with a British victory. In the Treaty of Paris, France gave up all of its claims on Canada.

1947: "New Look" fashion is born. In Paris, clothing designer Christian Dior unveiled his new line of clothing, which featured longer skirts with close-fitting waistlines and jackets with padded shoulders. The style would become a cornerstone of 1950s fashion in the West.

1962: Gary Powers is released. The pilot of a Central Intelligence Agency U-2 spy plane had been shot down over Soviet airspace two years before while conducting covert reconnaissance. The CIA initially denied that the U-2 was a spy plane, unaware that Powers had been captured alive. After several months of interrogation by the KGB, Powers made a public confession and apology for his part in the mission. He was convicted of espionage and sentenced to ten years in a Soviet labor camp. Powers was exchanged for Soviet spy Rudolf Abel (a.k.a. Vilyam Fisher) at the Glienicke Bridge in Berlin after less than two years. He received the CIA's Intelligence Star after his return. The negotiations leading up to the prisoner swap were the subject of the 2015 movie *Bridge of Spies* starring Tom Hanks.

1996: Checkmate! World chess champion and Grand Master Gary Kasparov is defeated in the first game of a six-game match against Deep Blue, a chess computer designed by IBM. The computer was capable of evaluating 200 million different chess moves per second and was specifically designed to play Kasparov in chess. Deep Blue was the first computer to win a chess match against a reigning chess champion. Kasparov used a Sicilian Defense in its opening, and 36 moves later, realizing he was cornered, he resigned the game. Following the defeat, the media wondered whether artificial intelligence had reached the point where it could finally outmaneuver the best human chess champions. Kasparov was undeterred, however, and went on to win the match with three wins, two draws, and a single loss. He later suggested that in the second game human chess players had intervened on behalf of the computer. IBM denied the allegations, and although Kasparov requested a rematch, the company declined and dismantled Deep Blue. Since the match, chess computing has shifted from developing computer hardware to writing software programs that use heuristics—a kind of problem-solving system—to play chess.

2009: A collision in space. The satellites Iridium 33, a United States civilian communications satellite, and Kosmos 2251, a defunct Russian military satellite, collided in orbit at an altitude of 490 miles, destroying both. It was the first accidental collision between two intact artificial satellites in history.

February 11

1573: Sir Francis Drake sees the Pacific. While on a voyage to the West Indies on a privateering ship, Drake climbed a mountain ridge at the Isthmus of Panama and saw the Pacific Ocean for the first time.

1809: Robert Fulton invents the steamboat ferry. The inventor patented the idea in New York, and two years later developed the first commercial steamboat, the North River Steamboat, to ferry passengers from New York City to Albany and back, a round trip of 300 miles, in just under three days. Fulton would later be commissioned to develop some of the first torpedoes ever used by the British Royal Navy.

1812: Gerrymandering is born. Massachusetts governor Elbridge Gerry oversaw the passage by the state legislature of a new map of state senate districts. Gerry was opposed to the new, highly partisan districting scheme, but signed the bill anyway. A journalist soon caught wind of the arrangement, and compared the shape of one of the districts to that of a salamander, coining the term "Gerry-mander." Gerry lost his reelection race, partly as a result of the negative coverage, and partisan redistricting has been called "gerrymandering" ever since.

1858: Marian apparition at Lourdes. A fourteen-year-old peasant girl named Marie Bernarde-Soubirous claimed she had had a vision of the Virgin Mary in a grotto near the town of

Lourdes in southern France. The grotto would become a major Catholic pilgrimage site, and water from the spring there is widely believed to have miraculous healing properties by devout believers. Marie was later canonized as St. Bernadette.

1928: The recliner is invented. The cousins Edward Knabusch and Edwin Shoemaker designed the chair to provide "nature's way" to relax. Using a yardstick and a sheet of plywood, they created a wood-slat porch chair with a reclining mechanism. They later added upholstery and marketed it as an indoor, year-round chair for relaxing. They held a contest to determine the chair's name, and "La-Z-Boy" was the winner. The company manufactured only La-Z-Boys for forty years, making the brand name synonymous with the recliner, before branching out to other products including reclining sofas and sofa-beds.

1975: Margaret Thatcher is elected leader of the Conservatives. Having beaten four male opponents, she became the first woman in history to lead a British political party. She would be elected the first Prime Minister when the Conservatives took power in 1979.

1990: Nelson Mandela is freed. After spending 27 years in prison for fighting the South African Apartheid government, African National Congress (ANC) leader Nelson Mandela walked out of Victor Verster Prison, holding his wife Winnie's hand and cheered by huge crowds. The event was broadcast live around the world. Mandela had been convicted of sabotage for his leadership of Umkhonto we Sizwe (MK), or "Spear of the Nation," the military arm of the ANC, which carried out acts of sabotage against power plants and government buildings. He had been variously imprisoned in Pollsmor Prison and the notorious Robben Island prison. During his imprisonment, Mandela became an international cause celèbre and the subject of protests worldwide. Upon his release, he stayed at the home of his friend and colleague Bishop Desmond Tutu, and gave a speech to an assembled crowd of over 100,000 people in Johannesburg.

February 12

1502: Spain outlaws Islam. Queen Isabella I issued a proclamation that made it illegal to practice Islam in Spain, forcing nearly all Muslims to convert to Catholicism. Muslims had previously been allowed to practice their religion openly, although they were treated as second-class citizens following the reconquest of the last Spanish caliphate. Isabella's edict started the widespread repression of the religion in the kingdom.

1541: Santiago, Chile, is founded. The Spanish conquistador Pedro de Valdiva marched across the coastal desert of northern Chile with a small force of men from Cuzco, Peru, having been sent by Francisco Pizarro. He named the city after St. James, the patron of Spain.

1865: Dr. Henry Highland Garnet addresses the House of Representatives. Dr. Garnet, a former slave, was the first African American to deliver an address in the United States Congress. His sermon praised Union victories at Gettysburg and other battle sites during the American Civil War and praised the country's deliverance from the evil of slavery.

1924: "Rhapsody in Blue" debuts. George Gershwin debuted the groundbreaking composition when he was only 26 years old. He had written it so hurriedly that at the premiere in New York City, Gershwin had to improvise much of the piano solo.

1986: Anatoly Sharansky is released from prison. The Russian-born Jewish human rights activist had been imprisoned by the Soviet government on charges of high treason. As the leader of the *refuseniks* movement, Sharansky had publicized the treatment of Jews in the Soviet Union, including the government's denial of permission to emigrate to Israel from the USSR and other Eastern bloc countries. He was sentenced to thirteen years of forced labor in 1977. Much of his imprisonment was spent in solitary confinement; Sharansky, who had been a chess prodigy as a child, kept himself sane by playing games of chess with himself in his head. His health deteriorated to the point where he nearly died. In the 1980s, Mikhail Gorbachev's policies of *glasnost* and *perestroika* greatly reformed the government's attitude toward political detainees. Sharansky was the first to be released by Gorbachev following intense pressure from the Reagan administration. He was exchanged with the United States along with three Western spies for several Eastern bloc spies. Sharansky immigrated to Israel, where he was elected to public office as the leader of Ba'Aliyah, a political party he founded. In 1996 he played chess with Grand Master Gary Kasparov, playing him to a draw.

1994: *The Scream* is stolen. Two thieves broke into the National Gallery of Norway and stole the iconic painting by Edvard Munch. The thieves left a note that read "Thanks for the poor security." The thieves were arrested during a sting operation in which British agents posed as underworld buyers, but after being convicted were later released on appeal.

1996: Peace rally in Belfast. More than 2,000 people brought paper doves to a rally for peace outside the Belfast city hall in Northern Ireland. The Irish Republican Army had set off a bomb in London two days earlier, killing two people and breaking a 17-month-old ceasefire with the British government. The rally was a pivotal point in the peace process that brought an end to the decades of violence known as the Troubles.

February 13

1429: Disguised as a man, Joan of Arc sets out across English-occupied France to seek an audience with the French crown prince. In the fateful meeting, she would request his permission to raise an army to fight the English.

1588: Danish astronomer Tycho Brae proposes a theory of the universe that combines the ideas of both Ptolemy and Copernicus. Ptolemy said that all heavenly bodies orbited the Earth; Copernicus argued that they all orbited the sun. In Brae's proposal, the sun and moon orbited the Earth, while the remaining planets orbited the sun.

1929: Antibiotics are discovered. Alexander Fleming, a Scottish bacteriologist, had left a culture of *staphylococci* on a bench in his laboratory while he went on vacation. When he returned, he saw that a fungus was growing in some of the dishes and destroying the colonies. He grew the mold in pure culture and found that it produced a substance that killed several strains of bacteria. He isolated the substance, initially calling it "mold juice" before renaming it penicillin. Fleming's publication of his results in the *Journal of Experimental Pathology* initially did not attract much attention from the scientific community, but his research was continued in the early 1940s by Howard Florey and Ernst Chain in Oxford, England. They began mass producing penicillin, and it was instrumental in saving countless lives during World War II and after. Fleming, Florey, and Chain were jointly awarded the 1945 Nobel Prize in Medicine for their discovery and production of the first antibiotic.

1935: Bruno Hauptmann is convicted of murder. Following a widely publicized trial, Hauptmann was convicted of the kidnapping and murder of Charles Lindbergh's infant son. The Lindbergh Baby Kidnapping, as it became known in the press, was highly sensationalized, with a nationwide search for the boy lasting over a year. Lindbergh paid a ransom of $50,000 through an intermediary, but the child was not returned. After several months, the body of Lindbergh's son was eventually found close to his family home in New Jersey. Hauptmann was arrested because he was found to be in possession of some of the ransom money. Although he protested his innocence, insisting he had received the money as payment from another man, enough evidence was presented to convict him. He was executed the following year.

1960: France becomes a nuclear power. The successful detonation of an atomic bomb, code-named "Gerboise Bleue," in the middle of the Algerian Sahara, was the largest first test bomb by any nation up to that date. With the detonation three times as powerful as the bomb dropped by the United States on Hiroshima, France became the fourth country to go nuclear.

2004: Astronomers discover the largest diamond in the universe. The white dwarf star BPM 37093 is composed of highly pressurized carbon atoms arrayed in a lattice formation. The star wass nicknamed "Lucy" after the Beatles' song "Lucy in the Sky with Diamonds."

2011: The Umatilla tribe hunts bison. For the first time in more than a century, the Native American tribe was allowed to hunt bison outside of Yellowstone National Park, a right that had been guaranteed by the U.S. government in 1855.

February 14

1849: James K. Polk is the first sitting President to be photographed. His picture was taken by Mathew Brady, who would go on to photograph other presidents and Civil War battlefields. Brady's use of a mobile darkroom during the War allowed him to make war photographs available to the public very quickly.

1876: Competing patents are filed for the telephone. Alexander Graham Bell and Elisha Gray both applied for telephone patents. Bell's patent was found to have an identical description of the variable transmitter, a crucial component of the telephone, leading to controversy over who was actually the first to invent the device.

1895: *The Importance of Being Earnest* premieres. Written by Irish playwright Oscar Wilde, it remains one of his most famous works. The Marquis of Queensbury was denied admission to the theater when he arrives to confront Wilde over the relationship the playwright had with his son.

1920: The League of Women Voters is founded in Chicago.
The League was organized to help women obtain a bigger role in public affairs in response to the forthcoming passage of the 19th Amendment, which granted the right to vote to women.

1929: The Saint Valentine's Day Massacre occurs. During Prohibition, George "Bugs" Moran's North Side Gang was fighting for control of bootleg liquor distribution in Chicago with South Side mobster Al Capone. Dressed as policemen, hit men working for Capone marched seven members of the North Side Gang into a garage, made them face the wall, and riddled them with bullets from Thompson sub-machine guns. When police arrived on the scene, they found Frank Gusenberg alive, but bleeding from fourteen bullet wounds. When they asked him who had shot him, Gusenberg replied "no one shot me," and died shortly after. The executions sparked public outrage about mob violence and helped boost Chicago's reputation during Prohibition as a gangland and bootlegging haven. No one was ever charged with the crime.

1989: A fatwa is issued against Salman Rushdie. Rushdie's publication of *The Satanic Verses*, in which the life of the Prophet Muhammed, the Angel Gabriel, and the Koran feature heavily in a tale of magical realism, sparked outrage among devout Muslims. The Ayatollah Khomeini, Supreme Leader of Iran, issued a fatwa calling for the assassination of Rushdie and his publishers. Rushdie went into hiding under the protection of British police. Several persons associated with the book's publishing have been killed or seriously injured in assassination attempts.

2005: YouTube is launched. A group of former employees at PayPal, the online money transfer site, started YouTube as a video-sharing website. One year later, they sold it to Google for $1.65 billion.

February 15

1764: St. Louis is founded. French fur traders Pierre Laclede and Auguste Choteau were commissioned to build a trading post near the confluence of the Mississippi and Missouri rivers by a New Orleans merchant. There, they built a fur trade with Native tribes living along the Missouri River.

1879: Women are allowed before the Court. The United States Congress passes legislation making it legal for female attorneys to argue cases before the Supreme Court.

1898: The U.S.S. *Maine* explodes and sinks in Cuba. The ship was anchored in Havana Harbor, sent there to protect United States interests during a revolt against Spain. While the cause of the explosion remained unclear, the press blamed Spain, and the event became a pivotal reason for the outbreak of the Spanish-American War later the same year.

1903: The first teddy bear goes on sale. The bear was created by Morris Michtom after he saw a political cartoon that showed Teddy Roosevelt refusing to shoot a small bear. The cartoon was based on a real-life event: Roosevelt had been presented with a tied-up black bear during a hunting trip, but refused to shoot it as he felt it was unsportsmanlike. Michtom sent one of his bears to the President and received permission to use his name, calling it "Teddy's bear."

1933: Giuseppe Zangara attempts to assassinate Franklin D. Roosevelt. The President-elect was delivering a speech from the back of an open car in Miami. Zangara, armed with a revolver that he had purchased at a local pawn shop, mixed with the crowd while waiting for a chance to shoot Roosevelt. He was only five feet tall, and could not get a clear view of Roosevelt, so he had to stand on a metal folding chair. The chair was wobbly, and Zangara's first shot went wide of its intended target, hitting Chicago mayor Anton Cermak, standing on the car's running board. Zangara was grabbed and subdued by the crowd as he fired several more shots wildly, hitting four more people. Roosevelt held Cermak in his arms as the car sped to the hospital. As he was taken into the hospital, Cermak reportedly said to Roosevelt "I'm glad it was me instead of you." He died nineteen days later, two days after Roosevelt's inauguration, and his last words to the President-elect were engraved on his tomb. Zangara confessed immediately to the shooting, saying "I kill kings and presidents first and next all capitalists." He was convicted and executed with the electric chair six weeks later.

1946: ENIAC is formally dedicated at the University of Pennsylvania. The Electronic Numerical Integrator and Computer (ENIAC) was one of the first electronic computers ever built. It was a thousand times faster than existing computers, and could perform calculations that took humans 20 hours in just 30 seconds.

2001: The Human Genome Project is published in *Nature*. The fifteen-year project to map the DNA sequence of the entire human genome was a major breakthrough in genetics. At the time it was the largest collaborative biology research project ever attempted by humans. It has since led to an explosion in medical and scientific advances.

2013: A meteor explodes in Russia. The Chelyabinsk meteor entered the Earth's atmosphere at a speed of 42,900 miles per hour and exploded over the Ural region of Russia. It was more brilliant than the sun, and visible up to 60 miles away. The explosion damaged over 7,000 buildings and injured 1,491 people.

February 16

1852: The Studebaker Company is founded. The American brothers and engineers Henry and Clement Studebaker founded a blacksmith and wagon-building business in South Bend, Indiana. The company would go on to become one of the largest independent auto manufacturers in the United States.

1874: The Silver Dollar becomes legal tender. The 1874 Trade Dollar was produced until 1885 and was 90% silver. It was the first silver coin placed into circulation by the United States Treasury.

1933: Prohibition is ended. The Blaine Act was passed by the Senate, initiating the repeal of the 18th Amendment to the Constitution, which had established Prohibition.

1937: Nylon is patented. Wallace Carothers, a researcher at the DuPont corporation, discovered the material during a nine-year research project experimenting with the production of synthetic and cellulose-based fibers. DuPont had previously invented rayon from its experiments with cellulose products. With the invention of nylon, originally called "polymer 6–6" because of its molecular structure, the company had hit upon a product that would revolutionize the fabric industry. In 1940 DuPont began marketing women's nylon stockings. All 4,000 pairs at the first store to carry them sold out in just under three hours. The company eventually hit upon producing a nylon blend to make an affordable, durable, and elastic fabric that continues to be used in a wide range of garments today.

1946: The first commercial helicopter flight. The four-seat, single-rotor helicopter had a range of 250 miles and a top speed of 100 miles per hour. Its creator, Igor Sikorsky, designed multiple helicopters for civilian and military use during his career and almost single-handedly revolutionized the aerospace industry with his work.

1959: Fidel Castro comes to power. Castro, who had led the revolution that brought him to power in Cuba, was sworn in as prime minister a little over a month after driving former dictator Fulgencio Batista from the island.

1978: The Bulletin Board System (BBS) is created. Its creators, Randy Suess and Ward Christensen, were stranded indoors during the Chicago Blizzard of 1978. The storm gave them the time and space they needed to write the program, which allowed users to send packets of information via modems. The BBS was the forerunner of other bulletin board systems on the early internet.

2005: Hockey is locked out! The National Hockey League canceled the entire 2004–2005 regular season and playoffs. The League's owners decided on a lockout after they could not come to an agreement with the players' union. This was the first year that the Stanley Cup was not awarded since 1919, and the first time a labor dispute resulted in a professional sports league missing an entire season since the 1994 Major League Baseball strike.

February 17

1600: Giordano Bruno is burned at the stake. The Italian philosopher and astronomer had developed a theory of an infinite universe. Bruno suggested that the stars were distant suns, like our own, and that they were surrounded by their own planets which themselves might harbor life. He argued that the universe could have no "center." All of these theories went against the doctrines of the Catholic Church, and Bruno was found guilty of heresy and burned at the stake. His ideas have since been found to be largely correct.

1673: Moliere collapses onstage. The French poet and playwright was giving a performance of the last play he ever wrote, *The Imaginary Invalid*. He was carried home, where he died from tuberculosis later that night.

1818: An early prototype of the bicycle is invented. German baron Karl von Drais de Sauerbrun patented the "draisine." It was a two-wheeled contraption with a steering handle and a saddle, but no pedals. Riders would propel themselves along by kicking at the ground.

1863: The Red Cross is founded. After witnessing the Battle of Solferino during the Italian War for Independence, Henry Dunant was stunned by the suffering of wounded soldiers and lack of medical care available to them. When he returned to his home in Geneva, he organized the International Committee for Relief to the Wounded with five colleagues. The committee held an international conference that same year with eighteen delegates from European governments to establish agreements concerning the welfare of wounded soldiers and civilians during wartime. The conference agreed that national volunteer forces to help the wounded would be established and that wounded soldiers should be considered neutral and protected.

1864: Torpedoes in the water! The H.L. *Hunley*, an early submarine developed by the Confederate navy during the U.S. Civil War, becomes the first submarine to engage and sink a warship. Its target, the sloop of war USS *Housatonic*, was anchored in Charleston harbor; five Union sailors were killed in the attack.

1933: *Newsweek* is first published. The magazine was founded by Thomas J.C. Martyn, formerly the foreign news editor at *TIME* magazine. The first issue featured stories about the foreclosure crisis hitting farmers in the American Midwest, Herbert Hoover's final address, and a speech by Adolf Hitler in Germany.

1959: The first weather satellite is launched. The *Vanguard 2* was designed to measure cloud-cover distribution for a period of nineteen days. It was a major advance by the United States in the ongoing space race with the Soviet Union. After the mission was completed, the satellite remained in orbit for decades, becoming one of the first pieces of space junk in low earth orbit.

1980: First winter ascent of Mount Everest. Climbers Krysztof Wielicki and Leszeck Cichy successfully climbed the 29,000-foot peak in the middle of winter.

2008: Kosovo gains independence. The country was one of the last ones to become independent as a result of the breakup of Yugoslavia. The tiny nation had been fighting for independence for nearly two decades.

February 18

1678: *Pilgrim's Progress* is published. John Bunyan's book would go on to become one of the most frequently reprinted book in the English language, second only to the Bible. The book is one of the most significant religiously-themed works in English literature, and is considered the first novel printed in English.

1878: The Lincoln County War begins. Outlaw Jesse Evans murdered John Tunstall, a rancher in Lincoln County, New Mexico, during a dispute over who would control dry goods distribution and cattle interests in the territory. The War lasted for five months and is remembered for the involvement of William Bonney, better known as Billy the Kid, and his posse of Regulators.

1943: The White Rose activists are arrested. The group was organized to nonviolently resist the rise of the Nazis. They distributed anti-Nazi pamphlets and painted anti-Hitler graffiti in and around Munich, Germany. They were sentenced to death and executed by guillotine.

1948: Eamon de Valera resigns as Prime Minister of Ireland. Originally from New York, New York, de Valera became one of the leaders of the 1916 Easter Rising, the War for Irish Independence, and the Anti-Treaty Forces during the Irish Civil War.

1954: Scientology opens its first Church. The Church of Scientology, founded by science fiction writer L. Ron Hubbard in 1953, would eventually grow to have about 8 million members worldwide, among them famous Hollywood celebrities such as Tom Cruise.

1970: The Chicago Seven are acquitted. The seven defendants—Rennie Davis, David Dellinger, John Froines, Tom Hayden, Abbie Hoffman, Jerry Rubin, and Lee Weiner— had been charged with conspiracy and inciting a riot in connection to antiwar protests that took place during the 1968 Democratic National Convention in Chicago.

1977: Make it so! NASA flight tested the first Space Shuttle, named the *Enterprise*, by flying it attached to a 747 jumbo jet. The spacecraft would go on four more test flights attached to the 747 before it was released for its first solo landing.

2001: Dale Earnhardt is killed. The legendary stock car racer was one of the most famous members of the NASCAR circuit, winning seven Winston Cup championships, including the 1998 Daytona 500. Earnhardt began his professional NASCAR career in 1975 and soon rose to prominence as one of the best drivers on the circuit. He won his first Winston Cup championship in 1980. He was one of the dominant competitors throughout the 1980s and 1990s, repeatedly leading NASCAR in wins and receiving many awards for his abilities. He was also wildly popular among fans, and was the face of the sport as it went from a relatively small local audience to one that had national prominence. During the final lap of the 2001 Daytona 500, Earnhardt's car collided with Ken Schrader's after Earnhardt made contact with Sterling Marlin. His car hit the outside wall head-on, killing Earnhardt instantly. His son, Dale Earnhardt Jr., took second place in the race just seconds later. GM Goodwrench retired his stock car number (3), and NASCAR placed a moratorium on other drivers using the number.

February 19

1600: The stratovolcano Huaynaputina erupts in Peru. The eruption was the largest in the known history of South America, and covered the surrounding area in a layer of volcanic ash ten inches deep. Flowing mixtures of mud and lava reached the Pacific Ocean, 75 miles away, and destroyed whole villages in their path. More than 1500 people were killed and ten villages were buried in volcanic ash.

1601: Henry Wriothesley goes on trial for treason. The Third Earl of Southampton was a patron of William Shakespeare. He had convinced the Globe Theatre to revive the Bard's play *Richard II*, which described the events leading up to the monarch's overthrow. The play was shown on the eve of a real-life rebellion in Essex that the Earl had taken part in. He was thrown in jail but later pardoned by James I.

1807: Aaron Burr is arrested for treason. The Vice President of the United States from 1801–1805 had conspired with a group of planters, Army officers, and politicians to create an independent nation in the American Southwest and Mexico. Several letters Burr had sent to other conspirators were shown as evidence of the plot. The trial was presided over by Chief Justice John Marshall, who narrowly defined the definition of treason, and the jury acquitted Burr. He left for Europe in disgrace.

1847: Rescuers finally reach the Donner Party. The group of American pioneers had become snowbound in the Sierra mountains the previous winter while trying to reach California. While trapped, they had been forced to resort to cannibalism to survive. Of the 83 pioneers who had started the journey to California, only 45 survived and were rescued.

1942: Franklin D. Roosevelt signs executive order 9066. Following the attack on Pearl Harbor, the U.S. military was concerned that the Japanese Empire was preparing to invade the West Coast of the United States. While public opinion initially considered Japanese-Americans to be staunchly loyal citizens, a letter written by the Joint Immigration Committee of the California Legislature and published in California newspapers soon changed that. Nationalist groups such as the Native Sons and Daughters of the Golden West and the American Legion began demanding that Japanese-Americans be placed in internment camps. Roosevelt's Executive Order 9066 authorized the military to create "exclusion zones" from which they could deport anyone at their discretion, but was clearly written to target Americans of Japanese ancestry. Over 110,000 Japanese-Americans, 62% of whom were American citizens, were eventually sent to internment camps. Many of them lost most of their personal property, homes, and farms when they were interned.

1963: *The Feminine Mystique* is published. The book, written by Betty Friedan, helped spark the second-wave feminist movement in the United States. Friedan interviewed suburban housewives and found that many were unhappy despite living in relative material comfort. She concluded that the restrictions placed on domestic life were the cause. The book was widely read and brought many white, middle-class women to the feminist movement for the first time. Friedan later helped found the National Organization for Women.

1985: William J. Schroeder leaves the hospital with an artificial heart. He was the first recipient of an artificial heart who was able to leave the hospital. After less than three weeks, he suffered a series of strokes and died in August 1987. At the time he had been the longest living person with an artificial heart.

February 20

1792: President George Washington signs the Postal Service Act. The legislation established the United States Post Office Department. The idea for the postal service was originally conceived by William Goddard, who was frustrated that his newspaper, the *Pennsylvania Chronicle*, was not delivered to readers by the colonial postal services. Benjamin Franklin took up Goddard's idea and became the first Postmaster General in 1775. The Act was passed to standardize routes and prices.

1872: The Met opens in NYC. The Metropolitan Museum of Art is the largest art museum in the United States, and after the Louvre, the second most-visited art museum in the world with over 7 million visitors each year. It was established by legislation passed by the state legislature and funded by artists, philanthropists and industrialists. The opening reception was in a picture gallery at 681 Fifth Avenue, and featured a Roman stone sarcophagus and 174 paintings, mostly from European painters. Luigi Palma de Cesnola, a former Union officer in the Civil War, was the Museum's first director. He oversaw the acquisition of many new paintings and artifacts, and the Cesnola collection of antiquities from Cyprus is named for him. The Museum soon outgrew its location and eventually moved to its current location at 1000 Fifth Avenue. It features art, armor, furniture, and sculpture from around the world displayed in galleries that take up a total of over half a million square feet—far larger than its humble beginnings!

1877: *Swan Lake* premieres in Moscow. The ballet, composed by Peter Tchaikovsky over a period of two years, was initially poorly received by audiences and critics alike. The production suffered from a number of replacements of dancers, and the musical score was overshadowed by the poor performance. Despite its inauspicious beginnings, it would go on to become one of the most celebrated ballets ever written.

1935: Caroline Mikkelsen arrives in Antarctica. Mikkelsen was a Norwegian explorer and became the first woman to set foot on Antarctica when she arrived at a coastal island there while accompanying her husband on an expedition searching for new land that could be annexed by Norway.

1942: Edward O'Hare becomes the first American flying ace in World War II. A Lieutenant Commander in the United States Navy, O'Hare was flying a patrol mission when he saw a formation of Japanese bombers approaching his aircraft carrier group. He attacked them single-handedly, shooting down or damaging several and preventing the attack.

1962: John Glenn orbits the earth. Glenn was the first American and the second human astronaut in space (Yuri Gagarin of the Soviet Union being first). He made three orbits of the Earth in just under five hours, reaching speeds of more than 17,000 miles per hour.

1986: *Mir* is launched. It was the first modular space station, and remained in orbit until 2001. Mir, the largest artificial satellite in orbit, housed an aerospace research laboratory, docking module, Earth research module, flight control, and living quarters. It was replaced by the International Space Station in 1998.

1977: Life is found at the bottom of the ocean. Scientists working near the coast of the Galapagos Islands discovered a hot water geothermal vent that was teeming with life. Far too deep to be reached by sunlight, the thriving ecosystem of worms, crabs, and other organisms revolutionized the perception of where life could exist. Scientists later determined that the ecosystem was dependent on a species of bacteria that used the hydrogen sulfide spewing from the vents.

February 21

1848: Karl Marx and Friedrich Engels publish *The Communist Manifesto*. The revolutionary pamphlet was one of the most influential political documents in history. In 1847 Marx was commissioned by the Communist League of London to write a manifesto summarizing the League's formative principles. He procrastinated for two months until the League sent him a firm deadline, and then he wrote the entire pamphlet in just under three weeks. The *Manifesto* was distributed through Europe during the revolutions of 1848, but following their defeats it fell into obscurity. It became popular again during the time of the 1871 Paris Commune, a radical socialist government that took power in France for three months.

Over the next few decades, hundreds of editions were published in thirty languages. When the Bolsheviks seized power in Russia in 1917, establishing the world's first socialist state, the *Manifesto* became a major feature of Soviet State ideology. It continued to influence revolutionaries the world over throughout the twentieth century and beyond.

1885: The Washington Monument is dedicated. The obelisk was erected to honor the first President of the United States, and stands 555 feet, 5 1/8 inches tall on the National Mall in Washington, D.C. At the time of its dedication it was the tallest such column in the world.

1916: The Battle of Verdun begins. The battle raged over nine months in northern France, and over 300,000 soldiers were killed. It was the longest and largest battle of World War I. The German forces initially captured strategic positions near Verdun, but were eventually pushed back by the French, leaving the status of the Western Front largely the same as it had been before the battle began.

1925: The first issue of *The New Yorker* is published. It was originally published as a humorous, cosmopolitan magazine, but later shifted its focus to serious journalism and literary fiction pieces.

1947: The Polaroid Land Camera is unveiled. The camera, named for its creator Edwin Land, was the first to produce "instant" photographs, which developed in about one minute. It was produced by Polaroid until 1983.

1965: Malcolm X is assassinated in New York City. The revolutionary African-American civil rights leader and Muslim minister was gunned down as he delivered a speech at the Audubon Ballroom. He had become disillusioned with the Nation of Islam and its leader Elijah Muhammad the previous year.

1995: Steve Fossett completes the first solo flight across the Pacific in a balloon. Fossett, an American member of the Explorers Club and Fellow of the Royal Geographic Society, had set off from Korea thirteen days before landing in Saskatchewan, Canada. The explorer disappeared in 2007 while flying a light aircraft over the Great Basin Desert in Nevada. Remains of the plane's wreckage were found the following year.

February 22

1371: Robert II is crowned King of Scotland. Robert's father-in-law was Robert the Bruce. Robert II established the royal House of Stewart and resumed the Scottish wars for independence. He ruled Scotland until his death in 1390.

1632: Galileo publishes his *Dialogue*. The book compared the Copernican theory of the universe, in which the sun was at the center, with the traditional Ptolemaic version, in which the Earth is at the center. The work was placed in the Inquisition's *Index of Forbidden Books*, and Galileo was tried for heresy for writing it.

1819: The Adams-Onìs Treaty is signed. Under the terms of the Treaty, the United States acquired Florida from Spain for five million U.S. dollars. Spain could no longer afford the cost of troops required to hold the territory or the settlers required to establish it.

1856: The Republican Party holds its first convention in Pittsburgh. The American political party was founded by Northern anti-slavery activists, many of whom were former members of the Free Soil and Whig parties. It elected John Fremont to run in the 1856 presidential election under the slogan "free labor, free land, free men."

1879: Frank Woolworth opens his first five-and-dime store in Utica, New York. Woolworth's would grow to be an international business and established the precedents for the modern retail store. It remained a fixture of American retail for much of the twentieth century, before declining in the 1980s and going out of business in 1997, its niche largely replaced by Wal-Mart.

1915: The Imperial German Navy begins the U-boat campaign in the Atlantic. The strategy called for unrestricted attacks in the Atlantic on Great Britain and its allies during World War I. German U-boats sank dozens of merchant and other civilian ships without warning. The campaign culminated with the sinking of the RMS *Lusitania*, a luxury passenger liner, causing international outrage and helping draw the United States into the war.

1980: Miracle on Ice. In the 1980 Winter Olympics in Lake Placid, New York, the United States Men's hockey team defeated the heavily-favored Soviet Union team 4–3. The upset victory was achieved by a team of amateur, mostly college-age players who took on a dominant team of professional hockey players that had won gold medals at six of the previous seven Winter Olympics. After counting down the final seconds of the game live on air, ABC television sports reporter Al Michaels famously quipped, "Do you believe in miracles? Yes!" The United States team went on to defeat Finland 4–2 and win the gold medal.

1994: Aldrich Ames is charged with spying for the Soviet Union and Russia. Ames, an officer in the Central Intelligence Agency (CIA), began selling secrets to contacts in the Soviet embassy in 1985. Over the next nine years he was paid $4.6 million for intelligence by the Soviets and later Russian Federation. Ames came under suspicion because of his lavish spending beyond what his salary could support. He was sentenced to life without parole.

February 23

1554: Mapuche warriors win. Mapuche warriors commanded by Lautaro overwhelm a force of Spanish conquistadors led by Francisco de Villagra at the Battle of Marihueñu in southern Chile during the Arauco War.

1778: Baron von Steuben arrives at Valley Forge. The Prussian military officer was recruited by Benjamin Franklin to help train and lead the newly formed Continental Army in the American War for Independence. Under his guidance the Army was transformed from a militia that relied mostly on ambush tactics into a professional fighting force capable of meeting and defeating the British on the field of battle.

1836: The Siege of the Alamo begins. A Mexican army 1,800 strong commanded by General Santa Anna arrived in San Antonio, Texas, and surrounded the small Alamo Mission. The Alamo was defended by a force of 156 soldiers commanded by William Barret Travis and James Bowie. Santa Anna offered the defenders a chance to surrender, and Travis replied by opening fire on the Mexican soldiers. Thirteen days later, the Mexican army launched an assault on the Mission that killed nearly all of the defenders.

1886: Charles and Julia Hall produce synthetic aluminum for the first time. The brother-sister team set up a coal-fired furnace behind their family home and began performing experiments to try to find a compound that could allow them to produce aluminum. They eventually hit upon passing electricity through a bath of dissolved aluminum oxide, which produced aluminum from the solution.

1917: The February Revolution begins in Russia. It was the first of two revolutions in Russia that year. The February Revolution led to the abdication of Tsar Nicholas II and the establishment of a Provisional Government, while the second, in October, swept Vladimir Lenin and the Bolsheviks into power.

1941: Glenn Seaborg discovers plutonium. He produced the element by bombarding uranium with neutrons. The discovery was a crucial step in the Manhattan Project's research into the development of nuclear weapons. Seaborg went on to discover nine more elements, one of which was named seaborgium.

1945: United States Marines raise the American flag on Mount Suribachi. The mountain was the highest point on the island of Iwo Jima, where Marines fought one of the fiercest battles in the Pacific Theater during World War II. During the bloody five-week battle, 110,000 Marines attacked a force of 21,000 Imperial Japanese Army that was firmly dug into a series of caves and tunnels on the island. The Marines suffered over 26,000 casualties, including 6,281 killed. Over 18,300 Japanese soldiers were killed. Joe Rosenthal photographed Marines raising the flag, and the image became one of the most iconic images of the battle. It became the only photograph to win a Pulitzer Prize for Photography the same year that it was published. It was later used by the sculptor Felix de Weldon in designing the Marine Corps War Memorial near Arlington National Cemetery. Three of the six Marines in the photograph were killed in battle over the next few days.

1954: The polio vaccine is administered to the first group of children. Jonas Salk's discovery stopped one of the worst public health problems in the world at the time, and rather than patent the vaccine, he chose to make it publicly available. Its use led to the disease going from causing annual epidemics to being almost completely eradicated worldwide.

February 24

1582: Pope Gregory XIII introduces the Gregorian calendar. The calendar was an update of the Julian calendar, which had been used for the past 15 centuries. The Gregorian calendar corrected the length of the year by just two thousandths of a single percentage, but doing so stopped the calendar from drifting around the solstices and equinoxes as the Julian one had. The calendar drift had resulted in the date of Easter, which fell after the vernal equinox, moving away from the spring date the early Catholic Church had established it on. The Gregorian calendar reduced the length of the year by 10 minutes 48 seconds, and also made every year that is exactly divisible by four into a leap year, except centurial years, which must be divisible by 400 to be a leap year. The Gregorian calendar was initially only adopted by predominantly Catholic countries in Europe, but Eastern Orthodox and Protestant countries followed suit over the next several centuries. In 1923, Greece became the last country to adopt the Gregorian calendar.

1607: The first opera premieres in Italy. *La favola d'Orfeo* was written by Claudio Monteverdi and Alessandro Striggio. It was based on the Greek myth of Orpheus descending into Hades to search for his bride Eurydice. It was performed at the court of the Duke of Mantua.

1809: Drury Lane Theatre burns down in London. It was the third incarnation of the theater, having been expanded in 1794. The theater's owner, R.B. Sheridan, was discovered on the street drinking a glass of wine while watching the theater burn. He famously said "A man may surely be allowed to take a glass of wine by his own fireside." Sheridan was financially ruined by the loss and resigned from the theater in 1811.

1854: The first perforated postage stamp is issued. The Penny Red was a British postage stamp initially issued in 1841 without perforations; stamps had to be cut from a sheet with scissors. Perforations were experimented with beginning in 1850 and came into use soon after. Originally costing a penny apiece, they are now collector's items worth hundreds of thousands of pounds sterling.

1868: Andrew Johnson is the first American President to be impeached. The U.S. House of Representatives voted to impeach Johnson because of his removal of Edward Stanton, the Secretary of War, and his attempt to replace him with Lorenzo Thomas. A Senate trial later acquitted Johnson with a vote of 35 to 19.

1920: The Nazi Party is founded. The party was founded in Munich by Adolf Hitler, Herman Göring, Anton Drexler, and Dietrich Eckhart. The fascist, anti-Semitic organization would eventually rise to power in Germany and enormously affect world events in the 20th century. By the end of World War II, all four of its founders would be dead.

1942: The MV *Struma* is torpedoed by a Soviet submarine. The ship was trying to take 781 Jewish refugees fleeing Nazi-allied Romania to Palestine for resettlement. It broke down outside Istanbul in December 1941 and was quarantined outside the city harbor. The Soviet submarine *Shch-213* sank the ship with torpedoes, killing all but one of the passengers and crew.

2008: Fidel Castro retires as President of Cuba. Castro had ruled the communist island since his guerilla army seized Havana nearly fifty years prior. He remained head of the Cuban Communist Party for three more years.

February 25

1336: The defenders of Pilėnai choose death over surrender. The fort was the site of a last stand by the pagan Grand Duchy of Lithuania against a crusade by the Teutonic Knights, who were attempting to convert him and his people to Catholicism. When it became clear the fort would fall, all four thousand people within committed mass suicide rather than be taken prisoner.

1848: The Paris Commune guarantees workers' rights. The short-lived revolutionary socialist government was the first in modern history to formally recognize the rights of labor.

1866: The Calaveras Skull is discovered. Miners working in California found the skull in mine 130 feet below the surface, and passed it on to Josiah Whitney, a Professor of Geology at Harvard University. Whitney declared the skull was proof that humans, mastodons, and elephants had coexisted in California. The skull was later shown to be a hoax.

1901: U.S. Steel is incorporated. Industrialist J.P. Morgan combined the Carnegie Steel Company, the Federal Steel Company, and the National Steel Company. For a period during the 20th century, it was the largest steel producer and corporation in the world. It was downsized considerably in the 1980s.

1928: The first television broadcast license is issued. The Federal Communications Commission granted a broadcast license for television to Charles Jenkins Laboratories.

1956: Khrushchev's secret speech. In a speech to the 20th Congress of the Communist Party of the USSR, Nikita Khrushchev denounced the cult of personality that Stalin had created, as well as purges he had carried out.

1964: Muhammad Ali shocks the world. The heavyweight boxer, still called Cassius Clay at the time, won a bout against Sonny Liston with a technical knock-out in the seventh round. Liston, who had beaten former heavyweight champ Floyd Patterson in the first round, was an eight-to-one favorite to win the fight. Ali had predicted his victory in interviews before the fight, declaring he would "float like a butterfly, sting like a bee." After six rounds of pummeling from Ali, Liston refused to come out of his corner for the seventh-round  bell. The new heavyweight champion of the world shouted to the press "I shocked the world!" He would go on to be one of the greatest heavyweight boxers in history, amassing a record of 56 wins, 37 of which were by knockout, and 5 losses. He also became an influential Civil Rights figure, military draft resistor, and humanitarian activist.

1986: Ferdinand Marcos flees the Philippines. The People Power Revolution, a series of nonviolent mass demonstrations in Manila, drove Marcos out of power after 20 years of iron-fisted rule and led to the restoration of democracy in the Philippines.

1991: The Warsaw Pact is dissolved. The Pact was a defense treaty signed in Warsaw, Poland, between the Soviet Union and seven states in Eastern and Central Europe during the Cold War. It was designed to counterbalance the North Atlantic Treaty Organization. The Warsaw Pact's biggest military conflict was the invasion of Czechoslovakia during the uprising of 1968.

February 26

1266: Charles defeats Manfred in battle. Charles, the Count of Anjou, led an army of German and Sicilian fighters against the Manfred, King of Sicily, in Benevento, Italy. Charles' army routed Manfred's, and the Sicilian king was killed in the battle. Pope Clement IV proclaimed Charles King of Sicily and Naples. He ruled until his death in 1285.

1815: Napoleon Bonaparte escapes from Elba. He had been exiled to the island, 12 miles off the coast of Tuscany, by the Allies in the Treaty of Fontainebleau. Napoleon initially attempted to commit suicide with a pill, but it was not strong enough to kill him. While he was there, he heard rumors that he was going to be banished to an island far out in the Atlantic Ocean. He escaped with 700 supporters and sailed to the French Mainland. When they were met on the road by a regiment of French troops, Napoleon famously declared "Here I am: kill your emperor, if you wish!" The soldiers cheered him instead, and Napoleon marched to Paris at the head of an army. He ruled France for just over three months, in a period known as the Hundred Days. The other European Allies declared him an outlaw and went to war. They defeated him at the Battle of Waterloo, and Napoleon abdicated the throne soon after.

1878: The word "microbe" is coined. The French lexicographer Paul Emile Littrè proposed the neologism to Louis Pasteur to describe the organisms Pasteur was observing under the microscope at the time. The word soon caught on and quickly became the general descriptor for microscopic organisms.

1909: The first color movie is shown. Kinemacolor was a process developed by George Albert Smith in Brighton, England, to develop colorized motion pictures. Black-and-white film was projected behind alternating red and green filters. Smith showed 21 short films in Kinemacolor to a crowd at the Palace Theatre in London.

1935: Hitler violates the Versailles Treaty. The treaty had specifically prohibited the development of military aviation by Germany. Hitler signed a secret order that established the German Reich's *Luftwaffe*, or air force. He placed Herman Göring, one of the founders of the Nazi Party and a World War I flying ace, in charge of the air force.

1936: Attempted coup d'état in Japan. A group of young officers in the Japanese Imperial Army attempted to assassinate Prime Minister Keisuke Okada and seize control of the government. They were unable to find Okada or take the Imperial Palace and surrendered soon after. Nineteen of the ringleaders were executed and 40 others were imprisoned.

1966: The Saturn rocket lifts off. The first flight of the Saturn IB launch vehicle lifted off Pad 34 at the Kennedy Space Center and flew to a height of 260 nautical miles before the Command Service Module separated and continued its ascent alone. The rocket splashed down 37 minutes after launch and was recovered by the U.S. Navy.

1971: Earth Day is created. U Thant, the Secretary-General of the United Nations, signed a proclamation making the vernal equinox international Earth Day.

1993: The World Trade Center is bombed. A truck packed with 1,336 pounds of explosives blew up in the parking garage of the North Tower of the World Trade Center. The bomb was intended to bring down both towers but failed to do so. Six people were killed in the blast and hundreds more were injured.

February 27

1594: Henry IV is crowned King of France. Known as "Good King Henry," he was the first king in the Royal House of Bourbon. The Catholic signed the Edict of Nantes, which granted substantial rights to Protestant Huguenots. Henry remained on the throne until his death in 1610.

1776: The North Carolina Militia defeats Loyalist forces during the American War for Independence. The victory helped foster public support for the cause of Independence and the Militia increased its recruitment of volunteers substantially as a result.

1801: Washington, D.C. is placed under the jurisdiction of the U.S. Congress. The Congress passed the District of Columbia Organic Act to formally organize the unincorporated territory in the District. As a result of the Act, residents of Washington, D.C., are not represented in the Senate and have only a non-voting Representative in the House, although they do vote in Presidential elections.

1844: The Dominican Republic declares independence from Haiti. Juan Pablo Duarte led a secret group called the *Trinitarios*. They organized backing from wealthy cattle ranchers and mounted an insurrection, seizing strategic locations in Santo Domingo. Haitian soldiers in the capital fled, and Duarte declared a provisional government.

1864: The first Union prisoners arrive at Andersonville prison. The infamous prison in southern Georgia would quickly become overcrowded with desperate, starving men. Disease spread rapidly due to unsanitary conditions, and of the roughly 45,000 Union prisoners held there, some 13,000 died from scurvy, diarrhea, and dysentery. The camp's commander, Henry Wirz, was the only Confederate official executed for war crimes following the war.

1951: U.S. Presidents are limited to two terms by law. The 22nd Amendment to the United States Constitution was ratified to limit Presidential terms. Before the administration of Franklin D. Roosevelt, presidents declined to seek a third term out of deference to a tradition established by George Washington. Washington had not run for a third term due to his old age, which he stated in his Farewell Address in 1796. The importance of limiting Presidential terms was acknowledged early in the nation's history: Thomas Jefferson, the third President, declined to seek a third term and wrote of the need for a Constitutional Amendment fixing term limits. The four presidents who followed him also served two terms only. Ulysses S. Grant sought nomination to a third term at the 1880 Republican National Convention, but lost to James Garfield. While the nation was in the midst of World War II, Roosevelt ran for and won an unprecedented four terms as President, dying in office in 1945. Although the Amendment did not apply to Harry S. Truman, he declined to seek a third term.

1991: George H.W. Bush announces "Kuwait is liberated." After less than a month of ground combat operations code-named Operation Desert Storm, and an intense aerial bombardment, the United States military and a coalition of its allies successfully expelled the Iraqi Army from the oil-rich Gulf State.

2015: Boris Nemstov is assassinated. The Russian politician was an opponent of President Vladimir Putin who spoke out against the 2015 invasion of the Ukraine. An unidentified assassin fired eight shots at Nemstov, killing him instantly. Five Chechen men were later found guilty of the murder after having agreed to kill Nemstov for hire by persons they were unable to identify.

February 28

1528: Hernan Cortes has Aztec king Cuauhtémoc executed. Cuauhtémoc had ascended the throne while the Aztec capital city, Tenochtitlan, was under siege by Cortes; conquistadors. He was captured and tortured to reveal the location of gold, but did not do so. Cortes initially allowed him to live, but later had him killed, fearing that Cuauhtémoc would attempt to kill him.

1784: John Wesley creates the Methodist Church. Wesley was an English Anglican theologian who preached his ministry outdoors. He believed that Christians could attain a state of grace by accepting God's love into their hearts, and was one of the first Protestant ministers to preach the concept of a personal relationship with Jesus Christ.

1827: The first combined passenger and freight railroad is incorporated. The Baltimore and Ohio (B&O) Railroad was founded to compete with the Erie Canal, which had opened ten years earlier and was drawing business away from the Port of Baltimore. The B&O was the first railroad to offer both passenger and freight service. It continued to operate until 1987.

1844: Disaster on the Potomac. A gun on the USS *Princeton* exploded, killing six people. The Secretary of the Navy Thomas Walker Gilmer and Secretary of State Abel P. Upshur were among the dead. Although President John Tyler was aboard, he was not injured.

1849: Steamboat service from New York to San Francisco begins. The SS *California* arrived in San Francisco harbor four months and 22 days after leaving New York Harbor. During the journey, word of the California Gold Rush spread, and the steamboat picked up many passengers on the trip. The *California* continued to operate until it sank in 1895.

1885: AT&T is incorporated. The American Telephone & Telegraph Company was established as a subsidiary to the American Bell Telephone company. AT&T initially established a long-distance telephone network from New York to Chicago before distributing it nationwide. It held a monopoly on telephone service in the United States for most of the 20th century, until it was broken up in 1974 in an antitrust effort.

1954: The first color TVs go on sale. The Westinghouse H840CK15 was the first television sold to the public that used National Television System Committee (NTSC) color coding. It was priced at the then-expensive $1,295. The more inexpensive RCA CT-100 went on sale two months later.

1983: The final episode of *M*A*S*H** airs. The two-hour episode, titled "Goodbye, Farewell and Amen," was the series' 251st and was directed by series star Alan Alda. It had a total audience of 125 million viewers, making it the highest-rated television series finale ever broadcast. It remained the highest television viewership until 2010, when it was surpassed by Super Bowl XLIV. *M*A*S*H** was developed from the 1970 feature film of the same name and aired from 1972–1983. The show followed the exploits of a team of doctors at a "Mobile Army Surgical Hospital" during the Korean War. While the show was a situation comedy, it occasionally took on dark or dramatic subject matter, and like the movie was an allegory for the Vietnam War, which was still ongoing when the show premiered.

February 29

1504: Columbus tricks Jamaican natives. While he was marooned in Jamaica with a crew that was threatening to mutiny over lack of supplies, Christopher Columbus used his knowledge of a predicted lunar eclipse to terrorize natives into providing them with food. He told them that if they did not give his men the supplies they needed he would cause the moon to lose its light. When the eclipse occurred as predicted, the frightened indigenous people gave him what he needed.

1704: The Deerfield Raid on the Massachusetts Bay Colony. In one of the bloodiest battles of Queen Anne's War, fought between England and France, a party of French and Native American forces attacked the Deerfield settlement just before dawn. They destroyed roughly 40 percent of the village's buildings, killed 47 villagers, and took more than one hundred captives. The raid sent shock waves through New England, and prompted frontier settlements to become fortified and heavily armed camps. The story of the raid was later publicized in *The Redeemed Captive*, an account by the Puritan minister John Williams. He and his family were among the prisoners the French and Mohawk soldiers took with them into Canada.

William's wife was killed along the way and his daughter Eunice was adopted by a Mohawk family. The account was published in 1707 and widely read throughout the colonies. It influenced later works including James Fenimore Cooper's book *The Last of the Mohicans*.

1796: The Jay Treaty is signed. Named for its chief American negotiator, John Jay, the treaty eased relations between the United States and Great Britain for the next ten years. Americans were intensely divided over the treaty, and France, which had supported the American cause of independence and was at war with England, was enraged.

1864: The Kilpatrick-Dahlgren raid is thwarted. The Union Army in the American Civil war sent a massive cavalry contingent of over 3,500 riders south into Virginia. They were tasked with the goal of freeing prisoners of war, publicizing the Union offer of amnesty to any Confederate soldier who unconditionally surrendered, and killing Confederate president Jefferson Davis. The raid was bungled and 335 Union troops were killed before they could make it back to federal lines.

1916: South Carolina passes child labor laws. The legislation provided for some of the first legal protection of child laborers, and raised the minimum age for factory, mill, and mine workers from twelve to fourteen years.

1940: Hattie McDaniel becomes the first African American to win an Oscar. McDaniel had portrayed the house slave "Mammy" in the film *Gone with the Wind*. She won the Academy Award for Best Supporting Actress for the role.

1988: Desmond Tutu is arrested. The South African Archbishop of Cape Town was participating in a five-day series of demonstrations against the apartheid regime. During a march toward Parliament, he led the demonstration in a recitation of the Lord's Prayer while police sprayed them with high-pressure hoses.

Chapter 3

March

March 1

1360: Geoffrey Chaucer is ransomed. During the Hundred Years' War, the young poet took part in the invasion of France led by King Edward III. He was captured during the siege of Rheims, and Edward paid sixteen pounds, a large sum at the time, for his release.

1565: Rio de Janeiro is founded. The city was established by Portuguese colonists and named after the large bay, which they had called the *River of January*.

1781: The Continental Congress adopts the Articles of Confederation. It was the first collective agreement by the 13 founding states of the United States of America. The Articles were intended to provide the states with sovereignty and established a very limited federal government. Over the next several years, the lack of a strong central government caused enough problems for the young nation that the states met again to draft the United States Constitution.

1872: Yellowstone, the world's first national park, is founded. The park was created by the United States Congress in legislation signed by President Ulysses S. Grant. The park's geothermal features, including geysers and hot springs, are the result of a massive supervolcano called the Yellowstone Caldera that lies beneath the Earth's crust there. The volcano last erupted approximately 640,000 years ago. The park was created after extensive lobbying by Ferdinand Hayden, a geologist who led several expeditions there in the 1860s. Yellowstone's first super-intendent was Nathaniel P. Langford, who served in a volunteer capacity for five years. In its early years, Yellowstone was plagued by poaching, vandalism, and lawlessness because Langford lacked the resources to properly administer the park. Harry Yount was appointed Yellowstone's gamekeeper to bring the park under control, making him the first national park ranger in history. In 1886, the park came under the Army's protection. In 1978 the park was designated a UNESCO World Heritage Site. More than three million visitors go to Yellowstone every year.

1896: An Ethiopian Army defeats an Italian force at Adwa. The battle ended the First Italo-Ethiopian War and secured Ethiopian independence from Italy for nearly forty years until Fascist Italy occupied it during World War II.

1936: The Hoover Dam is completed. The dam, in the Black Canyon of the Colorado River, was the largest hydroelectric-ity facility in the world when generators were installed. Over 3.25 million cubic yards of concrete was used in the construction of the dam, and another 1.1 million cubic yards in the dam's power plant.

1966: *Venera 3* lands on Venus. This made it the first spacecraft to land on another planet's surface. The probe's communications systems failed as it passed through Venus' atmosphere, so it was unable to send back data about the planet.

1974: Indictments are delivered in connection the Watergate burglary. Seven indictments were handed up by the grand jury to Judge John Sirica along with a sealed briefcase that would be delivered to the United States Congress' impeachment investigators. Although the jury declined to indict President Richard Nixon, it did name him as a co-conspirator in the case.

2003: The International Criminal Court opens in The Hague. The Court has 123 member-states, and it hears cases for geno-cide, crimes against humanity, and war crimes.

March 2

1657: The Great Fire of Meireki burns Edo, Japan. The fire burned for three days, destroying seventy percent of the capital city (now Tokyo). Over 100,000 people perished in the inferno. According to legend, the fire was accidentally started by a priest who was cremating a supposedly cursed kimono.

1919: The first Communist International meets in Moscow. The Comintern, as it was informally known, was organized after the success of the Bolshevik revolution in 1917 to spread socialist revolution worldwide. It operated until 1943, when the Soviet Union joined the Allies during World War II.

1943: American and British planes overwhelm and sink a Japanese convoy in the Battle of the Bismarck Sea. The convoy was attempting to reinforce Japanese troops in the South Pacific. An Allied attack force of over 100 bombers and more than 50 fighter planes attacked over the course of two days, destroying all eight troop transports and four accompanying destroyers. The Allied victory discouraged the Imperial Navy from attempting to resupply New Guinea by sea.

1949: James Gallagher completes the first nonstop flight around the world. The pilot and a three-man crew flew a United States Air Force Boeing B-50 Superfortress named *Lucky Lady II* in a 94-hour flight around the globe. It was refueled four times in air and flew at an average ground speed of 249 miles per hour.

1962: Wilt Chamberlain scores 100 points. The achievement set the record for most points scored in an NBA game by one player. Chamberlain broke his own single-game scoring record of 71 points, established earlier that season. The third-year center scored the hundredth point with only 46 seconds remaining on the clock. The celebration lasted nine minutes before play could resume and the game be completed.

1972: *Pioneer 10* is launched. The probe's primary mission was to explore the outer solar system and in particular, Jupiter. The spacecraft achieved a number of notable firsts in human space exploration: it was the first to pass through the asteroid belt, the first to encounter Jupiter, and the first to leave the vicinity of the solar system's major planets.

1983: Compact discs go on sale in America. The technology had been unveiled in Japan the year before with a rerelease of *52nd Street* by Billy Joel on compact disc (CD). CBS released sixteen titles in the United States and Canada, including albums by Journey, Barbra Streisand, Earth Wind & Fire, and Simon & Garfunkel. Compact discs quickly became popular among classical music enthusiasts and audiophiles, and spread to pop music genres as the price of CDs and CD players came down over the next few years. The technology remained widespread through the 1990s and early 2000s, until digital music players became common. CD sales dropped off precipitously with the advent of the iPad and later the iPhone.

1998: The *Galileo* spacecraft discovers a liquid ocean on Jupiter's moon Europa. The probe had been launched nine years earlier and had been studying Jupiter and its moons for three years when it found indications of a liquid ocean underneath the moon's icy surface.

March 3

1776: The United States Marines stage their first amphibious assault. The Marines attacked Fort Montagu at the port city of Nassau, Bahamas, to seize gunpowder and arms the British stored there. They quickly took the fort and remained in Nassau for two weeks before loading British officers they had taken prisoner and as much weaponry as they could carry onto accompanying ships and sailing to Connecticut.

1873: The Comstock Law is passed. The legislation was one of the first obscenity laws in the United States, and prohibited sending "obscene, lewd, or lascivious" materials through the mail. It included educational materials, contraceptives, and pornography in its definition of obscenity.

1875: Bizet's opera *Carmen* premieres in Paris. The opera was scandalous when it debuted because the lead character was not a woman of virtue: one critic decried her as "the very incarnation of vice." *Carmen* would go on to be one of the most acclaimed operas ever written.

1923: The first issue of *TIME* magazine is published. The news magazine was founded by Henry Luce, an influential publisher who also started *Life* and *Sports Illustrated*. *TIME* eventually became the most widely circulated weekly news magazine in the world, with a readership of over 25 million worldwide.

1931: "The Star-Spangled Banner" is adopted as the United States' national anthem. Its lyrics were taken from a poem written by Francis Scott Key after he witnessed the shelling of Fort McHenry, Baltimore, during the War of 1812. The melody is from a British song written by John Stafford Smith that was widely popular in the United States. It was played at Fourth of July celebrations and at military parades through the 19th century. Congressman John Linthicum introduced six bills to the House of Representatives to officially adopt "The Star-Spangled Banner" as the anthem, beginning in 1918, until it passed following a petition drive by the Veterans of Foreign Wars.

1951: Jackie Brenston records *Rocket 88*, considered the first rock and roll record. Brenston recorded the song in Memphis, Tennessee, with the backing of Ike Turner's band Kings of Rhythm. The song combined jump blues, fuzz guitar, and swing tempos to create the sound that would revolutionize popular music in the 20th century.

1986: Australia becomes fully independent of the United Kingdom. The country had become a federation of colonies and a member of the British Commonwealth in 1901, and most of the constitutional links between the U.K. and Australia were abolished in 1942. The Australia Act of 1986 severed all remaining legal ties between the nations.

1991: Rodney King is beaten by Los Angeles police officers. King was apprehended by police following a high-speed chase. Officers tasered King and struck him with their batons to subdue him, and continued to do so after he was incapacitated. A witness named George Holliday surreptitiously videotaped the beating from across the street. The footage was eventually aired on news reports worldwide, resulting in four of the officers being charged with assault and use of excessive force. When three were acquitted of all charges, Los Angeles erupted in riots that lasted six days, resulted in 63 deaths, over 2,000 injuries, and property damage of more than $1 billion.

March 4

1519: Hernan Cortes arrives in Mexico. Cortes was a Spanish conquistador whose expedition through Mexico would eventually lead to the fall of the Aztec Empire, which had ruled Mexico since 1428. His overthrow of the Aztecs was the first stage of the colonization of Central America by the Spanish.

1789: The first Congress of the United States meets in New York. The first Congress consisted of representatives of the original thirteen states. During its first session, the Congress passed several acts establishing the Departments of State, War, and the Treasury. It also established the federal judiciary and the office of Attorney General. The first Congress passed twelve amendments to the United States Constitution, including the Bill of Rights, which established guarantees of personal freedoms, freedom of the press and religion, limits on governmental power, and establishment of the rights of persons accused of a crime. The Bill of Rights was later ratified on December 15, 1791.

1837: The City of Chicago is incorporated. The first non-indigenous settler in the area, Jean Baptiste Point du Sable, a

Black French fur trader, had arrived in the 1780s. The U.S. Army built Fort Dearborn near his trading post in 1803. Native inhabitants initially resisted settlement, destroying the fort in 1812, but later ceded land in an 1816 treaty, and were forcibly removed in 1833. The city became a major transportation hub to the rest of the Midwest and eventually grew to be the third largest in the United States.

1882: The first electric trams in London begin service.
Following the invention of the storage battery, electric tram lines were built in east London. Electric tram service eventually spread to serve all of London until 1952, when they were replaced by double-decker buses.

1917: Jeannette Rankin becomes the first woman elected to the U.S. House of Representatives. She represented Montana from 1917 to 1919 and again from 1941 to 1943. She was a major force in the passage of the 19th Amendment to the Constitution, which granted universal suffrage to women, having opened Congressional debate on the issue. A lifelong pacifist, Rankin was the only member of Congress to vote against the United States entering both World Wars.

1957: The Standard & Poor's 500 is introduced. It is one of the primary American stock market indexes and is widely considered to be a barometer for the U.S. economy. The S&P 500 originally started as the Composite Index and only tracked a few dozen stocks. It expanded to the S&P 90 in 1926, and expanded again to 500 stocks in 1957. Originally published weekly, it eventually was able to disseminate stock information in real-time.

1974: *People* magazine makes its publishing debut. The weekly human-interest and celebrity news magazine was created by Andrew Heiskell, a former publisher of *Life* magazine who was CEO of Time Inc. Headquartered in New York City, the magazine is known for yearly issues that name its choices for the "most beautiful" and "best dressed" people.

1986: *Vega 1* begins sending back images of Halley's Comet. The Soviet space probe intercepted the comet after it had already explored Venus' atmosphere. It took more than 500 pictures of the comet over several days, and greatly expanded our understanding of the composition of comets.

March 5

1616: Copernicus' theory of heliocentrism is declared "false and erroneous" by the Catholic Church. His book *On the Revolutions of Heavenly Spheres*, originally published in 1543, was also added to the Church's Index of Forbidden Books. The Church vehemently opposed any theory about the universe that disagreed with Biblical scripture, and forbade the book so that heliocentrism "may not creep any further to the prejudice of Catholic truth."

1770: The Boston Massacre. A British sentry guarding the Custom House in Boston got into an argument with a wigmaker's apprentice and struck the boy with his musket. An angry crowd soon formed, shouting at the sentry, throwing snowballs and stones at him and challenging him to fire. Eight more soldiers came to the sentry's aid. As the crowd pressed closer, they suddenly opened fire, killing three men outright, including Crispus Attucks, who is considered the first American killed in the American Revolution. Although two of the eight soldiers were tried and found guilty of manslaughter, the event turned public opinion against the British Crown. Paul Revere and Samuel Adams used the Massacre as a propaganda device to stoke anti-British sentiment in the thirteen colonies.

1872: The air brake is patented. George Westinghouse designed a compressed air brake that relied on a failsafe mechanism to work. A reduction in air pressure causes the system to apply brakes, so any failure in the train's brakeline will cause it to engage. Brakes based on Westinghouse's design are still in use in modern trains.

1931: The Gandhi-Irwin Pact is signed. Mahatma Gandhi and Lord Irwin, British Viceroy of India, signed the agreement at the Round Table Conference in London. Under the terms of the agreement, the Indian National Congress agreed to discontinue the nationwide campaign of mass civil disobedience and the British agreed to release all prisoners arrested for civil disobedience and drop the charges against them, remove the tax on salt, effectively legalize the Congress, and invite it to the Conference. The Pact was a major victory in Gandhi's and the Congress' struggle for Indian independence.

1970: The Nuclear Non-Proliferation Treaty goes into effect. The treaty signatories agree not to share nuclear weapons or nuclear technology with other states, to engage in negotiations for disarmament, and to pursue nuclear technology for primarily peaceful means.

1979: Soft gamma repeaters are discovered. Receivers in several locations in the Solar System detected a strong gamma-ray burst almost simultaneously. By triangulating the signal, astronomers were able to determine it originated in the Large Magellanic Cloud. This was the first indication of a repeating source of gamma rays, believed to be a neutron star.

1981: The Z×81 home computer is launched by Timex. It was the first inexpensive home computer that could be bought at retail stores, and more than 1.5 million units were sold. The computer's video output was connected to a television rather than standalone monitor, and programs were loaded with audio tape cassettes. The Z×81 had 1 kilobyte of memory.

1982: *Venera 14* lands on Venus. The Soviet space probe took photographs of the planet and analyzed the composition of its rocky surface. After only 57 minutes, the incredible atmospheric pressure on Venus, which is 94 times that of the Earth, crushed the probe.

March 6

1857: The *Dred Scott vs Sanford* ruling is handed down.
Scott, born into slavery in Virginia, had been taken by his owners to free states, and attempted to sue them for his freedom. The Supreme Court held that no African-American who was descended from slaves could be an American citizen, and therefore had no standing to sue in federal court. It is widely considered the Supreme Court's worst decision.

1869: Dmitri Mendeleev unveils the first periodic table of the elements. Mendeleev organized the known elements by their atomic weight, and chemical properties, and correctly predicted that gaps in his table would eventually be filled as new elements were discovered.

1899: Aspirin is trademarked. Plants rich in the compound acetylsalicylic acid (ASA) were used to treat conditions related to pain and inflammation for centuries, and it was first synthetically produced by Charles Gerhardt in 1853. Bayer trademarked the drug and began selling it worldwide. It remained one of the most popular over-the-counter medicines until the development of acetaminophen in 1956.

1951: Ethel and Julius Rosenberg go on trial. The husband and wife were accused of stealing nuclear secrets and giving them to Soviet intelligence agents, as well as information about radar, sonar, and jet propulsion technology. Ethel's brother David Greenglass was the prosecution's primary witness against them. He testified that while he was working at Los Alamos National Laboratory, he gave sketches for nuclear bomb mechanisms to Julius and that Ethel had typed up notes describing the sketches. Based largely on his testimony, the Rosenbergs were convicted after a three-week trial and sentenced to death. Despite an international campaign for clemency, they were executed at Sing Sing Prison in New York on June 19, 1953.

1957: Ghana gains its independence. It was the first British colony in Sub-Saharan Africa to gain independence, led by Kwame Nkrumah, who became its first Prime Minister. Nkrumah advocated for Pan-Africanism and created the Organization of African Unity. He led Ghana until he was deposed in a coup in 1966.

1975: The Zapruder film is shown to a national audience for the first time. The film, shot by Abraham Zapruder, showed the assassination of President John F. Kennedy in Dallas, Texas, on November 22, 1963. It was shown on *Good Night America*, hosted by Geraldo Rivera. The broadcast led to the House Select Committee on Assassinations investigation and spawned countless conspiracy theories as to what "really" happened that day.

1992: The Michelangelo computer virus begins operating. The computer virus was so named because it went live on Michelangelo's birthday. Because the virus was dormant for most of its time, it could infect any floppy disk inserted into the computer and spread into any other computer the disk was inserted into. John McAfee, the founder of the anti-virus company that bears his name, claimed that millions of computers were potentially infected, even though the real number was somewhere in the hundreds.

March 7

1814: An army led by Napoleon I wins the Battle of Craonne. Napoleon's Imperial French army attacked a combined force of 22,000 Russian and Prussian soldiers of the Sixth Coalition near Craonne in northern France. While Napoleon's army succeeded in winning the field and driving the Coalition forces from the area, they suffered many casualties that affected their battle-readiness and led to Napoleon ultimately losing the war.

1933: The board game Monopoly is trademarked by Charles Darrow. The game was a variation on The Landlord's Game, which was patented by Elizabeth Magee in 1904. It expanded the original format to include features such as the Jail and Community Chest and Chance cards.

1936: Germany reoccupies the Rhineland. The post-World War I Treaty of Versailles expressly forbid Germany from remilitarizing the area west of the Rhine river. Doing so meant Germany was poised to invade Belgium and threaten France. The reoccupation allowed the Third Reich to begin constructing a series of defensive fortifications along the river and tipped the military balance of power in its favor.

1945: General George Patton leads American armor in capturing Cologne, Germany. The city spanned the Rhine River, and its capture established the first Allied bridgehead on the river's east bank. The capture of the city signaled the beginning of the end of World War II.

1965: Bloody Sunday in Selma, Alabama. The Southern Christian Leadership Conference (SCLC) had been organizing nonviolent resistance to voter suppression in Alabama in the months leading up to the march. During a march for voter rights in Selma, Jimmie Lee Jackson, a deacon in a local Baptist church, was beaten and shot by Alabama State troopers. He died eight days later. In protest of his killing, the SCLC attempted to march from Selma to the state capitol in Montgomery. The march was met by Alabama State troopers at Edmund Pettus Bridge. They attacked the peaceful marchers with tear gas and billy clubs, beating Amelia Boynton unconscious. Images of her lying on the Bridge were publicized worldwide. The ensuing public outcry directly led to President Lyndon B. Johnson presenting the Voting Rights Act to Congress later that month.

March 8

1817: The New York Stock Exchange (NYSE) is founded.
Stockbrokers in New York had been operating under the
Buttonwood Agreement of 1792 until then. After agreeing to
new rules and reforms regarding commission rates and manipu-
lative trading, traders founded the NYSE to exclusively trade
securities. It later grew to be one of the largest securities and
exchange trading houses in the world.

1862: The *Monitor* and *Virginia* battle one another. The battle
was the first fought between ironclad ships in history, and was
one of the most important naval battles of the war. The ironclad
Confederate ship CSS *Virginia* attacked several wooden-hulled
Union ships, sinking two, before the USS *Monitor* intercepted
her and the two fought to a stalemate. The battle attracted worldwide attention and changed the course of naval warfare forever.

1936: The first oval stock car race is held at Daytona Beach.
Stock car racer Sig Haugdahl organized the race along a 3.2-mile
course at the Beach, with a $5,000 first prize offered. Thousands
of fans arrived to watch the race, which had to be stopped after
the 75th lap as the sandy turns became too difficult to drive
through. Milt Marion was declared the winner.

**1949: Mildred Gillars, a.k.a. "Axis Sally," goes to prison for
treason.** Gillars, an American musician who was working as
an English teacher in Berlin in 1940, married Paul Karlson, a
naturalized German citizen. When war broke out, the State
Department advised Americans living in Germany to return
home, but Gillars' husband refused to leave and she stayed with
him. He joined the German Army and was later killed in action.
Gillars got a job at German State Radio and was working there
when Japan attacked Pearl Harbor. Although she initially condemned the attack to her coworkers, she realized that she could
be sent to prison or a concentration camp for disloyalty, and soon
signed an oath of allegiance to the Third Reich. In 1942 she was

cast in the program *Home Sweet Home*, which was broadcast to American troops and was designed to make them feel homesick. She also was the voice of three other propaganda shows directed at troops and at their families in the United States. After the war Gillars was captured, tried, and imprisoned until 1961.

1965: 3,500 Marines are the first American ground combat forces to go to Vietnam. The United States had already been engaged in an extensive bombing campaign and providing advice to the South Vietnamese military. Its commitment of ground troops would eventually swell to more than half a million and result in 58,318 casualties over the next ten years.

1971: Joe Frazier beats Muhammad Ali in the "Fight of the Century." Both heavyweights were undefeated prior to the bout, which Frazier won by unanimous decision in fifteen rounds. They fought three times, with Ali winning the next two fights.

1983: Ronald Reagan calls the Soviet Union the "evil empire." He characterized it as such in a speech before a national convention of Evangelical Christians, saying that the Cold War was part of the eternal struggle between good and evil. The speech indicated a new, hardline approach to dealing with the Soviet Union by his administration.

March 9

1796: Napoleon marries Josephine de Beauharnais. She became the first Empress of France, and while he was invading Italy Napoleon famously wrote her many love letters. But after news of her affair with a young cavalry officer reached him, his attitude changed. He divorced her in 1810 so that he could remarry and attempt to produce an heir.

1815: Francis Ronalds invents the battery-operated clock. Although the battery Ronalds used in the clock had a very long lifespan, it was affected by changes in the weather, and as a result the clock was not very precise and had to be regularly corrected.

1841: The Supreme Court rules in *United States v Amistad*. The Spanish slave ship *La Amistad* was carrying African captives near Cuba when the prisoners, led by a man named Sengbe Pieh, a.k.a. Joseph Cinque, took control of the ship and demanded to be returned to Sierra Leone. The crew instead sailed to the United States, and the Africans were tried for killing their captors. The Supreme Court found that they had been illegally captured and were justified in killing the slavers in self-defense.

1908: The soccer club Inter Milan is founded. The club was established following a schism with A.C, Milan, and has played in the Italian top tier of professional soccer since its inception, winning the league title more than 17 times.

1916: Pancho Villa attacks Columbus, New Mexico. What began as a cross-border raid to capture arms and supplies soon escalated into a full-scale battle between Villa's men and the United States Army. The U.S. 13th Cavalry Regiment soundly defeated Villa's force, driving them back to Mexico. The United States invaded Mexico to track down and capture Villa as a result of the raid, but were unable to apprehend the rebel leader.

1945: U.S. Warplanes firebomb Tokyo in Operation Meetinghouse. The air raid was the single most destructive bombing run conducted in human history, destroying over 16 square miles of central Tokyo and killing more than 100,000 civilians. In comparison, the atomic bomb dropped on Hiroshima caused a firestorm that was two miles in diameter and caused a similar number of deaths.

1982: "Krononauts" hold a meeting for time-travelers in Baltimore. The all-night party took place in a bookstore to welcome future time-travelers. None, as far as they could tell, attended the party.

2011: Space Shuttle *Discovery* makes its final landing. The Shuttle had been in service for over 27 years, launching and landing 39 separate missions, making it the spacecraft with the most flights at the time of its retirement. The Shuttle's notable missions included its carrying the Hubble telescope into orbit, the "Return to Flight" missions following the *Challenger* and *Columbia* disasters, and John Glenn's 1988 return to space at the age of 77, making him the oldest person to go into space at the time. Following its retirement, *Discovery* was put on display at the Smithsonian Institution's National Air and Space Museum.

March 10

1629: Charles I dissolves the English Parliament. This was the fourth time Charles had dissolved Parliament, which was his right under the Royal Prerogative. The dissolution of 1629 would be the longest period the United Kingdom would go without Parliament under his rule, and became known as the Eleven Years' Tyranny. Charles was eventually overthrown and executed by Oliver Cromwell.

1804: The Louisiana Purchase is formalized. The Purchase saw 828,000 square miles west of the Mississippi River transferred to the United States from France for the price of 11.25 million dollars and the cancellation of debts owed to France worth another 18.75 million dollars. Territory in fifteen states was acquired in the deal. The Purchase was formalized in a ceremony in St. Louis, Missouri, celebrated as Three Flags Day.

1848: The Treaty of Guadalupe Hidalgo is ratified. The U.S. Senate ratified the Treaty by a vote of 38 to 14 in a major step towards the formal end of the war.

1876: Alexander Graham Bell successfully tests the first telephone. It was the first successful test of two-way communication over the device. Bell famously spoke the words "Mr. Watson, come here, I want to see you" to his assistant over the electromagnetic transmitter, and Watson answered him. Bell would publicly demonstrate his telephone later that summer.

1952: Fulgencio Batista takes control of Cuba in a coup.
Batista had previously served as President of Cuba from 1940 to 1944. When he ran again in 1952, it was clear that he would lose the election badly. Rather than wait for the election, Batista overthrew president Carlos Prío Socarrás and cancelled the elections. He established business relationships with the American Mafia and ran a severely authoritarian regime until was overthrown in 1959 by Fidel Castro.

1969: James Earl Ray pleads guilty to assassinating Martin Luther King, Jr. He was found guilty and sentenced to life in prison, although he later recanted his guilty plea. Witnesses had seen him fleeing from the scene of the assassination, and a rifle was found near the site that had Ray's fingerprints on it. Dr. King's family later said they believed Ray was not the killer.

2000: The dot-com boom reaches its peak. The NASDAQ composite stock market reached a high on that day, driven by speculation on growth of internet and technology companies. The boom was so named because most of these companies had the internet-related '.com' suffix in their name. Many of the companies were wildly overvalued, leading to a huge stock market bubble that grew from approximately 1,000 points to over 5,000 in just five years. Companies that had never seen any profit whatsoever were able to raise money via initial public offerings, further fueling the bubble. As many of these companies went bankrupt, the NASDAQ declined nearly eighty percent over the next two years in what became known as the dot-com crash. Some companies were completely wiped out, while others, including Amazon.com, survived and eventually surpassed the stock values they attained during the boom.

2006: The Mars Reconnaissance Orbiter (MRO) reaches Mars. The MRO measured the water ice volume in the Red Planet's polar caps, discovered ice in craters on Mars, located chloride deposits, and observed what appeared to be flowing water on the surface.

March 11

1702: The first daily newspaper in England is published.
The *Daily Courant* was produced by the printer and bookseller
Elizabeth Mallet. It contained international news on the front
of a single page, with advertisements on the back page. Mallet
eventually sold the paper to Samuel Buckley, who continued to
publish it until 1735.

1824: The Bureau of Indian Affairs is created. Native
American relations were administered by the U.S. War
Department before the BIA was created. Throughout the 19th
and early 20th century, the BIA engaged in the controversial policy of educating Native American children at boarding schools
that emphasized assimilation to European-American culture.
In 1972, members of the American Indian Movement occupied
the BIA's offices in Washington, D.C. for a week in order to raise
awareness of issues affecting Natives.

1851: The opera *Rigoletto* debuts in Venice. The opera was
written by Giuseppe Verdi and based on a play by Victor Hugo.
It was one of Verdi's first masterpieces, and was a huge success
at its opening. The opera's famous aria, "La donna è mobile" was
sung in the streets of Venice the morning after the premiere.

1918: The first case of Spanish influenza is discovered. The
first victim, Private Albert Gitchell, was diagnosed at Fort Riley,
Kansas, where American troops were being trained for World
War I. The virus was carried by Allied sailors and soldiers to
the battlefields of Europe, and from there it spread around the
globe. Spanish influenza eventually infected 500 million people
around the world. Between 50 and 100 million people died
from the virus, which accounted for ten percent of the deaths in
1918 alone. While influenza typically only kills patients who are
very young, elderly, or have compromised immune systems, the
1918 flu primarily killed previously healthy adults. It remains
one of the deadliest epidemics in human history.

1945: The Japanese Navy launches a *kamikaze* attack on the U.S. Pacific Fleet. Before dawn, 24 long-range Japanese bombers loaded with bombs set off to attack the Fleet. The attack force lost multiple planes to mechanical failures on the way, and only one made it to its targets. That plane crashed into the USS *Randolph*, killing 27 men and wounding 105.

1985: Mikhail Gorbachev gains power in the Soviet Union. As General Secretary of the Communist Party, Gorbachev was the last leader of the Soviet Union. His policies of *glasnost* and *perestroika*, or openness and restructuring, respectively, led to the end of the Cold War and ultimately brought about the dissolution of the Soviet Union.

1993: Janet Reno becomes the first female U.S. Attorney General. Reno was the State's Attorney for Miami-Dade County, Florida, from 1978 to 1993, when she was appointed Attorney General by President Bill Clinton. During her tenure she oversaw the capture of Unabomber Ted Kaczynski, the disastrous siege of Branch Davidians at Waco, the capture and conviction of Oklahoma City bomber Timothy McVeigh, and the capture and conviction of World Trade Center bombing mastermind Omar Abdel-Rahman, and the seizure and return of Elian Gonzalez to Cuba. She was the second-longest serving Attorney General when she stepped down on January 20, 2001.

March 12

1894: Coca-Cola is bottled and sold for the first time. Before then, it was only available as a fountain drink. Joseph Biedenharn began bottling the soft drink at his candy factory in Vicksburg, Mississippi, at the request of the company.

1912: The Girl Guides, a forerunner of the Girl Scouts, are founded. At a rally of Boy Scouts in London, a group of girls asked their founder, Robert Baden-Powell, if they could join. He, in turn, asked his sister Agnes Baden-Powell to establish a separate organization for girls.

1918: Moscow officially becomes the capital of Russia again.
St. Petersburg had been the capital for the previous two centuries, but with the Bolshevik takeover and establishment of the Soviet Union, it was moved back to Moscow.

1930: Mahatma Gandhi begins the Salt March. As part of the campaign of nonviolent civil disobedience for independence from Great Britain, Gandhi led a march of 78 people from Ahmedabad nearly 240 miles to the coast. Hundreds of people joined the march on its way. There, he made salt from seawater in defiance of the British tax on salt, inspiring millions of Indians to do the same. The protest drew international recognition of the independence movement.

1933: Franklin D. Roosevelt addresses the nation in his first "fireside chat." In his address he discussed the Banking Crisis and the relief the government was providing for banks hit hard by the Great Depression. He would go on to deliver thirty chats over the course of his presidency, speaking directly to the American people on topics ranging from the Works Relief Program to the Declaration of War on Japan.

1947: The Truman Doctrine is declared. It was an American foreign policy designed to counter Soviet expansion of communism after World War II. The Doctrine led to the formation of NATO and shaped American policy for the duration of the Cold War.

1967: Suharto takes power in Indonesia. He was appointed acting president after he wrested power from President Sukarno, and went on to consolidate his power with brutal purges that saw 1.5 million Indonesians imprisoned and at least half a million killed. He remained in power until he was deposed in 1998.

1999: Former Warsaw Pact members join NATO. The Czech Republic, Hungary, and Poland had all been members of the Eastern bloc, a group of Communist-aligned countries in Eastern Europe dominated by the Soviet Union.

2011: The Fukushima Daiichi nuclear disaster. The Tohoku earthquake struck off the northeast coast of Japan on March 11 with a magnitude of 9. It was the most powerful earthquake ever recorded in Japan and the fourth most powerful in recorded history. Although the Fukushima nuclear reactors automatically shut down when the earthquake struck, a tsunami that followed the earthquake inundated the nuclear plant and disabled its cooling and control mechanisms. The plant suffered a nuclear meltdown as well as multiple explosions that resulted in the release of radioactive materials into the surrounding area. The first radioactive isotopes were detected on the 12th, and began to appear in Russia and on the west coast of the United States in the days that followed. Radiation contaminated a wide area and made it into water plants in Tokyo. An exclusion zone was established around the plant for up to twenty miles.

March 13

1567: The Eighty Years' War begins. The war began as a rebellion by the Dutch against rule by the Habsburgs of Austria. It ended with the Peace of Westphalia, with the Dutch Republic being formally recognized as an independent country.

1781: William Herschel discovers the planet Uranus. Herschel, a British astronomer, had constructed a large telescope in 1774 and spent the next nine years cataloguing heavenly bodies. When he realized Uranus was a planet, it made him instantly famous as the first person to discover a planet since antiquity. He became the Court Astronomer and was elected to the Royal Society as a result.

1862: The Union forbids Army officers from returning fugitive slaves. The order effectively invalidated the 1850 Fugitive Slave Act, which mandated the return of escaped slaves to their owners, even if they were in free states. It also set the stage for Lincoln's Emancipation Proclamation on New Year's Day, 1863.

1954: The Battle of Dien Bien Phu begins. During the First Indochina War, French forces established a base of operations in northwest Vietnam that was resupplied and reinforced entirely by air. The French command thought they would be able to destroy the Viet Minh there using superior firepower. The Viet Minh, however, were able to take advantage of terrain to besiege the French with ongoing artillery and ground assaults for nearly two months. The French defeat there resulted in them abandoning Indochina and the division of Vietnam at the 17th Parallel.

1962: The Chairman of the Joint Chiefs of staff proposes Operation Northwoods. The proposal, made by General Lyman Lemnitzer, called for false flag terrorist attacks to be carried out against American targets and blamed on the Cuban government of Fidel Castro. President John F. Kennedy rejected the proposal and had Lemnitzer relieved as Chairman of the Joint Chiefs.

1979: Maurice Bishop takes power in Grenada. The leader of the New Jewel Movement (NJM), a left-wing political party, Bishop had run unsuccessfully in national elections in the years leading up to the revolution. When Eric Gairy, the Prime Minister of Grenada, was out of the country, the NJM's National Liberation Army seized government buildings and police stations in a bloodless coup. Although Bishop sought recognition and aid from the United States, he was rebuffed by President Ronald Reagan, and turned to Cuba for support.

1988: The Seikan Tunnel opens. At 33 miles long, it was the longest tunnel with an undersea segment ever constructed when it was opened. The tunnel serves the Kaikyo Railway line in northern Japan and took five years to complete.

2003: Ancient human footprints are verified in Europe. The science journal *Nature* reported on 350,000-year-old footprints in Italy. The fossilized footprints were first found in the 19th century. Archaeologists determined their age to be some of the oldest human footprints discovered outside of Africa.

March 14

1794: Eli Whitney patents the cotton gin. The machine made it much easier to remove cotton seeds from the plant's fibers, a process that had been done by hand until the invention. Eli patented two devices: one could be operated by a hand crank, and the other, an industrial-sized version, was driven by horses or waterwheels. The device was one of the major technological advances that brought about the Industrial Revolution and completely changed the economy of the American South. The cotton gin greatly improved cotton production in the Antebellum South, with yields of raw cotton doubling every decade in the 19th century. The result of increased demand for cotton directly leading to a huge escalation in the American slave trade. As a result, the invention is widely credited with being a causative factor in the events leading up to the American Civil War.

1885: *The Mikado* premieres in London. It was the ninth opera by W.S. Gilbert and Arthur Sullivan, and had 672 performances in its first run at the Savoy Opera. Although the opera was set in Japan, it satirized British politics of the day.

1900: The Gold Standard Act is ratified. The Act established gold as the only standard for which paper money could be redeemed in the United States. Prior to the Act, money could also be redeemed for silver. The United States remained on the gold standard until 1933.

1903: Pelican Island National Wildlife Refuge is created. President Theodore Roosevelt established the United States' first National Wildlife Refuge to protect birds from extinction due to hunting. The American Ornithologists' Union and Florida Audubon Society were instrumental in lobbying for the Refuge's creation.

1943: The Nazis liquidate the Jewish ghetto in Krakow.
The ghetto was established after the Nazis invaded Poland in 1939. Beginning in 1940, the Nazis deported nearly 70,000 Jews from Krakow and forced the remaining 15,000 into the ghetto. Beginning in May 1942, they systematically deported thousands more to concentration camps. On March 13, 8,000 Jews were taken to Plaszów concentration camp. The following day, some 2,000 remaining Jews who had been deemed unfit to work were killed in the streets of the ghetto.

1994: Linux kernel 1.0.0 is released. It became one of the most popular open-source operating systems, having been built with the contributions of thousands of programmers and more than 1,200 companies. Linux was originally created by Linus Torvalds, a Finnish software developer.

1995: Norman Thagard is the first American to go to space on a Russian spacecraft. Thagard traveled to space on the Russian *Soyuz* TM-21 and performed research on the space station *Mir* until his return on July 7, 1995. The cooperation was a major event in the space program, which had originally been a major element of the Cold War between the two nations.

March 15

44 BC: Julius Caesar is assassinated. A group of Roman senators led by Gaius Cassius Longinus and Marcus Brutus feared that Caesar was planning to overthrow the Senate and rule as dictator. As many as sixty senators attacked Caesar, stabbing him to death at the base of the Curia in the Theatre of Pompey. The leaders of the conspiracy attempted to restore the Roman Republic but failed to do so. Brutus and Longinus eventually were overthrown by Mark Antony

1783: George Washington thwarts the Newburgh Conspiracy. Officers in the Continental Army, angry that they had not received any pay for nearly a year, were considering overthrowing Congress in a military coup. General Washington addressed his officers in an emotional plea that moved some of them to tears and immediately put an end to any ideas of a coup.

1819: Augustin-Jean Fresnel proves that light behaves like a wave. Challenged to do so by the French Academy of Sciences, he devised a series of proofs called the Fresnel integrals, which definitively showed that Isaac Newton's theories that light was a wave were true.

1877: The first cricket test match is played. The term "test match" refers to the length of the games, which can last several days and be a demanding "test" of the teams' relative strengths. The first game of the match was played between Australia and England over the course of four days in Melbourne. Australia won the game 245–196. England came back to win the second game, played over five days, and the match ended in a draw.

1927: The first Women's Boat Race at Oxford. The half-mile race was held between students at Oxford and Cambridge and became a tradition that took place every year thereafter. The Oxford racing team won the first race by fifteen seconds.

1943: The Third Battle of Kharkov ends. The city of Kharkov changed hands several times during World War II, and in the third battle the German Army's 1st SS Panzer Division recaptured it. The Soviet Red Army suffered more than 86,000 casualties in the battle, with some 45,000 killed. The German Army lost 4,500 men in the battle.

1955: "Fats" Domino records "Ain't That a Shame." The recording was a hit that catapulted Domino to fame, reaching number 1 on the *Billboard* R&B chart and selling a million copies. "Ain't That a Shame" was one of the songs that inspired the Beatles, and was the first song John Lennon learned to play.

1985: The first Internet domain name is registered. Symbolics, Inc. was a computer manufacturing company that was founded by a group of computer hackers and members of the Massachusetts Institute of Technology Artificial Intelligence

Lab. The company developed one of the first computer algebra systems, and also produced operating systems, software, and computers known as Lisp machines. It notably developed the first workstations that could process high-definition video. It has been repeatedly referenced in popular culture: the company was mentioned in the Michael Crichton novel *Jurassic Park*, and one of its computers, the Symbolics 3600, was shown in the film *Real Genius*. It registered the domain name symbolics.com, which was later sold to XF.com in 2009.

March 16

1621: Samoset greets the settlers of Plymouth Colony. Samoset was a member of the Abenaki tribe that lived in the area, and he had learned to speak English from English fishermen. He had been visiting Massasoit, the chief of the Wampanoag people, when he learned of the Pilgrim's settlement. Samoset entered their camp and greeted the Pilgrims in English, saying "Welcome, Englishmen! My name is Samoset." He then asked if they had any beer. After staying in the camp with the Pilgrims, he brought five other Abenaki men to meet the Pilgrims the following day, and eventually introduced them to Squanto, who showed the Pilgrims how to grow native crops and trap game for the fur trade.

1802: The U.S. Army Corps of Engineers is formally established. President Thomas Jefferson established the Corps with the Military Peace Establishment Act, which stationed them at West Point. The Corps engaged in many public works, building roads, dams, and canals throughout the United States. It also played a major role in the Civil War, building bridges and forts for the Union Army.

1865: Union forces fight the Confederates in the Battle of Averasborough. A Confederate force led by Joseph Johnston attacked a Union army commanded by Major General William Tecumseh Sherman as it was marching through North Carolina.

While each side lost about 700 men in the battle, the Union could afford such losses, while the Confederates could not. The battle was a prelude to the decisive Battle of Bentonville, which occurred three days later and was a major defeat for the Confederate army.

1926: Robert H. Goddard launches the first liquid-fueled rocket. Goddard, an American engineer and physicist, launched 34 rockets over the next fifteen years, ushering in the era of space flight. He was one of the first people to assert that rockets could eventually reach the moon, and that rockets could travel in the vacuum of space. NASA's Goddard Space Flight Center is named in his honor, as is a crater on the Moon.

1945: British bombers level Wüzburg, Germany in 20 minutes. The city was targeted because it was a major traffic hub. The Royal Air Force dropped incendiary bombs that set fire to the city, left roughly 80 percent of its buildings in rubble and killed 5,000 civilians. The rubble was not completely cleared until 1964.

1984: CIA agent William Buckley is kidnapped during the Lebanese Civil War. Buckley was the CIA station chief in Beirut while on his way to work. He was killed by Islamic Jihad after more than a year in captivity. He is buried in Arlington National Cemetery.

1988: Oliver North is indicted for conspiracy. North had been a member of the National Security Council during the Iran-Contra Affair. The Affair involved several members of the Reagan administration in the sale of arms to Iran and the use of proceeds from the sale to arm Contra rebel groups in Nicaragua. He testified before Congressional hearings on the scandal, during which he defended his actions. North was eventually convicted of three felonies, but the convictions were later vacated after the ACLU helped him appeal them.

March 17

1776: The Siege of Boston ends. The British Army held
Boston and was resupplied by the British Navy, which secured
the harbor. The Continental Army blocked land access to
Boston, but could not drive the British from the city and lacked
the heavy artillery needed to bombard their positions. General
Washington sent Henry Knox, a former bookseller, to Fort
Ticonderoga to bring cannons to Boston. When the cannons
were delivered to Dorchester Heights, a position overlooking
Boston, the British realized they could not hold the city
and withdrew.

1861: The Kingdom of Italy is established.
The Kingdom was the first unified country
that included the entire Italian peninsula. It
was founded largely by Giuseppe Garibaldi
and King Victor Emmanuel II. Italy became
a republic in 1946.

**1941: The National Gallery of Art is officially
opened in Washington, D.C.** The gallery was established
by Congress and initially comprised a collection of art gathered by Andrew Mellon. The building was designed by John
Russell Pope. Mellon and Pope both died before the Gallery was
completed.

1943: The Battle of the Atlantic reaches a climax. In what
would be the largest convoy attack during World War II, twenty
German submarines attacked a heavily defended convoy of merchant marine ships. The Germans sank twenty-one merchant
ships and lost only one U-boat over a five-day rout. The attack
was the peak of German dominance in the Battle of the Atlantic,
in which Allied navies fought German submarines for control
of shipping lanes between Europe and North America for the
duration of the war. The German Navy had refined the U-boat
Enigma code system, which prevented Allied codebreakers from

determining what their U-boats were doing, allowing them to attack shipping lanes virtually at will. After ten days of attacks the Allies were finally able to break the codes, thanks in part to the computing power arrayed at Bletchley Park by British mathematician Alan Turing's team. Once the code was broken, U-boat kills fell considerably, and fifteen were destroyed the following month.

1958: *Vanguard 1* is launched by NASA. It was the first solar-powered satellite, and measured the shape of the Earth. The satellite remained in communication until 1964, when its mission ended, but remained in orbit indefinitely.

1969: Golda Meir becomes the first female Prime Minister of Israel. Meir was born in Kiev, the Russian Empire, in 1898 and emigrated to the United States with her family when she was a child. She became active in the Zionist movement in Milwaukee, and moved with her husband Morris to the British Mandate of Palestine in 1921. She was a principal negotiator between Jewish settlers in Palestine and the British authorities in 1946, and later raised $50 million in the United States to purchase arms for the new Israeli nation. She served in various roles in government before being elected Prime Minister. She resigned in 1974. Meir died in 1978 and was buried in Jerusalem.

1973: The photograph "Burst of Joy" is taken. The photograph depicts a family rushing to greet an American prisoner of war returning from Vietnam. The image became an iconic representation of the end of the United States' involvement in the Vietnam War.

1992: A suicide bomber attacks the Israeli embassy in Buenos Aires. The attack killed 29 civilians and injured 242. The attack was carried out by Islamic Jihad, who said the attack was revenge for the assassination of the Secretary General of Hezbollah the month before.

March 18

1850: American Express is founded. The company was originally an express mail carrier founded by John Butterfield, Henry Wells, and William Fargo. Wells and Fargo would later create Wells Fargo & Co. when they were prevented from extending American Express mail service to the west coast by Butterfield. American Express expanded into money order and traveler's checking businesses in the 1890s, making it an international company. In 1958 American Express began offering charge cards to customers, and was the first company to issue plastic cards that became the industry standard. It also invented the tiered system of charge cards with its Platinum Card, as well as the first card that did not require a balance to be fully paid off each month. Largely on the success of its charge card business, American Express would become one of the most successful multinational financial companies in the world.

1892: Lord Stanley offers a silver challenge cup to the best Canadian hockey team. Stanley was appointed to Governor General of Canada in 1888, and he soon became an enthusiastic fan of the sport. The cup that he first presented in 1892 became the championship Stanley Cup of the National Hockey League, awarded annually to the winner of the playoffs.

1915: Three Allied battleships are sunk at Gallipoli. The battleships were attempting to seize the straits of Dardanelles and open a supply route to Russia. They struck submerged mines and sank with hundreds of casualties. The Allies failure of the assault led to the dismissal of the First Lord of the Admiralty, Winston Churchill.

1940: Hitler and Mussolini meet at Brenner Pass. In the meeting, the two fascist dictators agreed to form a military alliance against France and the United Kingdom. The Pact of Steel, as it became known, established the Axis powers, which Japan would later join.

1962: The Évian Accords formally end the Algerian War of Independence. The agreement provided for amnesty for Algerians who had collaborated with the French occupation forces and citizenship for the French community in Algeria. In spite of this, nearly all French nationals left Algeria almost immediately, and Algerian nationalists carried out reprisals against collaborators.

1965: Alexey Leonov walks in space. Upon exiting the *Voskhod 2* space capsule, Leonov became the first human to perform extravehicular activity in space. He remained outside the capsule for twelve minutes. He nearly could not reenter the capsule because his space suit had inflated in the vacuum of space, but was able to do so after releasing some of the suit's pressure.

1990: Thieves steal twelve paintings from the Gardner Museum in Boston. The thieves were disguised as Boston police officers and made off with art works worth an estimated $500 million. The thieves were never apprehended, and empty frames hang in the gallery where the paintings once were on display.

1994: The Washington Agreement is signed in Vienna. The ceasefire agreement ended the war between the Croatian and Bosnian forces. It established the Federation of Bosnia and Herzegovina, a loose confederation of states designed to prevent either ethnic group from having too much influence over the other.

March 19

1863: The Confederate cruiser *Georgiana* is sunk on her maiden voyage. The cruiser was said to be the most powerful cruiser in the Confederate fleet. It had an iron hull, was steam-powered, and carried sixteen large cannons. While trying to run the Union naval blockade outside Charleston, North Carolina, the ship was raked by Union ships' cannons and sank.

1895: The Lumière brothers record the first footage with their cinematograph. They filmed a scene of workers leaving their factory in Lyon, France. The 46-second footage is considered the first real motion picture ever filmed.

1918: Standard and daylight saving times are standardized. The Standard Time Act established modern time zones in the United States and formalized the practice of setting clocks forward during daylight saving time, from the second Sunday in March to the first Sunday in November.

1931: Nevada legalizes gambling. Gambling had been a feature of unregulated mining towns in Nevada during the 19th century, but the state had outlawed it in 1909. The state legislature legalized gambling during a local recession that followed declines in mining and agriculture. Legalized gambling soon made the sleepy roadside town of Las Vegas into one of the centers of casino gambling and entertainment in the world.

1941: The Tuskegee Airmen are established. Nicknamed the Red Tails and officially known as the 332nd Fighter Group and 477th Bomber Group of the United States Air Force, they were a group of African-American pilots during World War II. The NAACP lobbied Congress for over twenty years to allow African Americans to join the military, and in 1939 Senator Harry Schwartz amended an appropriations bill to fund training of African-American pilots. First Lady Eleanor Roosevelt boosted the program when she visited it in 1941, going on a test flight with the program's chief instructor, C. Alfred Anderson. The Airmen flew 1,578 combat missions during the war, destroying 112 enemy aircraft and damaging 148 more. They became widely known for their tenacity during bomber escort missions, with one of the best records of protection of any squadron in World War II, losing only seven bombers on 179 missions. Members of the squadron were awarded 96 Distinguished Flying Crosses, 14 Bronze Stars, and three Distinguished Unit Citations, and in 2007 collectively received the Congressional Gold Medal.

1962: Bob Dylan releases his first album. The self-titled record was a mix of country blues and folk music, including two songs written by Dylan and eleven covers. *Bob Dylan* initially received little praise, and the singer-songwriter did not become widely popular until the release of his second album the following year.

1979: C-SPAN begins broadcasting from the U.S. House of Representatives. It was the first televised session of Congress, and began with a floor speech by Representative Al Gore of Tennessee. C-SPAN expanded to C-SPAN2, which broadcast from the Senate, in 1986.

2011: The French Air Force begins bombing Libya. The air campaign supported rebels who were fighting Muammar Gaddafi's forces. It lasted two weeks and destroyed multiple ground units. The rebels won the war and executed Gaddafi a few months later.

March 20

1602: The Dutch East India Company is founded. The company was given a twenty-year monopoly on the spice trade by the Dutch government at its inception. It is considered the first true multinational corporation in the world, and was the first to issue bonds and stocks to the general public be listed on a stock exchange. It had the powers to wage war, imprison and execute, negotiate treaties, mint coins, and found colonies. It operated until 1799.

1760: A fire destroys over 300 buildings in Boston. The fire broke out in the center of downtown Boston, and spread east and south toward the harbor, destroying everything it its path. More than a thousand residents were left homeless by the blaze.

1852: *Uncle Tom's Cabin* is published by Harriet Beecher Stowe. It became the best-selling novel of the 19th century, and sold 300,000 copies in the United States in its first year alone. Its anti-slavery themes greatly helped the abolitionist cause, and was roundly criticized by Southern slavery supporters. The novel was one of the major cultural factors that helped contribute to the outbreak of the Civil War. When President Lincoln met Stowe in 1862, he reportedly said "so this is the little lady who started this great war." Many abolitionists later credited the novel with inspiring them to join the movement.

1915: Einstein publishes his theory of general relativity. Along with his theory of special relativity, published in 1905, Einstein's general theory completely revolutionized human understanding of classical physics, the shape and behavior of space and time, and our perception of the universe. General relativity proposed that space and time are one entity, called spacetime. Spacetime becomes curved whenever gravity, energy, or matter interact with it: very massive objects, such as the Sun, cause dimples in spacetime that smaller objects, such as the Earth, fall into, resulting in planetary orbits. Einstein demonstrated the links between these forces with a set of proofs known as the Einstein field equations. Research of physics was largely concerned with studying Einstein's theories for much of the twentieth century, and later with reconciling them with quantum theory. General relativity has been used in practical applications such as global positioning systems, television cathode rays, and space travel. Einstein was awarded the Nobel Prize in Physics in 1921 for his work.

1985: Libby Riddles wins the Iditarod. In doing so, she became the first woman to win the race. Iditarod competitors race over a

938-mile course with teams of sled dogs. She later wrote several books about her experience.

1987: AZT is approved for treatment of HIV/AIDS. The anti-retroviral drug was the first medication approved for treatment of the disease, but until a generic version was made available, its high cost—nearly $8,000 per year—prevented many patients from accessing it. AZT is now prescribed as part of a multidrug mixture of antiretroviral drugs.

1995: A cult releases sarin gas on the Tokyo subway. The Aum Shinrikyo doomsday cult believed it could overthrow the Japanese government and hasten the end of the world with the gas attack. The cult members released gas on three train lines during rush hour. Twelve people were killed and nearly 5,000 more were affected by the gas. The attackers were quickly caught.

2003: The United States invades Iraq. American and coalition forces attacked multiple sites along the southern Iraqi border and quickly penetrated the country. Air raids on Baghdad and military targets around the country were carried out as the ground force carried out its invasion. In less than a month, major combat operations were declared over, but the war would drag on for years as a determined insurgency resisted the occupation.

March 21

1556: Thomas Cranmer is burned at the stake. Cranmer had been the Archibishop of Canterbury under Henry VIII and Edward VI, and helped Henry divorce his wife Catherine of Aragon. He was the founder of the Church of England and established much of its doctrine and liturgy. Cranmer was imprisoned when the Roman Catholic Mary I ascended the throne and executed for heresy.

1844: The first Baha'i New Year is celebrated. The Baha'i Faith preaches the equality of all people and value of all religions, growing out of the Bábí religion. Its calendar is comprised of nineteen months that are nineteen days long, with four days between each year, and begins on the Spring equinox.

1925: The Butler Act outlaws teaching evolution in Tennessee.

The law specifically prohibited public school teachers from contradicting the Biblical account of the origin of the world. The law was challenged by the Scopes trial, in which William Jennings Bryan argued for the prosecution of a local teacher, and Clarence Darrow defended him. The law remained on the books until it was repealed in 1967.

1946: Kenny Washington is signed by the L.A. Rams.

Washington was the first African-American football player to be signed to the National Football League. He had been a star running back on the UCLA football team. George Halas tried to sign him to the Chicago Bears in 1940, but was prevented by other teams' owners. He played two seasons with the Rams.

1952: The first rock and roll concert is held in Cleveland.

The Moondog Coronation Ball was organized by Alan Freed, a disc jockey who popularized the term "rock and roll." Musicians Tiny Grimes, Paul Williams, the Dominoes, and Varetta Dillard all played.

1960: The Sharpeville Massacre takes place in South Africa.

The Apartheid government of South Africa maintained control of the population largely with pass laws. The laws prevented black Africans and other minorities from freely traveling around the country, and forced them to carry pass books at all times. About 10,000 people assembled at the Sharpeville police station without their pass books, demanding to be arrested in a non-violent action to protest the laws. The police attempted to disperse the crowd with tear gas and baton charges, but the crowd remained. The police then opened fire on the crowd, killing 69 unarmed demonstrators and wounding another 180. Many of the demonstrators were shot in the back as they tried to run away. The massacre caused a national outcry, with unrest spreading throughout the country the next day. Thousands were arrested in protests and strikes over the following weeks, and the African National Congress was banned by the government in response to the disorder. The event is now commemorated as Human Rights Day in South Africa.

2006: Twitter is founded. Jack Dorsey and other members of the company Odeo came up with the idea to create a text-based system for posting group messages and status updates. Dorsey published the first Twitter message, which read "just setting up my twttr," referring to the original name of the domain. Twitter eventually became one of the most popular online news and social networking sites in the world, with more than 100 million users per day.

March 22

1508: Amerigo Vespucci becomes Chief Navigator of the Spanish Empire. Vespucci made repeated voyages to the Americas on behalf of the Spanish and Portuguese. He was the first European to discover the coast of Guyana and the mouth of the Amazon, and was the first modern European to map the Southern Cross. The name "America" is derived from the Latin for Vespucci's first name, *Americus*.

1765: Parliament passes the Stamp Act. The act was the first to directly tax the colonies of British America, and it required that any printed material in the colonies use special paper that was produced in London and marked with the tax stamp. The tax was designed to raise revenue to pay for the cost of the French and Indian War and of keeping a sizeable British military presence in the colonies. The American colonists saw the tax as an insult, and felt that they were largely responsible for their own defense on the frontier. They also believed that the tax violated their rights, because it was levied without the consent of colonial legislatures. The slogan "No taxation without representation" spread through the colonies and led to the creation of patriotic clandestine organizations such as the Sons of Liberty. Although the tax was repealed the following year, it remained one of the reasons the colonists eventually rebelled against the Crown, culminating in the 1776 Declaration of Independence.

1906: The first rugby match between England and France is held. England defeated France 35–8 in the match, which was held at the Parc des Princes in Paris. It became a traditional annual rivalry between the two nations, and is referred to by both sides as *Le Crunch*.

1933: Low-alcohol beer is made legal during Prohibition. President Franklin D. Roosevelt signed an amendment to the Volstead Act that made wine and beer with less than 4% alcohol by volume legal. He famously said after signing it, "I think this would be a good time for a beer." The 18th Amendment outlawing the sale of alcohol was repealed later that year.

1960: The first laser is patented. Physicists Arthur Leonard Schawlow and Charles Townes were the first to use visible, rather than infrared, light in designing their laser. Schawlow hit upon the use of dual mirrors to make the laser work with visible light. He won the 1981 Nobel Prize in Physics for his work.

1972: The Equal Rights Amendment is sent to the states for ratification. The amendment was intended to guarantee equal rights for men and women before the law, and was applicable to situations such as divorce, property rights, and employment. Congress passed the Amendment, but following a campaign by conservatives led by Phyllis Schlafly, only 36 of the required 38 states ratified it, and it was not adopted.

1993: The first Pentium chips are introduced. The P5 chip was one of the first publicly available microprocessor cores, and incorporated all the central processing unit's functions on a single integrated circuit. The P5 contained 3.1 million transistors and was smaller than two centimeters square. Its introduction allowed computers to be added to just about any home appliance.

March 23

1540: Henry VIII completes his Dissolution of the Abbeys.
In his suppression of the Catholic Church in England, Henry closed monasteries, convents, abbeys, hospitals, and other Catholic institutions. Their property, including thousands of books from libraries hundreds of years old, was sold off and used to fund his military campaigns. The campaign was widely unpopular and sparked small rebellions in some areas. When Mary became Queen in 1553, her attempts to restore the Church failed largely because the Dissolution completely destroyed its infrastructure.

1775: Patrick Henry declares "Give me liberty or give me death!" Henry delivered the speech at the Second Virginia Convention in Richmond during debate over whether Virginia should send troops to join the Revolutionary War effort. Following the fiery speech, the delegates to the Convention reportedly sat in stunned silence for several minutes. A man named Edward Carrington was listening to the speech outside the Convention, and later requested to be buried where he had heard it. In 1810, he was. The immediate effect of the speech was the passage of an amendment by the Convention declaring the American colonies to be independent of Great Britain. Henry was appointed to lead the effort to raise a militia to join the War. The phrase has become an iconic symbol of the American Revolutionary War. Patrick Henry would go on to be the first Governor of Virginia following the war.

1806: Lewis and Clark begin their journey home. The pair had led a group of explorers from Saint Louis across the territory acquired in the Louisiana Purchase, over the continental divide, and to the West Coast over the previous two years. Their journey home was much faster, aided by their knowledge of the Columbia and Missouri rivers, and they reached Saint Louis in September of that year.

1857: The first safety elevator is installed in New York City. The safety elevator was designed by Elisha Otis, founder of the Otis Elevator Company. Otis installed a braking mechanism on the elevator that would engage if the hoist cables should snap. He demonstrated his invention at the 1854 World's Fair, and it soon became widely used. The Otis elevator made the first skyscrapers possible and greatly contributed to building booms in New York and elsewhere in the 19th century.

1909: Theodore Roosevelt goes on safari in Africa. The trip was sponsored by the Smithsonian Institution and was intended to collect specimens for the Institution's newly opened Natural History museum. Roosevelt led the safari through present-day Kenya, the Congo, and Sudan, and collected over 23,000 specimens for the museum's collection.

1940: The Lahore Resolution is passed. The resolution called for an independent state and homeland for Muslims to be established in British India. It later formed the basis for the Partition of India in 1947 and the establishment of the Islamic Republic of Pakistan.

1965: Gus Grissom and John Young lift off in *Gemini 3*. It was the first manned mission of the Gemini program, the first mission to carry more than one American astronaut, and the final space flight directed from the Cape Kennedy Air Force Base. *Gemini 3* made three orbits of the Earth in just under five hours.

2001: *Mir* is scrapped in a controlled de-orbit. The Russian space station reentered the Earth's atmosphere in three stages, eventually breaking up over Fiji in the southern Pacific Ocean.

2003: The Battle of Nasiriyah begins in Iraq. It was the first major battle in the invasion of Iraq, and PFC Jessica Lynch was captured during it, becoming the first American prisoner of war in the conflict. She was rescued by Special Operations Forces the following week.

March 24

1721: Johann Sebastian Bach dedicates the Brandenburg Concertos. The collection of six concertos was dedicated to Christian Ludwig, the ruler of Brandenburg. They are generally considered some of the greatest Baroque era pieces written for orchestra.

1765: The British Parliament passes the Quartering Acts. The legislation required American colonists to provide British soldiers with food and shelter on demand. They became one of the colonists' grievances against the Crown, and contributed to discontent leading up to the American Revolution.

1832: Joseph Smith is tarred and feathered in Hiram, Ohio. Smith was the founder of the Mormon Church of Latter-Day Saints. Residents who were angry that his church was gaining political power attacked him and left him for dead. Smith left Ohio soon after.

1944: Allied prisoners break out of a German POW camp. Seventy-six prisoners tunneled out of Stalag Luft II during World War II. Of them, 73 were captured. The breakout later inspired the 1963 movie *The Great Escape*, starring Steve McQueen.

1958: Elvis Presley is drafted into the U.S. Army. He had been granted a deferment to finish filming the movie *King Creole*, and was drafted after it was complete. Presley joined the Army at

Fort Chaffee, Arkansas, in a media circus. He served in the Army until 1960, stationed for most of that period in West Germany. While there he met his future wife Priscilla Beaulieu. He also became dependent on amphetamines while in the Army, which contributed to his early death in 1977.

1980: Òscar Romero is assassinated. Romero was Archbishop of San Salvador during the Salvadoran Civil War. He became an outspoken critic of the Salvadoran Army's human rights abuses and repression of dissidents. He was shot after delivering a sermon while celebrating the Catholic Mass in a small church. His funeral was also attacked, and 31 people were killed.

1989: The Exxon Valdez oil spill occurs. The oil spill released 10.8 million gallons of crude oil into the Prince William Sound in Alaska. It was one of the most destructive oil spills in history, with the oil eventually covering more than 1,300 miles of coastline and 11,000 square miles of ocean. Hundreds of thousands of marine and coastal animals were affected by the oil spill, and the total environmental impact could never be determined.

1993: Comet Shoemaker-Levy 9 is discovered. Carol and Gene Shoemaker and David Levy discovered the comet while they were observing Jupiter at the Palomar Observatory in California. The comet broke apart and collided with Jupiter in 1994.

March 25

1199: Richard the Lionheart is wounded in France. The English King had ruled for ten years and led multiple Crusades in the Middle East, where he had been captured and ransomed. He began a conquest of Normandy in 1196. While besieging a castle there, he was shot by a French crossbowman. The wound quickly became infected, and Richard died two weeks later.

1584: Sir Walter Raleigh is granted permission to colonize Virginia. He was given seven years to establish a settlement before the charter, issued by Queen Elizabeth, expired. Elizabeth hoped Raleigh would not only discover riches in the New World, but that his settlement could serve as a base of naval operations to harass Spanish galleons. Raleigh never went to Virginia himself, but sent settlers to found the ill-fated colony of Roanoke.

1807: The slave trade is abolished in the British Empire. The Slave Trade Act only outlawed the capture and sale of enslaved people in the Empire. Slavery itself, however, was not abolished by the British for another 26 years.

1911: The Triangle Shirtwaist Factory Fire kills 146 workers. At the time, it was the worst industrial disaster in the United States. The garment factory was on the upper floors of a building in Manhattan, and the stairwell doors were locked by management. When the fire broke out, the workers could not escape, and either died in the fire or by jumping from windows. The fire led to the growth of the International Ladies' Garment Workers Union and improved safety standards in factories.

1931: The Scottsboro Boys are arrested in Alabama. The nine African-American teenagers were accused of rape and sentenced to death in rushed trials where they did not have adequate legal representation. The NAACP and Communist Party USA took up the appeals of the case, and eventually charges against four were dropped. The case was a *cause célèbre* in civil-rights circles and highlighted the unjust treatment African-Americans received in Southern courts.

1948: The first tornado forecast is made. U.S. Air Force officers Robert Miller and Ernest Fawbush issued a tornado warning after analyzing weather conditions near Tinker, Oklahoma. The warning was the forerunner of modern severe weather warnings issued by forecasters.

1957: Allen Ginsberg's poem "Howl" is seized on obscenity grounds. The poem's references to drug use and sexuality led the U.S. Customs to seize 520 copies that were shipped from

London. The seizure led to an obscenity trial that the defense eventually won with the help of the ACLU.

1969: John Lennon and Yoko Ono begin their Bed-In for Peace. The pair had just gotten married, and used the related media attention to protest the ongoing Vietnam War. Inspired by the sit-ins of the Civil Rights Movement, Lennon and Ono adorned their hotel suite with signs reading "Hair Piece" and "Bed Peace" and invited the media to their room to hear them promote the cause of world peace every day for a week. They also sent acorns to world leaders, asking them to plant them for the cause of world peace. They then flew to Montreal, where they held a second bed-in. The press was largely dismissive of their events, to Lennon's frustration, although he later insisted the event was held in the spirit of fun and meant to be taken lightly. The events of the bed-in were later mentioned in the Beatles' song "The Ballad of John and Yoko."

March 26

1344: The Siege of Algeciras ends. During the Spanish *Reconquista* that ousted the Muslim caliphate, a Castilian army commanded by Alfonso XI besieged the city for 21 months. The city was strategically important because it controlled the Strait of Gibraltar. It was the first siege in Europe in which gunpowder was used in the battle.

1484: William Caxton publishes *Aesop's Fables*. Caxton introduced the printing press to England and became the first seller of printed books there. Because the printing was rushed, it contained several inaccuracies.

1830: The Book of Mormon is published. It is the sacred book of the Mormon Church of Latter-Day Saints, founded by Joseph Smith. According to Smith, an angel had appeared to him and told him he would find golden plates written by ancient prophets buried in upstate New York. He claimed the text of those plates formed the basis of the Book.

1917: A Turkish army halts a British advance at the Battle of Gaza. The British attempted to invade Palestine with 31,000 troops, and were stopped by an entrenched and determined Ottoman force of 19,000. The British press falsely reported the British as victorious.

1954: Nuclear detonation at Bikini Atoll. Operation Castle Romeo tested a nuclear bomb at the Atoll that had an explosive yield of 11 megatons—nearly three times the predicted yield. It was the third-largest test of a nuclear weapon conducted by the United States during the Cold War. A photograph of the explosion's fireball and mushroom cloud became an iconic image of the nuclear arms race.

1967: The Central Park be-in attracts 10,000 people. It was the first of several be-ins that took place in the Park between 1967 and 1970. The crowd, consisting of thousands of hippies and curious onlookers, gathered to protest the Vietnam War and share flowers with one another.

1979: Egypt and Israel sign a peace treaty. The treaty, signed by Egyptian President Anwar Sadat and Israeli Prime Minister Menachem Begin, was negotiated by United States President Jimmy Carter at the Camp David Accords. The treaty ended a state of war that had existed between the two nations since 1948 and resulted in the withdrawal of Israeli armed forces from the Sinai Peninsula. Sadat and Begin shared the Nobel Peace Prize for their efforts toward reconciliation.

1997: The Heaven's Gate mass suicide is discovered. Heaven's Gate was a millenarian cult whose leader, Marshall Applewhite, preached that the earth was going to be "recycled" in an apocalypse that would wipe it clean and renew it. He said that an alien spacecraft was following the Hale-Bopp Comet to rescue survivors of the apocalypse, and that he was its emissary to Earth. He convinced 38 members of the cult that mass suicide would allow their souls to leave their bodily "containers" and be taken aboard the spacecraft. The spacecraft, he said, would then take the cult members to an "existence above human." The cult members committed suicide over the course of three days. One of the cult members who committed suicide was the brother of Nichelle Nichols, who had portrayed Lieutenant Uhura in the television show *Star Trek*.

March 27

1513: Juan Ponce de León arrives in the Bahamas. The Spanish conquistador initially sailed to the Caribbean in 1493 with Christopher Columbus. After serving as the first governor of Puerto Rico, he explored the area, supposedly searching for the Fountain of Youth.

1794: The United States Navy is founded. President George Washington established the Navy, stating that "we could do nothing decisive without it," and authorized the building of six warships, including the USS *Constitution*. The *Constitution*, also called "Old Ironsides" is the oldest commissioned naval vessel still in operation, although its duties are wholly ceremonial.

1814: Andrew Jackson is victorious in the Battle of Horseshoe Bend. The Battle was between U.S. forces and the Red Sticks, a faction of the Creek tribe who were resisting U.S. expansion during the War of 1812. The victory effectively ended the war with the Creeks and contributed to Jackson's popularity.

1836: The Mexican Army kills Texan prisoners in the Goliad massacre. Following the Battle of Goliad, nearly 450 Texan prisoners of war were killed under the orders of Mexican General Santa Anna. The massacre is mentioned in Walt Whitman's poem *Leaves of Grass*.

1886: Geronimo surrenders. He was the leader of a band of the Chiricahua Apaches who engaged in a guerilla campaign against the United States Army and settlers for two decades before his final surrender. When the U.S. government forced Chiricahua off of their traditional lands in Arizona and New

Mexico, Geronimo and a group of followers refused to comply, escaping instead to Mexico. From there, they launched raids on mining camps and settlements in the region until Geronimo was captured in 1877. He was taken to the reservation, but once there he gathered 700 more followers and fled again. He was convinced to return in 1883 by General George Crook, who made several concessions to Geronimo regarding the administration of the reservation. Geronimo left again in 1885, pursued by more than 5,000 U.S. soldiers under General Nelson Miles. Geronimo surrendered for the last time and was imprisoned in Florida. He performed at the 1904 World's Fair in St. Louis, but never returned to his home, and died in Oklahoma in 1909.

1964: The Good Friday earthquake strikes Alaska. The 9.2-magnitude earthquake hit near Anchorage, Alaska, causing landslides and ground fissures, collapsing buildings, and killing more than 130 people. It was the most powerful earthquake in North American recorded history.

1977: The Tenerife airport disaster kills 583 passengers. Thick fog was obscuring the runways at the airport in the Canary Islands when two Boeing 747 passenger jets collided on the runway. It was the deadliest aviation accident ever to occur.

1998: Viagra is approved by the FDA. Pfizer developed the drug initially to treat hypertension and heart disease, but it was discovered to have additional applications. It became extremely successful, with sales climaxing in 2008 at nearly $2 billion.

1999: An American stealth bomber is shot down. The bomber was participating in the NATO bombing of Yugoslavia during the Kosovo War in an effort to force the Yugoslav armed forces to withdraw from the region. Stealth bombers were believed to be invisible to radar, but Yugoslav forces had discovered a method of modifying their radar to detect them and shot down the plane with surface-to-air missiles. The stealth pilot, Lieutenant Colonel Dale Zelko, and the commander of the anti-aircraft crew, Colonel Zoltan Dani, became friends after the war.

March 28

1814: Two American ships are captured in the Battle of Valparaiso. The battle was one of the first naval actions fought between the United States and Great Britain during the War of 1812. Although the captain of the American ships had been ordered to avoid contact with the Royal Navy, he sought them out to achieve personal glory.

1842: The Vienna Philharmonic Orchestra gives its first concert. The Orchestra was formed by composer Otto Nicolai and nine of his colleagues. Its members were drawn from the Vienna State Opera. The Orchestra would go on to become one of the finest concert orchestras in the world.

1854: France and Britain declare war on Russia. The Ottoman Empire had already been at war with Russia since October 1853, and its allies joined the war in support. The Crimean War raged across Eastern Europe until the Allies won it 1856. A charge by British light cavalry against Russian forces during the war was immortalized as the poem "The Charge of the Light Brigade" by Alfred, Lord Tennyson.

1939: Franco's fascist troops capture Madrid. Franco had besieged the Spanish capital for more than two years. A coalition of forces loyal to the Spanish Republic courageously defended it against Fascist assaults and aerial bombing. Thousands died during the siege, either from outright battles or disease and starvation. As the Republican cause faltered elsewhere in Spain, the defense of the city became untenable, and its defenders were forced to surrender.

1942: The St Nazaire Raid destroys a major port in Normandy. St Nazaire was a port and dry dock that the Nazis used for repairing and building large warships for the Battle of the Atlantic. Destroying the dock would compel the Germans to send any warship that needed repairs back to Germany through the English Channel or around the northern coast of England, both of which were heavily defended by the British Navy. The British turned the destroyer HMS *Campbeltown* into a floating bomb by packing it with explosives with time-delay charges. Accompanied by 18 other ships, the *Campbeltown* crossed the Channel and crashed into the dock at St Nazaire while a force of 265 commandos came ashore to destroy machinery and buildings. The German defenders were able to sink most of the accompanying ships, and following the raid the commandos were forced to flee through the countryside, but were surrounded and captured when they ran out of ammunition. Five commandos were able to escape and return to England. At noon, a huge explosion ripped through the hull of the *Campbeltown*, completely destroying the dry dock, which remained out of commission for the rest of the war.

1979: The Three Mile Island nuclear meltdown occurs. It was the worst nuclear accident in United States history. It caused public perception of nuclear power to grow very negative in the following years. As a result of the accident, anti-nuclear demonstrations spread across the country, with over 200,000 people marching in New York City alone.

2005: A massive earthquake hits northern Sumatra. The 8.6 magnitude earthquake killed more than 1,000 people and destroyed hundreds of buildings on the Sumatran island of Nias. It also generated a small tsunami that hit the coasts of Thailand and Sri Lanka.

March 29

1806: Construction of the Cumberland Road is authorized. Also called the National Road, it was the first federal highway in the United States, and was built between 1811 and 1837. The highway was 620 miles long and connected the Ohio River to the Potomac. It was eventually extended west to Southern Illinois, but Congressional funding dried up before it could reach St. Louis, its intended destination.

1847: General Winfield Scott captures Veracruz during the Mexican-American War. Following a 20-day siege of the city that included bombardment by gunboats anchored in the harbor, the Mexican garrison surrendered. The capture of Veracruz was an important step in preparing for a full-scale invasion of Mexico and advance on the capital later that year.

1865: The Appomattox Campaign of the U.S. Civil War begins. It was the final campaign of the Eastern theater of the American Civil War. The campaign consisted of a series of battles fought between the armies commanded by Ulysses S. Grant and Robert E. Lee. Lee's army was forced to retreat across Virginia, with Grant winning battle after battle, until the campaign concluded with Lee's unconditional surrender

1882: The Knights of Columbus are founded. The group is the largest Catholic charity and service organization, with nearly 2 million members around the world. It was founded by Michael McGivney, a Catholic priest in New Haven, Connecticut, who organized the men of his parish to form the group's first chapter.

1886: John Pemberton invents Coca-Cola. Pemberton had been wounded during the American Civil War and addicted to morphine as a result. He was searching for a painkiller that he could use to wean himself off morphine and began experimenting with coca leaves and wine. He eventually hit upon using kola nuts, which contain caffeine, cocaine, and wine. His creation, Pemberton's French Wine Coca, would become the forerunner of Coca-Cola.

1927: The land speed record is broken at Daytona Beach. The Sunbeam 1000 HP *Mystery* was the first car to travel more than 200 miles per hour. It was powered by dual aircraft engines mounted in the front and rear of the car. Henry Segrave achieved the land speed record of 203.79 mph with the *Mystery*.

1961: The 23rd Amendment is ratified. The Amendment extended universal suffrage for presidential elections to the residents of the District of Columbia. It granted the district representation in the Electoral College for the first time.

1974: *Mariner 10* flies by Mercury. The NASA space probe was the first human-made object to fly past the planet, and the final spacecraft in the Mariner program. It made three passes of Mercury while orbiting the Sun, and discovered evidence of clouds, a magnetic field, and mapped the planet's surface.

1984: The Baltimore Colts move to Indianapolis. The NFL team had been based in Baltimore since 1953, and began the 1984 NFL season in Baltimore. The team's owner, Robert Isray, had been threatening to move the team since at least 1969 if the city of Baltimore did not help finance the building of a new stadium. He became embroiled in an ongoing feud with the Baltimore press, who were critical of his management of the team's finances and organization as well as the fact that the team had not made a playoff appearance since 1977. Indianapolis built a new stadium, called the Hoosier Dome, in 1982 in the hopes of attracting a major sports team to the city. The mayor of Indianapolis offered use of the stadium and a $12.5 million loan to the team, and Isray agreed. In the middle of the night, the team's belongings were loaded onto trucks and driven to Indianapolis. The move was unannounced and caught the entire city of Baltimore by surprise.

March 30

1842: Ether is used as anesthesia for the first time. Crawford Long, an American surgeon and pharmacist, discovered that inhaling ether had the same effects as nitrous oxide, which had been used as an anesthetic since 1794. When removing a tumor from the neck of a patient, he first had him inhale sulfuric ether, which rendered him safely unconscious. After experimenting with ether for several years, Long published his findings in 1849.

1857: The U.S. purchases Alaska from Russia. The Empire of Russia was looking for a buyer for the Alaska territory because it was too far-flung to be easily defended if a war should break out. American Secretary of State William Seward negotiated the purchase, for which the United States paid $7.2 million, by check.

1944: The Royal Air Force suffers its worst loss of World War II. A detachment of 795 bombers and fighter escorts was sent to bomb the city of Nuremberg. Of these, 95 were shot down while attacking the city or while attempting to return to Britain.

1945: Soviet forces liberate Danzig. The city, on the Baltic coast in Poland, had been captured by the Nazis in their 1939 invasion as an important strategic port. As the Nazis began losing the war on the Eastern Front, Danzig became a point of escape for hundreds of thousands of German refugees, many of whom became trapped in the city as the Soviet Red Army advanced.

1972: North Vietnamese forces begin the Easter Offensive. More than 250,000 soldiers of the People's Army of North Vietnam invaded South Vietnam in order to take control of as much territory as possible before the end of the Paris Peace Accords. The South Vietnamese military and their American allies were unprepared for the sheer size of the offensive or the fact that the North Vietnamese attacked in three separate areas at once. The North Vietnamese made significant gains before being repulsed later that fall.

1981: President Ronald Reagan is shot. As Reagan was leaving an event at the Hilton Hotel in Washington, D.C., John Hinckley Jr. stepped out of a crowd and shot at him six times. Although all six shots initially missed, the last bullet ricocheted off of the armor-plated limousine and hit Reagan in the chest, collapsing his lung and missing his heart by inches. Three others were also wounded in the shooting, including White House Press Secretary James Brady, a police officer, and a secret service agent. Brady was left paralyzed as a result of the shooting. Hinckley had become obsessed with and began stalking the actress Jodi Foster after seeing the movie *Taxi Driver*. He believed that by shooting Reagan he would get the actress' attention. He was eventually found not guilty by reason of insanity and confined to a mental hospital. Reagan survived the attempt and was taken to George Washington University Hospital, where he insisted on walking into the Emergency Room under his own power. In the operating room, the President joked with the surgeons, famously saying "I hope you are all Republicans." He returned to work at the Oval Office within a few weeks.

2017: SpaceX conducts the first relaunch of a reusable rocket system. The company, founded by Elon Musk, had been attempting to perfect relaunch technology to lower the cost of space travel for fifteen years. Its Falcon9 orbital rocket was the first time it was successful.

March 31

1774: Great Britain closes the port of Boston. The action was a retaliation for the Boston Tea Party the previous winter, when colonists had dumped stores of British tea into the harbor in protest of taxes. The closure was one of the so-called Intolerable Acts cited by colonists as justification for the later war.

1854: Commodore Perry opens Japan to American trade. The treaty opening trade was signed by the Japanese Shogunate under threat of invasion by Commodore Perry. It ended Japan's policy of national seclusion.

1889: The Eiffel Tower is opened. The Tower had taken two years to construct and would later serve as the entrance to the 1889 World's Fair. At 1,063 feet tall, it was the tallest human-made structure in the world when it opened. The first ascent of the Tower had to be made by foot, and took visitors more than an hour.

1930: The Motion Picture Production Code is established.
Also known as the Hays Code, it established a set of guidelines
regarding what was acceptable to be shown in motion pictures.
The Code prohibited profanity, nudity, reference to sex, mixed-
race relationships, ridicule of the clergy, and deliberate offense. In
1968 the MPAA film rating system replaced the Code.

1933: The Civilian Conservation Corps is founded. The work
relief program was designed to ease unemployment in the United
States during the Great Depression as part of Franklin D.
Roosevelt's New Deal. It hired young men between the ages of
17 and 28 to perform unskilled labor to conservation and development projects on federal lands. The workers planted billions
of trees, constructed trails and lodges in state and national parks
nationwide, and built roads in remote locations.

**1966: *Luna 10*, the first space probe to orbit the Moon, is
launched.** The Soviet probe entered lunar orbit three days later
and took measurements of the Moon's magnetic field, cosmic
radiation and surface.

1995: Selena is murdered. The American crossover singer was
one of the most famous Mexican-American entertainers of the
1990s. The music and fashion icon was often referred to as the
"Tejano Madonna" by the media. She is considered one of the
most influential Latin music artists of all time, and her records
sold millions of copies and won multiple Grammy awards.
Selena was killed by Yolanda Saldívar, the president of her fan
club and manager of the singer's retail stores. She had come
under suspicion of embezzling tens of thousands of dollars from
the retail stores and club. Selena's family confronted Saldívar and
fired her, but did not go to police. She then bought a gun and
lured Selena to a hotel room. An argument ensued, and as Selena
tried to leave, Saldívar shot her in the back. Selena died within
an hour. Over 30,000 people attended her wake, and 600 friends
and family attended her funeral.

Chapter 4

April

April 1

1789: Frederick Muhlenberg is elected the first Speaker of the House. Muhlenberg was a German-American colonist who lived in Pennsylvania. He had been a member of the Continental Congress and had been the first signer of the Bill of Rights. He served in the state House of Representatives before being elected to the U.S. House of Representatives.

1826: Samuel Morey invents the internal combustion engine. Morey had discovered the explosive properties of turpentine vapor, and designed a two-cylinder engine that was fueled by it. Although he successfully demonstrated its potential on a boat and a prototype of a car, he was unable to sell his idea, and it fell into obscurity.

1891: The Wrigley Company is founded. The company initially sold baking powder, and began packaging chewing gum as a bonus to buyers. The gum soon became more popular than the baking powder, and the company switched to manufacturing gum full time. It eventually became one of the largest chewing gum and candy companies in the world.

1948: The Berlin Blockade begins. After World War II, Germany was divided into zones of control by the Allies, with Britain, France, and the United States occupying areas of West Germany and the Soviet Union occupying East Germany. The capital Berlin was located in the East, with the Allies each controlling a sector of the city. Stalin, wishing to wrest control of Berlin from the Western Powers as the Cold War began, ordered the halt of all passenger trains and traffic to West Berlin and stopped resupply of the city from the West. The West responded by blockading East Berlin. West Berlin only had three weeks of food supplies, and the Allies were faced with a dilemma. They still had three 20-mile-wide air corridors that provided access to Berlin by plane, and eventually decided to airlift supplies to Berlin. It was a massive undertaking: at its height the Airlift was bringing more than 8,000 tons of cargo on 1,500 flights per day. In May of 1949, the Soviets lifted the blockade. The Airlift ended the following September, having delivered more than 2.3 million tons of supplies on over 278,000 flights.

1976: Steve Jobs, Steve Wozniak, and Ronald Wayne found Apple Inc. Wayne sold his shares in the company back to Jobs and Wozniak after a few weeks for $800. Apple's first computer, the Apple I, was merely a motherboard with CPU and RAM; it did not have a case, keyboard, or monitor. The company would go on to be one of the leading developers of computers and mobile devices, and produced multiple devices that revolutionized home computing, mobile phones, smart phones, and tablet computing. It eventually grew to 123,000 employees and annual revenues of more than $200 billion.

1999: Nunavut is established. It is an autonomous territory of Canada, and was separated from the Canadian Northwest Territories by the government to provide a separate territory for native Inuit Tapiriit Kanatami people.

2004: Google publicly launches Gmail. The project began as Google's internal email server, and was made public in a limited beta release that required users to obtain an invitation to the email system. It went on to become one of the most popular email systems in the world, and Gmail passed 1 billion active users in February 2016.

April 2

1800: Beethoven premieres his First Symphony in Vienna.
The concert also included works by Joseph Haydn and Wolfgang
Mozart. Beethoven's prominent use of wind instruments was
novel at the time, and the Symphony established him as one of
the top composers in Vienna at the time.

1865: The Confederate Army abandons Richmond. A massive Union assault of more than 100,000 troops broke the
Confederate defenses, and the army fled the capital city. The
Union Army captured Richmond the next day.

1900: The Foraker Act is passed. The legislation established
an independent civilian government in Puerto Rico, which the
United States had just seized during the Spanish-American war.
Puerto Rican residents were also granted U.S. citizenship by the
Act, and federal laws were extended to the island.

1930: Haile Selassie is proclaimed the emperor of Ethiopia.
Following the death of the Ethiopian empress Zewditu, Selassie,
who was then regent, proclaimed himself Emperor. He brought
Ethiopia into the United Nations and attempted to institute land
reforms during his rule. Selassie was overthrown in a coup d'état
in 1974 and died the following year.

1956: *As the World Turns* debuts on CBS. It was the first daytime soap opera to run in a half-hour format. The show was set
in the fictional town of Oakdale, Illinois, and featured the daily
dramas surrounding the life of the Nancy Hughes McClosky,
played by Helen Wagner. It had the second-longest run of any
daytime soap opera, airing for 54 years. It was cancelled following
Wagner's death in 2010.

1973: LexisNexis is launched. The company developed the concept of electronic database research for legal and public records.
It originally offered records of legal cases in Ohio and New York
before expanding to archive all federal and state cases in the U.S.

1982: Argentina invades the Falkland Islands. The Islands are located off the coast of Argentina but are a territory of Great Britain. Argentina believed the British would not go to war to defend the islands, but were mistaken. The British responded by dispatching a task force that recaptured the Islands after three months of conflict.

1992: John Gotti is convicted of murder and racketeering. Gotti, the boss of the Gambino mafia family, had earned the nickname "The Teflon Don" after three attempts to convict him for various crimes had all ended in acquittal. The flamboyant and opinionated mobster never shied from media attention, and his public image made him a major target of law enforcement, which eventually contributed to his downfall. Gotti had risen through the ranks of the Gambino crime family, and arranged the murder of the mafia boss Paul Castellano to consolidate his power in the organization. Although the FBI gathered evidence that tied him to extortion, murder, and assaults, jury tampering and witness intimidation prevented the government from obtaining a conviction. When his underboss, Sammy "The Bull" Gravano turned state's evidence, his testimony was damning enough to convict Gotti of five murders and a host of other crimes. He was sentenced to life in prison without parole, and died in 2002.

April 3

1860: The Pony Express completes its first successful run. Either Billy Richardson or Johnny Fry was the first rider to make the run, from St. Joseph, Missouri, to San Francisco, California. The first mail carried by the Pony Express reached San Francisco in just eleven days. The mail service was innovative in that it replaced stagecoaches with single horse riders traveling between Missouri and California. The route had 184 stations at ten-mile intervals. Riders would switch to fresh horses at each station, and the practice allowed them to make the entire 1,900-mile trip in record time. They carried 20 pounds of mail at a time in saddlebags, and were told that the mail pouches were to be guarded with their lives. Notable riders include "Buffalo" Bill Cody, who joined when he was only fifteen years old, and "Pony" Bob Haslam, who carried Lincoln's Inaugural address over 120 miles in 8 hours while wounded. The route the Pony Express travelled through Missouri later became U.S. Highway 36. With the invention of the telegraph, rapid mail service ceased to operate.

1882: Robert Ford assassinates Jesse James. Ford, a member of the James Gang, killed the outlaw to collect a $10,000 reward offered by the Governor of Missouri. He shot James in the back of the head while James was standing on a chair to clean a dusty picture. Ford was later shot to death by Edward O'Kelley in retaliation for the assassination.

1885: Gottlieb Daimler patents the vertical engine. Daimler's air-cooled flywheel engine had a running speed of 600 rpm, which was three times as fast as the average engine of the day. He later installed the engine on a two-wheeled car, thus inventing the first motorcycle with an internal combustion engine.

1895: Oscar Wilde brings a libel case against the Marquess of Queensbury. The Marquess had left a calling card at Wilde's club accusing him of committing the crime of sodomy, and Wilde retaliated by accusing him of libel. Queensbury hired a team of public detectives to dig up evidence against Wilde. The trial became a public sensation and Wilde was forced to drop the prosecution, tried for homosexuality, and later imprisoned.

1922: Joseph Stalin becomes General Secretary of the Soviet Communist Party. When Lenin was incapacitated by a stroke, Stalin was able to establish firm control over the Party, and as General Secretary he became the sole ruler of the Soviet Union. He ruthlessly purged his political enemies and ruled the Soviet Union until 1952, guiding the country through World War II but also committing numerous human rights abuses and leaving a legacy that was ultimately repudiated by his successor, Nikita Khrushchev.

1973: Martin Cooper makes the first mobile phone call.
Cooper, an executive and researcher for Motorola, called his rival, Dr. Joel Engel of Bell labs. The mobile phone he used weighed 2 pounds, took ten hours to charge and could be used for only thirty minutes.

1981: The Osborne 1 is unveiled. It was the first commercially successful portable computer that was commercially successful. The computer weighed 24 pounds and cost $1,795. It had a 64-kilobyte memory and a 5-inch display screen.

1996: "Unabomber" Ted Kaczynski is captured. Kaczynski had terrorized the United States in a 17-year-long bombing campaign that killed three people and injured 23 more. He espoused an anarchist ideology that advocated for the destruction of industrialization. Kaczynski was captured when his manifesto was published in the *New York Times* and his brother recognized similarities between it and letters Kaczynski had written. He was sentenced to life in prison.

2010: The iPad goes on sale. The touch-screen operated tablet computer's release had been anticipated since the company released its iPhone in 2007. It was wi-fi enabled, had 256 megabytes of RAM, was half an inch thick and weighed 1.5 pounds. It retailed at $499.

April 4

1721: Sir Robert Walpole becomes the first Prime Minister of Britain. Walpole, the First Earl of Oxford, is the longest-serving Prime Minister in British history, having remained in office until 1742. Walpole controlled the government with his brother-in-law and political ally Charles Townshed, the Secretary of State.

1796: Georges Cuvier gives the first lecture on paleontology. Cuvier, a French naturalist, presented an analysis of fossils of wooly mammoths along with the skeletal remains of African and Indian elephants. He showed that African and Indian elephants were separate species, and that the mammoth was neither and was therefore extinct. This finding was a milestone in the history of paleontology and definitively established that extinction does occur.

1841: President William Henry Harrison dies in office.

Harrison, the ninth President of the United States, died of pneumonia after only a month in office. His death prompted a minor crisis because the Constitution was ambiguous as to whether the Vice President should become acting President or become President. John Tyler, Harrison's Vice President, took the Oath of Office and was confirmed by the Congress as President for the remainder of Harrison's term. This established a precedent that remained until the 25th Amendment was ratified in 1967.

1887: Susanna Salter becomes the first female mayor in the U.S.

She was mayor of Argonia, Kansas, for one year. Her name had been placed on the ballot by a group of men attempting to humiliate women and intimidate them from running for office. The Women's Christian Temperance Union and the Republican Party switched their votes to Salter at the last minute. Salter served for one term.

1949: The North Atlantic Treaty Organization is founded.

The alliance originally included the United States, Canada, Great Britain and several European states as a mutual protection pact to counter the military prowess of the Soviet Union. It eventually expanded to include more than 40 states, including former satellite states of the USSR following its dissolution.

1968: Martin Luther King, Jr., is assassinated.

The American Civil Rights leader rose to prominence in his efforts to dismantle racist Jim Crow segregation, as well as lobbying for the passage of the Civil Rights Act of 1964, fighting for extension of universal suffrage in the South, and opposition of the Vietnam War. He had been awarded the Nobel Peace Prize for his use of nonviolent protest to call attention to the injustices African Americans faced. King was in Memphis in support of striking garbage workers as part of the Southern Christian Leadership Conference's "Poor People's Campaign," which was raising awareness about economic injustice in the U.S. While he was standing on the balcony of his motel, a sniper shot him from a building across the street. He died within an hour. His death was mourned around the world, and major riots broke out in cities across the U.S. He was buried in Atlanta, Georgia, and his grave is a National Historic Site.

1975: Bill Gates and Paul Allen found Microsoft. The pair had worked together to develop computers for several years before developing software—a BASIC interpreter for the Micro Instrumentation & Telemetry Systems (MITS) corporation's Altair 8800. MITS agreed to distribute the software, and Gates and Allen named their company after the venture. Microsoft began producing operating systems in 1980, and its development of Windows in 1985 catapulted it to becoming one of the most successful software companies in history.

1994: Netscape is founded. The company was the first to profit from the emerging World Wide Web. It created the JavaScript web scripting language, developed the secure sockets layer (SSL) security protocols for computer networks, and several web browsers. It was acquired by AOL in 1998 and later disbanded.

April 5

1614: Pocahontas marries John Rolfe. Pocahontas was a Native American woman and daughter of Powhattan, the chief of a confederation of tribes. She supposedly saved the life of John Smith, an English captive of her father's, when she was a young girl. Smith had established a fort in the region controlled by Powhattan's confederation. When Smith was captured by a group of confederation hunters, he was brought to Powhattan. He later recounted that Pocahontas prevented her father from executing him, although historians are varied as to the accuracy of the account. Pocahontas was captured by the English in 1613, and when she was given the chance to return to her tribe, she converted to Christianity and married Rolfe instead, taking the name Rebecca and giving birth to a son named Thomas. The family traveled to London, where she was presented as a "civilized savage" and enjoyed minor celebrity. They boarded a ship to return to America in 1617, but Pocahontas became ill and died in Gravesend before they could leave England.

1722: Jacob Roggeveen discovers Easter Island. He was sent by the Dutch West India Company to find the *Terra Australis*, a fabled continent believed to exist in the Southern Hemisphere. He found Easter Island instead, and was the first European to reach it. He reported seeing an estimated 3,000 native inhabitants, but did not land on the island.

1792: President George Washington issues the first veto. He issued the veto on a bill that allocated seats in the House of Representatives, believing that it gave Northern states disproportionate power in the Congress.

1900: The Linear B tablet is discovered. The British archaeologist Arthur Evans found the tablet while excavating a site at Crete. The tablet was written in an early form of Greek writing called Linear B by linguists, and was used during the Mycenaean Civilization, about 1300 B.C.E.

1922: The American Birth Control League is founded. Margaret Sanger, a nurse and educator, founded the organization, which opened the first birth control clinic in the United States. She popularized the concept of women controlling their fertility and planning intentional families. In 1942 the League changed its name to Planned Parenthood.

1969: Antiwar demonstrations take place in cities across the U.S. Opposition to the Vietnam War had grown to include a large segment of the American population by this time, and the antiwar demonstrations were some of the largest to date, involving hundreds of thousands of people.

1986: The La Belle discotheque is bombed in West Berlin.
Libyan agents bombed the disco, which was popular with American servicemen, killing three people and injuring over 200. The United States bombed Libya in retaliation, killing between 15 and 30 civilians but missing Muammar Gaddafi, who was the intended target.

April 6

1712: A slave revolt begins in New York City. When the British took control of New Amsterdam in 1664, renaming it New York, they instituted legal restrictions on the lives of enslaved Africans living there, and expanded the slave trade in the city. By 1712, roughly 20 percent of the city's population was in bondage. A group of 23 enslaved men set fire to a building near Broadway in downtown Manhattan to draw the attention of white colonists. When whites arrived to put out the fire, the men attacked them, killing nine and wounding six more. In retaliation, seventy black men were arrested and 21 were executed.

1830: Joseph Smith founds the Church of Christ. It was the forerunner of the Church of Latter Day Saints and the first to use the Book of Mormon as its holy scripture. About 30 members of Smith's Church met in New York State and formally organized the Church. The Church would eventually grow to include more than 14 million members.

1869: John Wesley Hyatt patents celluloid.
He was originally trying to find a synthetic sub-stitute for ivory in manufacturing billiard balls when he made the discovery, which was later widely used in photography.

1896: The first modern Olympics opens.
The French academic Pierre de Coubertin established the International Olympic Committee to organized the revival of the Olympics, which had not been held since antiquity. The 1896 Summer Olympics were held in Athens, Greece, as the

nation had been the site of the ancient Olympics. Fourteen nations participated in the 1896 games, with cycling, fencing, tennis, weightlifting and wrestling among the ten events. The original medals in the first modern Olympics were silver for first place and copper for runners-up. Ten nations received medals,  with the United States winning the most first-place medals and Greece winning the most overall. Over 200 nations would compete in later Olympic games, which have expanded to include 35 sports and over 400 events. Beginning in 1964, a medal named for Pierre de Coubertin has been awarded to Olympians who personify the spirit of sportsmanship in Olympic events.

1929: Impeachment hearings begin against Huey P. Long.
The opinionated populist governor of Louisiana attacked bankers with fierce rhetoric and advocated for the redistribution of wealth via progressive taxation. His opponents attempted to impeach him when he enacted a tax on oil production. During debate of the impeachment proceedings, a brawl erupted in the State Legislature between his supporters and opponents. Long survived the impeachment attempt and went on to serve in the U.S. Senate until his assassination in 1935 by the family member of a political rival.

1973: The designated hitter debuts in the American League.
The rule allows a player to take the place of the pitcher in the batting lineup. The National League does not have a designated hitter.

1994: The Rwandan genocide begins.
The Rwandan dictator and ethnic Hutu Juvénal Hayarimana was assassinated when his plane was shot down. The Tutsi minority was blamed for the assassination, and over the next 100 days close to 1 million were systematically massacred.

April 7

1141: Matilda becomes the first female ruler of England. She was the daughter of Henry I and moved to Italy while he was still King of England with her husband, the Holy Roman Emperor Henry V. Following her husband's death, she remarried a French Count and was named Henry I's heir. She was declared Lady of the English while warring with her cousin Stephen of Blois, who had also made a claim to the throne. Her son eventually led a campaign to overthrow Stephen and ascended the throne in 1154. Matilda died in 1167.

1798: The Mississippi Territory is established. It was acquired from Spain in the Treaty of Madrid. It remained a territory until 1817, when it was divided into the states of Alabama and Mississippi.

1827: The first friction match is sold. The English chemist John Walker developed the match, which was a stick of cardboard coated with sulphur. Walker sold about 160 boxes of matches. They were banned in 1829 because of safety concerns, as they often resulted in flaming balls of sulphur dropping to the floor when they were ignited.

1945: The Allies sink the *Yamato*. The Japanese ship was the largest and most heavily armed battleship ever constructed. It had nine 18-inch guns, the biggest ever mounted on a battleship, and was more than 270 yards long. The ship had a total of 33 artillery guns and another 162 antiaircraft guns. The *Yamato* was the crown jewel of the Japanese navy, and was designed to be able to attack several enemy battleships at once. In 1942 it was made the flagship of the Japanese Combined Fleet and took part in several naval engagements. In 1945 the *Yamato* was sent to Okinawa to engage in a suicide mission to delay Allied forces. Aircraft sent from U.S. carriers intercepted the ship before it reached Okinawa and sank it.

1948: The World Health Organization is founded. The United Nations agency manages public health issues around the world, and has its headquarters in Geneva, Switzerland. The WHO issues a yearly World Health Report and works to prevent pandemics, completely eradicate diseases such as polio and smallpox, and promote nutrition and hygiene.

1949: *South Pacific* debuts on Broadway. The musical, by Richard Rodgers and Oscar Hammerstein, was based on a book by James Michener and told the story of an American nurse stationed in the Pacific during World War II. It was an instant success and ran for 1,925 shows. It won a Pulitzer Prize for Drama and three Tony Awards, and was adapted to film in 1958.

1969: The first Request for Comments (RFC) is sent. The RFC was developed as part of the Advanced Research Projects Agency Network (ARPANET), the forerunner of the modern Internet. The RFC was invented by Steve Crocker to record notes about the development of ARPANET, and the publication of RFC1 is considered the symbolic "birth" of the Internet.

2003: U.S. troops capture Baghdad. Following weeks of bombing runs, U.S. forces took control strategically important crossroads on highways leading into the city, defending them against eighteen hours of Republican Guard assaults, captured a presidential palace, and ordered Iraqi forces in Baghdad to surrender. Much of the city remained unsecured, and civilians engaged in widespread looting over the next few days.

April 8

1730: The first synagogue in New York City is dedicated. The Congregation Shearith Israel was built to serve a Jewish community that had been in lower Manhattan since 1655. It was the only synagogue in New York until 1825 and is the oldest Jewish congregation in the United States.

1820: The *Venus de Milo* statue is discovered. A French naval officer named Olivier Voutier found the statue in two pieces on the Greek island of Milos while he was searching for ancient artifacts on shore leave. The marble statue, which depicts the Greek goddess Aphrodite, was taken to the Louvre Museum in Paris. It is one of the most famous statues of antiquity ever found.

1895: The Supreme Court finds income tax to be unconstitutional. In *Pollock v. Farmers' Loan & Trust Company*, the Court ruled that income taxes were effectively direct taxes that were not based on the population of the state, which the Constitution said that direct taxes had to be. The ruling was superseded in 1913 by the passage of the 16th Amendment, which exempted income tax from it.

1904: Longacre Square is renamed Times Square. The square was originally named after the Long Acre of London because, like its British counterpart, it was the center of horse and carriage trading in New York. When *The New York Times* moved its headquarters to the square, it was renamed in honor of the newspaper.

1911: Heike K. Onnes discovers superconductivity. Onnes was investigating the electrical resistance of mercury at very low temperatures when he found the resistance completely disappeared at 4.2 degrees Kelvin (-452.11 degrees Fahrenheit). He received the 1913 Nobel Prize in Physics for his discovery.

1935: The Works Progress Administration is established. The New Deal agency was organized to put millions of unemployed Americans to work during the Great Depression. WPA workers built public works all over the United States, engaged in music, theater ,and arts projects, and archived historical records.

1959: Grace Hopper organizes the creation of COBOL. The computer programming language was designed to be a portable language for data processing. Hopper went on to become a rear admiral in the U.S. Navy and was awarded honorary degrees from 40 universities around the world. In 1991 she was awarded the National Medal of Technology, and in 2016 she was posthumously awarded the Presidential Medal of Freedom.

1964: Gemini 1 is launched. It was the first test flight of the Gemini program and tested the Titan II expendable rocket launch system. Its success initiated the remainder of the United States' second manned space program.

1992: Arthur Ashe announces he has AIDS. Ashe was an American tennis player who was dominant in the 1960s and 70s. He was the first African-American to play for the U.S. Davis Cup team and the first to win three Grand Slam titles. He was ranked Number 2 in the world in in 1976, and retired from playing tennis in 1980. Ashe most likely contracted HIV when he received a blood transfusion while undergoing heart bypass surgery in the 1980s. It was not discovered until 1988. Ashe originally kept the discovery private, but when *USA Today* found out he was ill and began planning to run a story about it, he decided to preempt the story. He then established the Arthur Ashe Foundation and worked to raise awareness about the disease. Ashe died on February 6, 1993, of complications related to AIDS. He was posthumously awarded the Presidential Medal of Freedom later that year.

April 9

1413: Henry V is crowned King of England. He would later be immortalized by Shakespeare's play about him. In the play, Henry delivers a famous speech to his troops before the Battle of Agincourt, in which he says "we happy few, we band of brothers" will be remembered forever for what they did in the battle. In the actual battle, an outnumbered English Army defeated a much larger French force. Henry ruled England until his death in 1422.

1860: Éduoard-Léon Scott de Martinville records the human voice on his phonautograph. Martinville was a printer and book dealer from France who was interested in finding a way to automatically transcribe human speech. He based his design of the device on the actual human ear, with an elastic sheet substituting for the eardrum, a series of levers replacing the ear bones, and a stylus that would record sound vibrations on a treated surface. The device recorded visual images of sound waves generated when a person spoke near it.

1865: Robert E. Lee surrenders to Ulysses S. Grant at Appomattox. After his Army of Northern Virginia was defeated at the Battle of Appomattox, Lee realized he would not be able to continue his retreat and made the decision to surrender. All but one of his officers agreed that it was the only course of action left to them, and Lee agreed to meet Grant in the small town. The two generals met at the home of Wilmer McLean, whose first home was in Manassas during the First Battle of Bull Run at the beginning of the war. They briefly discussed the last time they had met, during the Mexican-American War nearly twenty years before. Grant then offered Lee his terms of surrender, and Lee accepted them. Lee's officers were allowed to keep their sidearms, and the rebels were also permitted to take their horses and mules home for the spring planting. As Lee rode away from the McLean home, the Union army began cheering, but Grant ordered them to stop. After four years and more than 750,000 deaths, the American Civil War was over.

1940: Vidkun Quisling takes power in Norway. Quisling, the founder of the Norwegian fascist party, seized power as the Nazis were invading Norway. He headed a puppet government during World War II, allowing a Nazi SS brigade to be permanently installed in Norway, and working to exile Jews. He was executed for treason following the war, and his name—*quisling*—is now a synonym for "traitor."

1942: The Bataan death march begins. After Allied forces in the Philippines surrendered to the Japanese Imperial Army on Bataan, they were forced to march more than 60 miles to prisoner of war camps. Tens of thousands of prisoners died during the march from exhaustion, disease, starvation, and summary executions by Japanese troops. The Japanese commanders responsible for the march were tried for war crimes and executed after the war.

1965: The Houston Astrodome opens. It was the first domed sports stadium ever constructed and the first major sports venue to feature artificial turf. It was the home of the Houston Astros until 1999, the Oilers until 1996, and the site of the Houston Rodeo until 2002.

2005: Prince Charles weds Camilla Parker Bowles. They had maintained an affair through both of their former marriages, and the British tabloids broke the news of the relationship in 1993. They made their first public appearance together following the scandal in 1999, and Parker Bowles first met the queen the following year. She is officially the Princess Consort and Duchess of Rothesay.

April 10

1606: The Virginia Company of London is founded. Chartered by King James I to establish colonies in America, it was granted territory on the east coast from Cape Fear, North Carolina, to Long Island, New York. The company established Jamestown, Virginia, the first permanent English settlement in the New World, in 1607.

1864: Maximillian is proclaimed Emperor of Mexico. The Austrian nobleman was granted the title by Napoleon III. Although European powers recognized his rule, the United States did not, and forces loyal to Mexican President Benito Juárez captured and executed him in 1867.

1912: The RMS _Titanic_ departs from England. It carried more than 2,000 passengers and was the largest ship in the world when it went into service. It also had state-of-the-art safety features, including remotely operated watertight hatches, but did not have enough lifeboats for all of its passengers. When the supposedly "unsinkable" ship struck an iceberg after five days at sea, it sank and more than 1,500 passengers and crew died.

1970: Paul McCartney leaves the Beatles. The band had been pursuing individual projects for over a year, with John Lennon embarking on a pacifist tour with Yoko Ono, and George Harrison focusing on expanding the Apple Records label's artists. But McCartney's announcement that he was leaving the band was the final blow that sealed its breakup. McCartney had quietly been working on his own self-titled album, and his unhappiness grew when Ringo and George attempted to prevent him from releasing it. The band had been together for just over ten years, and had completely revolutionized pop music during their tenure.

1972: The Yinqueshan slips are discovered. The slips were an ancient copy of *The Art of War*. These military instructions, written by Sun Tzu around 450 BC, were sealed in a tomb in the Shandong Province in China. The book is composed of thirteen chapters, with each dealing with a different aspect of military strategy, such as planning, maneuvering, or use of spies. The book has been used by military leaders throughout history ranging from Mao to Norman Schwarzkopf. It has also provided inspiration and guidance to business and political thinkers throughout history. The Yinqueshan slips were found in tombs dated to about 120 BC, and their discovery made it clear that Sun Tzu's classic was far older than what historians had previously believed. They were put on display at the Shandong Museum.

1988: The Good Friday Agreement is signed. The Agreement, signed by the United Kingdom and the Republic of Ireland, was a major step in the peace process that ended the Troubles, a period of significant civil unrest and violence. It established a system of government for Northern Ireland, required paramilitary groups such as the IRA to give up their arms, and normalized security from military occupation to civilian police forces.

2010: Polish President Lech Kaczynski is killed in a plane crash. Prior to becoming president, he was the mayor of Warsaw from 2000–2001 and a senator from 2001–2003. His plane crashed while carrying his delegation to Smolensk, Russia, killing everyone aboard.

April 11

1689: William and Mary begin their rule of Great Britain. The Protestant monarchs invaded England and overthrew the Catholic King James II in what became known as the Glorious Revolution. They were crowned at Westminster Abbey by the Bishop of London after the Archbishop of Canterbury, who traditionally performed coronations, refused to recognized them as sovereigns. William and Mary reigned until William's death in 1702.

1881: Spelman College is founded. It was the first liberal arts college for African-American women founded in the United States. Spelman, located in Atlanta, Georgia, has been the alma mater of many notable African American women in its tenure.

1909: Tel Aviv is established. Jewish settlers had established communities in the area beginning in 1884, and during the Second Aliyah, or immigration to Israel, between 1904–1914, many more settled there. The land was divided between 66 Jewish families on this date, and the settlement was named Tel Aviv the following year.

1945: Allied forces liberate Buchenwald. The Nazi concentration camp was established near Weimar, Germany, in 1937. Prisoners from Europe and the Soviet Union including Jewish, Polish, and Slavic people, Jehovah's witnesses, political dissidents, and homosexual people, were sent to the concentration camp, where more than 50,000 were killed. The US 9th Armored Infantry Battalion arrived at the camp at 3:15 P.M. and liberated over 20,000 prisoners.

1951: President Truman fires General Douglas MacArthur.
MacArthur was commanding Allied forces in the Korean War and had achieved several victories. But he had begun publicly contradicting the Truman Administration's policies regarding the war, and Truman had him relieved, and in so doing asserted civilian control of the military in the post-WWII era.

1968: LBJ signs the Civil Rights Act. The legislation was also called the Fair Housing Act, and was meant to reinforce the Civil Rights Act of 1964 by prohibiting discrimination in housing and real estate. The assassination of Dr. Martin Luther King, Jr., a week before the Act prompted Congress to pass it by a wide margin.

1970: Apollo 13 is launched. It was the Apollo program's seventh manned mission and the third in which the goal was to land astronauts on the Moon. James Lovell, John Swigert, and Fred Haise crewed the spacecraft; Swigert had replaced Ken Mattingly after he was exposed to rubella during training. The spacecraft was two days into its mission and had covered more than three quarters of the distance to the Moon when an oxygen tank aboard the Service Module exploded. The explosion crippled power to the Command Module, rendering multiple onboard systems inoperable and leaving the astronauts in the dark and without heat. More critically, the carbon dioxide removal system on the Lunar Module was damaged, and they had to adapt the Command Module's square CO_2 cartridges to fit the Lunar Module's round assembly. They then had to restart the Command Module while in flight and separate the two Modules before reentry to Earth's atmosphere, neither of which had ever been attempted before. They succeeded on all counts, returning safely to Earth six days after liftoff.

1976: The Apple I is invented. Steve Wozniak designed and built the desktop computer by hand, and Steve Jobs used it as the template for the first Apple computers sold. The pair sold about 175 units before introducing the Apple II in 1976.

April 12

1204: The Crusaders sack Constantinople. The sacking of the Christian capital of the Byzantine Empire by mutinous European troops continued for three days. The Crusaders destroyed ancient works of art and holy sites and burned the Library of Constantinople. Pope John Paul II formally apologized for the sacking nearly 800 years later.

1861: The American Civil War begins. Confederate artillery batteries in Charleston, South Carolina, opened fire on the Union garrison at Fort Sumter in the city's harbor. The troops at Fort Sumter returned fire and withstood the bombardment for 34 hours before surrendering. The attack prompted President Lincoln to call for 75,000 volunteers to join the Union Army and led four southern states to secede and join the Confederacy. The war would rage for four years and result in more than a million casualties, making it the deadliest war the nation ever engaged in.

1864: Confederates massacre African-American soldiers at Fort Pillow. Union forces had occupied the fort, located north of Memphis on the Mississippi River, since June 1862. African-American soldiers were recruited by the Union beginning in 1863, and a detachment was defending Fort Pillow when Confederates under the command of Nathan Bedford Forrest attacked it. After capturing the fort with relatively little bloodshed, Forrest's troops massacred the surrendering soldiers. Between 200–350 men were killed in cold blood. Forrest later served as the first Grand Wizard of the Ku Klux Klan.

1945: Franklin D. Roosevelt dies, and Harry Truman replaces him as President. Roosevelt was serving an unprecedented fourth term as President even as his health was declining. He died of a cerebral hemorrhage while resting at the Little White House in Warm Springs, Georgia, following the Yalta Conference. Harry Truman was sworn in as President that day.

1961: Yuri Gagarin becomes the first human to travel to outer space. The Soviet cosmonaut completed an orbit of the Earth in a Vostok Spacecraft, also becoming the first human to do so. His space flight was a major step in the advent of the Space Age in human history. Vostok 1 was launched from the Baikonur Cosmodrome in Kazakhstan. The mission lasted for one hour and 48 minutes, during which time Gagarin completed a single orbit of Earth. While reentering Earth's atmosphere, he experienced capsule gyrations that caused him to undergo g-forces of 8G. Gagarin ejected from the space capsule at 23,000 feet and parachuted to the ground in southwest Russia. Upon landing, he encountered a farmer and her daughter, who were shocked by his sudden appearance from the sky. He said to them "don't be afraid, I am a Soviet citizen…and I must find a telephone to call Moscow!" Gagarin was praised worldwide for his mission.

1981: The first Space Shuttle mission is launched. *Columbia* launched from the Kennedy Space Center with two crew members, John Young and Robert Crippen, aboard. The Shuttle orbited the Earth 37 times over two days before landing at Edwards Air Force Base. The Shuttle program continued until 2011, flying 133 successful missions and two notable failures.

1992: Euro Disney opens in France. The amusement park was modeled after Disneyland in America, and initially was a financial failure, with attendance falling far short of expectations. Disney made significant changes and was able to turn the park's fortunes around by 1995.

April 13

1742: Handel's *Messiah* makes its debut in Dublin. The oratorio was the composer's sixth work in the English language and is a reflection on the life of Jesus in 53 movements. It was immediately acclaimed in the press, and remains a widely popular choral piece.

1873: The Colfax Massacre occurs. White Southern Democrats in Colfax, Louisiana, angry at the extension of voting rights to African Americans during Reconstruction, attacked the Colfax County Courthouse. The Courthouse was defended by African-American Republicans, who surrendered after it was set on fire. The white mob then massacred 150 of them. The incident, and a related Supreme court ruling that the Federal government could not prosecute the perpetrators of it, led to widespread growth of white terrorist organizations in the South.

1902: J.C. Penney opens his first store. Penney had been working for a chain of retail stores when he was offered a partnership in a new store by the owners. He opened his first store in Kemmerer, Wyoming, and expanded the chain through the American West over the next decade. By 1929, he had over 1,400 stores nationwide. The company expanded into mall outlets in the 1980s and launched an online store in 1998.

1941: The USSR and Japan sign a neutrality pact. After a border war that was fought between 1939–1941, the two nations agreed not to attack one another during World War II. The pact lasted until August 9, 1945, when the Soviet Union declared war on Japan and joined the Allies' call for unconditional Japanese surrender.

1953: The CIA initiates Project MK Ultra. The program was a series of human experiments designed to identify interrogation techniques using "mind control" techniques including hypnosis, administration of drugs such as LSD, sensory deprivation and psychological torture. The clandestine program was carried out at hospitals, prisons, and universities, with the CIA hiding its involvement by using front companies to perform the experiments. The program was revealed in a 1984 report by the U.S. General Accounting Office.

1964: Sidney Poitier wins an Oscar for Best Actor. Poitier, who won for his performance in *Lilies of the Field*, was the first African-American man to win an Academy Award.

1974: Western Union launches a satellite. The *Westar 1* was the first commercial geostationary communications satellite. It was launched from Cape Canaveral and designed to send internal telegrams and mailgrams to Western Union offices. Other organizations, including PBS, NPR, and HBO also paid Western Union to use the satellite during the next decade. The satellite was decommissioned in 1983.

1992: The Great Chicago Flood begins. A wall of a utility tunnel under the Chicago River collapsed, flooding basements throughout the city's central business district, the Loop. Over 250 million gallons of water flooded the district, causing power outages and leading to an evacuation of the area. The flood continued for three days before city crews were able to locate and plug the leak, and cost nearly $2 billion in estimated damage and loss of business.

1997: Tiger Woods wins the Masters Tournament. At 21 years old, he was the youngest player ever to win the Masters. Woods won the tournament by twelve shots, the widest victory margin ever, and set a 72-hole record that stood for over three decades.

April 14

1561: Strange lights appear in the sky above Nuremberg. The residents of the city said they saw what appeared to be a battle in the sky, with hundreds of lights, spheres, and strange-looking objects flying through the sky. The apparition ended with a large black triangular object appearing, followed by a crash outside the city. The phenomenon has been variously explained as a UFO sighting or a naturally occurring solar event.

1775: The first antislavery organization in America is founded. The Pennsylvania Abolition Society was founded in Philadelphia by a group of Quakers that included Thomas Paine. In 1785, Benjamin Franklin was elected the Society's President, and he later petitioned the U.S. Congress to outlaw slavery.

1828: Noah Webster copyrights a dictionary. His *American Dictionary of the English Language* was the most extensive published at the time, containing seventy thousand entries. Its first edition sold for $15, a relatively large sum at the time. Webster's dictionary would be acquired by George Merriam in 1847 and republished as the *Merriam-Webster's Dictionary*.

1865: John Wilkes Booth assassinates Abraham Lincoln. Lincoln was attending the play *Our American Cousin* at Ford's Theater in Washington, D.C., when Booth entered his viewing box and shot him in the back of the head. Booth, a Confederate sympathizer, had initially planned to kidnap Lincoln and force the Union to release Confederate prisoners during the U.S. Civil War. When the Confederacy surrendered, Booth decided to assassinate the President instead. He conspired with three others to strike several blows at the Executive Branch at once: George Atzerodt was assigned to assassinate Vice President and Lewis Powell was tasked with killing Secretary of State William H. Seward. Atzerodt lost his nerve, and Powell only wounded Seward. Booth shot Lincoln with a Deringer pistol and leapt from the viewing box onto the stage, breaking his ankle as he landed, and shouted "*Sic semper tyrannis!* The South is avenged!" He escaped the theater and made his way to Virginia, where he was found hiding in a barn and killed in a shoot-out. His co-conspirators were hanged for their part in the assassination plot. Abraham Lincoln died the morning after he was shot and was mourned by the nation.

1906: The Azusa Street Revival launches the Pentecostal religion. William J. Seymour, an African-American Pentecostal preacher in Los Angeles, began the revival. It lasted until 1915 and prompted many attendees to start their own Pentecostal churches around the United States. The religion eventually grew to more than 500 million members.

1935: A severe dust storm hits Oklahoma and Texas. Cattle and sheep ranching had left much of the American West devoid of grass to anchor topsoil, and a massive drought combined with existing conditions to cause the Dust Bowl, a series of severe regional dust storms. In 1935, the "Black Sunday" dust storm displaced an estimated 300 million tons of topsoil and caused thousands of people to relocate.

1939: John Steinbeck publishes *The Grapes of Wrath*. The book, which tells the story of poor tenant farmers driven from their home in Oklahoma during the Dust Bowl and Great Depression, won the National Book Award and Pulitzer Prize for fiction.

2014: Boko Haram kidnaps 276 female students in Nigeria. The extremist organization kidnapped the young women from their school in northeast Nigeria, prompting an international outcry and "Bring Back Our Girls" campaign.

April 15

1755: Samuel Johnson publishes his dictionary. It is considered one of the most influential dictionaries of the English Language and took him seven years to complete. The dictionary was the preeminent English Dictionary until the *Oxford English Dictionary* was published in 1928.

1817: The American School for the Deaf is founded. It was the second attempt to found a school for the deaf in the United States, and the first successful one. Established in West Hartford, Connecticut, by Laurent Clerc, Thomas Gallaudet, and Dr. Mason Cogswell, it is the oldest institution for the deaf in America.

1892: General Electric is created. Thomas Edison's General Electric Company and the Thomson-Houston Electric Company merged to form the General Electric Company, headquartered in Schenectady, New York. It grew to become a multinational corporation with interests in television, power generation, and computing, worth several hundred billion dollars.

1924: Rand McNally publishes its first road atlas. The company, founded by William Rand and Andrew McNally, originally was a printing shop that did most of its business printing the *Chicago Tribune*. It began producing maps in 1872 and publishing a Business Atlas with maps and corporate data in 1876. Rand McNally began publishing road maps in 1904, and published its first comprehensive road atlas as the black-and-white *Rand McNally Auto Chum*.

1947: Jackie Robinson debuts for the Brooklyn Dodgers. In doing so, he became the first African American to play in Major League Baseball in the modern era. His signing brought about the end of racial segregation in professional baseball, and also spelled the beginning of the end of the Negro Leagues, which had operated since the 1880s. He endured death threats, racist taunts, and violence to do so. Robinson played for ten years, winning the Rookie of the Year in 1947 and National League MVP in 1949, playing in six World Series, and winning one in 1955.

1955: Ray Kroc opens his first McDonald's. The restaurant chain had existed in California since the 1940s, but Kroc's influence made it into a major brand. A former milk shake machine salesman, Kroc joined the company when he saw a McDonald's in San Bernardino, California. He invented the franchise model of fast food restaurants and grew the company into a national and eventually a global corporation.

1960: The Student Nonviolent Coordinating Committee is founded. Ella Josephine Baker, a civil rights activist, organized a student meeting at Shaw University in North Carolina to plan protests against segregation. At the meeting, SNCC was born. It would lead direct-action protests across the south, including sit-ins of lunch counters and public venues.

2013: Bombs explode at the Boston Marathon. The brothers Dzhokhar and Tamerlan Tsarnaev planted bombs near the marathon's finish line, which killed three people and seriously injured hundreds.

April 16

73: The fortress of Masada falls. The fortress was defended by the Sicarii, a group of Jewish rebels fighting the Roman occupation. They had fled to the fortress atop a high plateau, after defeating the Roman garrison at Jerusalem. They held out for three months while the Romans built a siege ramp to the plateau and began breaching the walls. Rather than be captured and enslaved, the 960 Sicarii committed mass suicide.

1746: The Battle of Culloden ends the Jacobite rebellion. The Highland Jacobites had rebelled in an attempt to seize the British Throne from Queen Anne for Charles Stuart. Following the defeat at Culloden, the British cleared the Scottish Highlands of inhabitants.

1862: Slavery is abolished in the District of Columbia. The Compensated Emancipation Act provided for payment of up to $300 to slave owners by the government for freeing their slaves. More than 3,000 slaves were freed as a result of the Act. Nine months later, President Abraham Lincoln signed the Emancipation Proclamation.

1943: Albert Hofmann accidentally discovers the effects of LSD. Hofmann, a Swiss chemist, was working at a pharmaceutical company when he made the discovery. He was researching new stimulants made from derivatives of lysergic acid when he synthesized LSD in 1938. He did not find any use for it and set the compound aside for five years. While working with it again in 1943, he accidentally touched his hand to his mouth, ingesting a small amount. He found himself later experiencing hallucinations and a dreamlike state for about two hours. A few days later, Hofmann deliberately ingested LSD, and experienced the effects of the drug while riding home on his bicycle. He was later critical of the hippie counterculture's use of LSD, arguing that it should be primarily applied to psychoanalytical pursuits and therapeutic applications.

1945: The Battle of the Seelow Heights. More than a million Soviet Red Army soldiers assaulted entrenched German positions outside Berlin. The Heights were the last defensive line protecting the city from being captured. After two days of heavy fighting, the Soviets broke through the German lines and began a two-week battle to take Berlin.

1961: Fidel Castro declares that he is a Marxist. The Cuban leader had carefully concealed his Marxism during the revolution to overthrow Fulgencio Batista, in order to gain support of less radical organizations. By 1961, however, this was no longer necessary, and in a televised speech he declared his political ideology and called Latin America to rise up in communist revolution.

1963: Dr. Martin Luther King Jr. writes "Letter from Birmingham Jail." The political essay argued for the moral responsibility of using nonviolent resistance to protest unjust laws and defended the strategy in overthrowing Jim Crow. He wrote it in response to "A Call for Unity," written by a group of white clergymen that condemned his methods. .

1990: Jack Kevorkian performs his first assisted suicide.
Kevorkian publicly championed the right of terminally ill
patients to die by euthanasia. He claimed to have participated
in 130 assisted suicides during his career. He was arrested in
1999 for his role in a physician-assisted suicide and served eight
years in prison.

April 17

1397: Chaucer performs *The Canterbury Tales*. The English
poet was also a civil servant of King Richard II, and he first
performed his collection of stories for the King's royal court. *The
Canterbury Tales* are written as a group of stories told by pilgrims
while they journey to a religious shrine in Canterbury Cathedral.
It was the first major instance of literature written in English,
rather than French or Latin, and Chaucer is considered the creator of English literature because of it.

1521: The trial of Martin Luther begins at the Diet of Worms.
The Diet of Worms was an imperial congress of the Holy
Roman Empire in the city of Worms, in what is now Germany.
Luther was asked whether he stood by his *Ninety-Five Theses*,
which were highly critical of the Catholic Church. He responded
the following day, refusing to recant his work. The Diet issued
a decree that declared Luther a heretic and ordered his arrest.
Luther escaped with the help of Frederick III, a Holy Roman
elector, who concealed him in his castle, where Luther
continued to develop doctrines that became the basis of the
Protestant Reformation.

1942: Henri Giraud escapes. Giraud was a French general who
had been captured by the Germans in the Ardennes while leading a reconnaissance patrol. He was taken to the high-security
prison at Königstein Castle, where he spent the next two years
plotting his escape. He memorized maps of the area and made
a 150-foot rope that he used to escape. He made his way to
Switzerland and then to France.

1961: The Bay of Pigs invasion begins. The operation was undertaken by Cuban exiles who were sponsored by the U.S. Central Intelligence Agency to try to overthrow the Cuban government of Fidel Castro. In the wake of the Cuban Revolution of 1959, Castro had grown increasingly friendly with the Soviet Union, and was showing signs of commitment to worldwide Marxist revolution. President Dwight D. Eisenhower approved a $13 million operation to recruit and train anti-Castro exiles in 1960, and President John F. Kennedy signed off on the invasion plans the following year. The Cuban government knew about the impending investigation, having been warned by their own intelligence agents in Miami and the Soviets. In the early morning of April 17, five infantry brigades made an amphibious landing at the Bay of Pigs, but were immediately overpowered by a local revolutionary militia and became trapped on the beach. Castro took personal control of the defenses, and after Kennedy declined to provide the invaders with American air support, the attack collapsed.

1975: The Khmer Rouge captures Phnom Penh. The Cambodian communist rebels had waged a war against the corrupt military dictatorship of the Khmer Republic for eight years. The Viet Cong and North Vietnamese Army supported the Khmer Rouge, while South Vietnam and the United States supported the Republic. Once in power, the Khmer Rouge evacuated the city and forced the residents to move into agricultural communes and labor camps. Intellectuals, Buddhists, Muslims, and ethnic minorities were massacred, and many others died of starvation, with the death toll estimated to be as high as 2.2 million.

2014: Kepler-186f is discovered. It was the first Earth-sized planet discovered in the "habitable zone" of another star—the region within which a planet can have liquid water with the right atmospheric conditions.

April 18

1506: Construction of St. Peter's Basilica begins. The Vatican City cathedral was designed by Michelangelo, Carlo Maderno, Donato Bramante, and Gian Bernini, and is the largest church in the world. The construction was financed by the selling of indulgences by the Catholic Church. The practice was central to Martin Luther's criticism in the 95 *Theses*, which led to the Reformation.

1775: Paul Revere makes his midnight ride. At the beginning of the American Revolutionary War, a detachment of 700 British troops was sent from Boston to seize and destroy an arms cache belonging to the Massachusetts militia at Concord. An intelligence network organized by Patriots learned of the British plans weeks in advance and prepared to ride ahead of the British column when it departed to warn militias. The arrival of the British troops was signaled to the riders from the belfry of Boston's Old North Church: one lantern communicated that they were coming on an overland route, and two lanterns signaled they were coming by boat across the Charles River. Revere and William Dawes rode through the countryside alerting colonial militias that the British regulars were coming. Revere passed the warning to militias in Somerville, Medford, and Arlington, triggering a network of other riders to spread the news. He was captured and briefly detained by a British patrol before being released without his horse. He helped John Hancock escape from Lexington while Minutemen fought the British on the town green in the opening battle of the war.

1847: American forces win the Battle of Cerro Gordo. The Americans, under General Winfield Scott, attacked a larger force of Mexican troops commanded by General Santa Anna. Although the Mexicans had a defensive advantage, the Americans were able to outflank them. Ulysses S. Grant and Robert E. Lee were notably among the American officers who took part in the battle.

1906: A massive earthquake strikes San Francisco. The earthquake, which hit with an estimated magnitude of 7.8, caused fires to break out all over the city, which burned it to the ground. As many as 3,000 people were killed in the earthquake and subsequent fires, making it one of the deadliest in the history of the United States. More than 80% of the city was destroyed, and $400 million worth of property was lost.

1923: Yankee Stadium opens. It was called "The House that Ruth Built" in commemoration of Babe Ruth's dominant years with the New York baseball club. The Yankees played there until 2008, when they moved to a newer stadium across the street.

1943: U.S. fighter planes shoot down Admiral Yamamoto. The Japanese naval commander had planned and led the attack on Pearl Harbor, and the operation was code-named Operation Vengeance because of it. U.S. Army Signals Intelligence decrypted the Admiral's itinerary and discovered he would be flying to the Solomon Islands in New Guinea. Eighteen U.S. fighters intercepted his plane over the Bougainville Island and shot it down, killing Yamamoto.

1983: The U.S. Embassy in Beirut is bombed. A suicide bomber drove a delivery van packed with one ton of explosives into the Embassy compound, killing 63 employees and civilians in the blast. The attack was in response to U.S. intervention in the Lebanese Civil War, and at the time was the deadliest attack on a U.S. diplomatic mission overseas.

April 19

1770: Marie Antoinette marries Louis XVI. Antionette was the daughter of the Holy Roman Emperor Francis I. Upon marrying Louis, she became the Dauphine of France, and was crowned Queen of France in 1774 when he ascended the throne. Antoinette was arrested for treason during the French Revolution along with the royal family. She was then executed in 1793.

1861: A pro-secession mob riots in Baltimore. When a detachment of Union soldiers made their way through the city by train, Confederate sympathizers attacked the train cars and blocked the route. The soldiers left the train and marched through the city, attacked by rioters wielding bricks and pistols. They opened fire on the mob, fighting them in a running battle through the city. Four soldiers and 12 civilians were killed; the incident was the first blood spilled in the Civil War.

1927: Mae West is jailed for obscenity charges. The actress wrote and starred in a Broadway play entitled *Sex*, which enjoyed multiple sold-out performances before religious groups complained to authorities. The theater was raided during a performance, and West was arrested along with the rest of the cast. She served eight days in jail. West wrote several other plays that dealt with risqué issues and societal norms, and went on to be a Hollywood star in a career that spanned six decades.

1957: Grace Kelly marries the Prince of Monaco. Prince Rainer III and Kelly met at the Cannes Film Festival in 1955. They courted for a year before marrying, and she was made Princess of Monaco in what the press called "The Wedding of the Century." The couple had three children.

1971: The first space station is launched. The Salyut 1 was designed by the Soviet Union to support a crew of three cosmonauts in low Earth orbit. Its first crew made contact with the space station aboard the Soyuz 10 on April 23 but were unable to enter it and were forced to abort. Its second crew successfully docked the Soyuz 11 with the station in June and stayed aboard for 23 days, orbiting the Earth 362 times. Upon reentry of the Earth's atmosphere, their space capsule malfunctioned, losing cabin pressure, and all three cosmonauts were killed.

1987: *The Simpsons* debuts on *The Tracey Ullman Show*. The animated short "Good Night Simpsons" had a two-minute run time. Each of the three children has a crisis at bed time that results in the entire family sleeping in their parents' bed.

The show would go on to become the most successful animated television show in history, running for more than 30 seasons and winning dozens of Primetime Emmy Awards.

1995: Timothy McVeigh bombs Oklahoma City. McVeigh, a disaffected Gulf War veteran, committed the terrorist attack as revenge for the federal government's siege of the Branch Davidian compound in Waco, Texas, that left 76 people dead. With the help of Terry Nichols, McVeigh detonated a 5,000-pound truck bomb parked in front of the Alfred P. Murrah Federal Building in Oklahoma City, killing 168 people and injuring more than 680. He stated before his execution in 2001 that his only regret was having failed to completely destroy the building.

2011: Fidel Castro resigns as Secretary of the Cuban Communist Party. He had stepped back from his duties as President of Cuba in 2008 due to declining health. His brother Raúl succeeded him as President, and was his successor as General Secretary as well. He continued to meet with foreign dignitaries and addressed the Communist Party several times before his death in 2016.

April 20

1534: Jacques Cartier sets out for Canada. Cartier was commissioned by the French King Francis I to seek a Northwest Passage by which ships could sail above the Americas to reach Asia. He was the first European to explore Newfoundland, Prince Edward Island, and the Gulf of St. Lawrence, and claimed Canada as a French possession.

1817: The last trial by combat occurs in England. Abraham Thornton was being tried a second time for the murder of Mary Ashford after having been acquitted the first time. Ashford's brother William appealed for a retrial, claiming that the evidence against Thornton was overwhelming. Thornton demanded the right to trial by combat, a practice that had been common in the medieval period and fallen out of usage, but never formally repeated by Parliament. The court granted his request. Ashford declined to meet Thornton in battle, and Thornton was set free. As a result of the trial, the Parliament hastily passed a bill abolishing trial by combat. Thornton, who was widely reviled by those who thought he was guilty, sailed to America, where he lived until he died in Baltimore in 1860.

1898: The Spanish-American War begins. After the USS *Maine* exploded in the harbor of Havana during the Cuban War for Independence, the United States entered the war. The war lasted four months and ended with the United States gaining control of Cuba, Puerto Rico, Guam and the Philippine Islands for $20 million.

1916: The Chicago Cubs play their first game at Wrigley Field. At the time the ballpark was called Weeghman Park after its then-owner. The Cubs beat the Cincinnati Reds 7–6 in eleven innings.

1946: The League of Nations is dissolved. The international body was established after World War I to mediate disputes between nations and prevent the outbreak of wars. It failed in that purpose due its inability to enforce its mandates, and its weakness allowed the Second World War to break out. Following the War, the Allies organized the United Nations to replace it.

1999: The Columbine school massacre. Two students at Columbine High School brought an arsenal of weapons to the school and murdered twelve students and one teacher. At the time it was one of the worst mass shootings in American history. The perpetrators also planted explosives that failed to detonate, and committed suicide following the shooting spree.

2008: Danica Patrick wins the Indy Japan 300. In doing so she became the first woman to win an Indy Car Series race, finishing nearly six seconds ahead of the next driver. She went on to become the first female driver to win a NASCAR pole, at the Daytona 500, and first to lead an Indianapolis 500 race.

2010: Deepwater Horizon explodes. The explosion of the oil drilling rig in the Gulf of Mexico killed 11 workers and resulted in the largest environmental disaster in U.S. history at the time.

April 21

753 BC: Romulus founds Rome. The legendary first king of Rome is said to have laid out the boundaries of the city with a plough and fortified the Palatine hill on this day. The founding of the city was commemorated by ancient Romans every year with the festival of the Parilia.

1509: Henry VIII ascends the throne of England. In doing so he became the second King of England of the House of Tudor, after his father. He would marry six times, clash with the Pope on his right to annul his marriage to Catherine of Aragon, and have two of his wives executed. His inability to have the Pope annul his marriage led him to split with the Catholic Church, founding the Church of England and sparking the English Reformation.

1836: Texas rebels win the Battle of San Jacinto. The Texans, led by General Sam Houston, attacked the Mexican Army under General Antonio López de Santa Anna. The Texans famously shouted "Remember the Alamo!" and "Remember Goliad!" as they attacked, referring to previous battles in the war that had seen the Mexican Army massacre Texan prisoners. The Mexicans, outnumbered, on unfamiliar terrain, and exhausted from a 24-hour march the day before, were routed in fewer than eighteen minutes. The Texans slaughtered Mexican infantry even after they surrendered, killing more than 600 in total. Santa Anna hid in a nearby marsh but was captured the following day and brought before Houston. He negotiated treaties that allowed the Mexican Army to withdraw from Texas, but the treaties were never ratified. But the defeat at San Jacinto had effectively ended the war and established the Texas Republic as a sovereign nation.

1918: The Red Baron is shot down. Manfred von Richthofen was a German pilot who was credited with 80 aerial combat victories during World War I. He was considered the top flying ace of the war. Richthofen was shot down by Canadian pilots over the Somme River while engaging them in a dogfight.

1934: A photo of the Loch Ness Monster is published. It was reportedly taken by Robert Kenneth Wilson and was called the "surgeon's photograph" after his occupation. The photo was widely reproduced, fueling many theories about its origin. It was later found to be a hoax.

1966: Haile Selassie visits Jamaica. The Ethiopian Emperor was believed to be a messiah of God or an incarnation of God himself by the Rastafarian religion, whose largest population is on the island. About 100,000 Rastafari met him at the Kingston airport. The day is celebrated as Grounation Day, the second most holy day of the calendar, by Rastafari.

1982: Rollie Fingers becomes the first pitcher to record 300 saves in baseball. He had already won the Cy Young Award and American League MVP Award in 1981, and was later just the second relief pitcher elected to the Baseball Hall of Fame.

1992: The first discovery of planets outside the solar system is announced. The planets, named "Draugr," "Poltergeist," and "Phobetor," orbit a pulsar located 2,300 light years from the Solar System in the constellation of Virgo.

April 22

1529: The Treaty of Zaragoza is signed. The treaty divided the Pacific into "spheres of influence" controlled by Spain or Portugal. The treaty gave Portugal control over the region including present-day Indonesia, and maintained Spain's control of the Americas west of the Tordesillas line, a point 370 leagues west of the Cape Verde Islands.

1864: "In God We Trust" is first minted on U.S. coins. In the Coinage Act, Congress authorized the use of the phrase on the two-cent coin. The coin continued to be minted until 1873, when it was replaced by the nickel. The phrase was officially adopted as the national motto, replacing *E Pluribus Unum* ("Out of Many, One"), in 1956.

1876: The first National League baseball game is played. This was the beginning of American Major League Baseball. The game was played between the Philadelphia Athletics and the Boston Red Stockings, and Boston won the game 6–5.

1889: The Land Rush of 1889 begins. The United States government had made two million acres of public lands in Oklahoma available for settlement with the passage of the Indian Appropriations Act of 1889. About 50,000 people lined up for the land rush, and at high noon they streamed into the area, staking claims to homesteads. Oklahoma City and Guthrie became cities of more than 10,000 inhabitants in less than a day.

1915: Chlorine gas is used in the Battle of Ypres. It was the first successful use of the deadly chemical gas in warfare. The German Army released over 170 tons of chlorine gas over four miles of trenches occupied by French troops. The French sustained more than 6,000 casualties within minutes. By July the Allies were using gas masks all along the front lines.

1944: Combat helicopters are first used in battle. The 1st Special Operations Wing was established to drop commandos in the Pacific theater of World War II. Six Sikorsky R-4 helicopters performed a combat rescue, retrieving a downed British pilot and crew, two men at a time, in Burma.

1970: The first Earth Day is celebrated. Roughly 20 million Americans attended festivals, demonstrations, and speeches supporting environmental reform in communities across the country. It spread to over 190 countries around the globe and is coordinated by the Earth Day Network.

2016: The Paris Agreement is signed. The agreement was signed at the United Nations headquarters in New York by 175 representatives from around the world. Signatory nations voluntarily agreed to take measures to reduce greenhouse gas emissions in an attempt to keep the global temperature rise below 2 degrees Celsius above preindustrial levels.

April 23

1516: The Reinheitsgebot codifies the recipe for beer. The law was adopted in Bavaria to limit the acceptable ingredients in beer to only water, barley and hops. When Germany unified in 1871, Bavaria demanded the law be nationalized, and other ingredients were heavily taxed as a compromise. The law continued to influence German beer-making well into the 20th century.

1849: Fyodor Dostoyevsky is arrested. Dostoyevsky, who would later become one of the most acclaimed novelists in Russian literature, was a member of a revolutionary group called the Petrashevsky Circle. They spent eight months in prison and endured a mock execution in front of a firing squad before being released at the last moment.

1867: William Lincoln of Providence, Rhode Island, patents the Zoetrope on behalf of Milton Bradley. The machine showed animated films by showing the viewer a series of pictures through a narrow slit, which made them appear to be moving. Milton Bradley released twelve Zoetrope strips for public viewing.

1940: The Rhythm Club fire. The dance hall in Natchez, Mississippi, was hosting a performance by a jazz orchestra from Chicago when the club caught fire. People became trapped as the only exit was through the single front door, and 209 died from smoke inhalation or by being crushed by the crowd. The orchestra continued playing as the fire spread to try to keep the crowd calm, and the band leader, Walter Barnes, died in the blaze.

1945: Hitler replaces Hermann Göring. Göring was Hitler's presumed successor in the event of his death. He sent Hitler a telegram requesting permission to assume leadership as Hitler sought refuge from the advancing Soviet Red Army in an underground bunker in Berlin. Fearing a coup was taking place, Hitler became enraged and ordered Göring to resign, replacing him with Joseph Goebbels. Three days later, Hitler committed suicide as the Red Army closed in.

1968: Students at Columbia University take over administration buildings. Members of the Students for a Democratic Society (SDS), an antiwar group, had discovered that the university had a secret institutional affiliation with a think tank associated with the Department of Defense. Combined with bitterness over the assassination of Martin Luther King, Jr. and the university's plans for a segregated gymnasium, the affiliation sparked an uprising that led hundreds of students to occupy the campus in a student strike for a week.

1985: Coca-Cola releases New Coke. The company had been losing market share for years to diet drinks and its main rival, Pepsi. In an attempt to rebrand itself with a new formula, it introduced New Coke, but the public reaction to it was overwhelmingly negative. When it reissued the original formula as "Coca-Cola Classic," it enjoyed a huge market rebound.

2005: The first YouTube video is uploaded. The nineteen-second video, titled "Me at the zoo," showed Jawed Karim, the cofounder of the website, standing in front of a group of elephants at the San Diego Zoo. Although the video had relatively low production quality, it was a successful proof of concept that ushered in the era of internet videos. Since it first aired, "Me at the zoo" has received tens of millions of views. Youtube was purchased by Google in 2006 for $1.65 billion. While Karim owned only a small stake of the company, the purchase was significant enough to net him more than $60 million worth of shares.

April 24

1558: Mary, Queen of Scots, weds the French Dauphin. Mary was the queen consort of France while the Dauphin was on the throne, but he died only two years into their marriage and she went home to Scotland.

1800: The Library of Congress is established. Founding Father James Madison came up with the proposal for a congressional library in 1783, and an act of Congress signed by President John Adams made it a reality nearly a quarter century later. The original collection had 740 books. The Library was destroyed when the British burned the Capitol during the War of 1812, and all but one of its books were lost. The collection was replaced when Thomas Jefferson sold his library of 6,487 books to Congress later that year. It would grow to eventually become the largest library in the world, with more than 16 million books and over 120 million other items.

1885: Annie Oakley is hired by Buffalo Bill's Wild West. Oakley, an American sharpshooter, was discovered by Frank E. Butler when she beat him in a shooting contest at the age of fifteen. They later married. The couple joined the Wild West show and toured the country for several years.

1915: The Armenian Genocide begins. Ottoman authorities arrested roughly 250 Armenian intellectuals and community leaders in Constantinople (now Istanbul). They were deported to Ankara and eventually killed. The Genocide continued for nine years, during which time 1.5 million Armenians were systematically killed by the Ottoman and Turkish governments. The word *genocide* was coined by Raphael Lemkin to describe the extermination of Armenians.

1916: Irish rebels launch an uprising in Dublin. Led by Patrick Pearse, James Connolly, and the Irish Republican Brotherhood, hundreds of Irish Volunteers seized strategic positions around the city and declared the Irish Republic. Although the rising was put down in under a week and its leaders executed, it was the beginning of the revolution that eventually led to Irish independence from Britain.

1918: First tank vs. tank combat. Although tanks had been in action in World War I for two years, they had mainly been used to breach gaps in infantry lines and overcome trenches, but had not encountered one another. Three British Mark IV tanks were surprised by three German A7V tanks at Villers-Bretonneux. They fired their machine guns at one another until artillery fire damaged the British tanks, forcing them to withdraw.

1980: Operation Eagle Claw fails. The operation was an attempt to free prisoners during the Iran hostage crisis. The mission was doomed from the start, with several helicopters rendered useless by a sand storm. One of the helicopters crashed, killing eight commandos. The failure significantly contributed to Jimmy Carter's loss in the 1980 presidential campaign.

2013: The Savar Building collapses in Bangladesh. The five-story building collapsed in the middle of the day, killing 1,134 people and injuring 2,500 more. The disaster was the worst modern building collapse and the deadliest accident in a garment factory in history.

April 25

1792: Nicolas Pelletier is the first person executed by guillotine. Pelletier was a French bandit who killed a man during a robbery in October 1791. He was captured almost immediately, tried, and sentenced to death. The French Assembly had just legalized the use of the guillotine, and Pelletier waited in prison for three months while one was built. He was executed in front of the Paris city hall while a crowd watched the novel form of execution be carried out.

1846: The Thornton Affair sparks the Mexican-American War. A force of 1,600 Mexican soldiers had crossed the Rio

Grande, where a U.S. cavalry detachment of eighty riders encountered them. The cavalry was immediately routed, with six men killed and the rest all captured. U.S. President Polk immediately asked Congress to declare war, which they did.

1859: Construction of the Suez Canal begins. It would take 1.5 million laborers nearly ten years to complete, and thousands died during its construction, mainly from disease. Many of the workers were unpaid, forced laborers called corvées. The canal was opened in November 1869.

1901: New York mandates automobile license plates. The first license plates were only required to display the initials of the driver. In 1903 the state changed to plates showing black numerals on a white background. New York did not issue the license plates at first; instead, drivers were expected to make their own.

1945: American and Soviet troops meet at the Elbe River. The Soviet Red Army was advancing through Germany from the east, and elements of the American First Army from the west, when they first made contact at the river. Photos commemorating the meeting were hastily arranged and publicized to spread the word that the war was nearly over in Europe.

1953: DNA's double helix is announced. Francis Crick and James D. Watson published a paper describing the molecular structure of DNA in *Nature*. Along with Maurice Wilkins and Rosalind Franklin, they had determined the structure by photographing DNA using X-ray diffraction. The photograph, known as "Photo 51," clearly shows a diffraction pattern that indicates a double-stranded structure. Watson and Crick developed a model of the DNA chain based on their observations of the photo. In 1962 they shared the Nobel Prize in Physiology or Medicine with Wilkins. Franklin had died in 1958, and the award is only given to living persons. The discovery led to understanding how DNA is copied and repaired, as well as the development of genetics as a field of research and to the discovery of how to sequence genes. .

1960: The USS *Triton* circumnavigates the Earth. Captained by Edward L. Beach, the nuclear-powered submarine was the first to circumnavigate the globe while submerged. It covered more than 30,000 miles in just sixty days without ever breaching the surface.

2004: The March for Women's Lives. An estimated 800,000 to 1 million women marched on the National Mall in Washington, D.C. to demonstrate against legislation restricting access to abortion. At the time, it was one of the largest protests in history.

April 26

1777: Sybil Ludington alerts patriots to the approach of British troops. The sixteen-year-old rode forty miles through New York and Connecticut to alert Patriot militias that British regulars were planning to attack a Continental Army arms cache at Danbury, Connecticut. She began her ride at 9 P.M. and continued until dawn the following day.

1805: Marines capture Derne during the First Barbary War. This was the first land battle fought by the United States after the American Revolutionary War, and the first victory in the Barbary War. Marines marched 500 miles through desert from Egypt to Libya, where they attacked and defeated a much larger force.

1865: Joseph E. Johnston surrenders to William Tecumseh Sherman. Johnston had learned of Lee's surrender to Ulysses S. Grant seventeen days before, but was able to negotiate marginally better terms with Sherman. Johnston's surrender was the largest of a Confederate force during the Civil War.

1933: The Gestapo is founded. The organization was the official secret police of the Nazi regime, and combined several existing police agencies into one. The Gestapo was established by Hermann Göring, Hitler's number two man, and was a parallel to the intelligence apparatus in the Nazi SS.

1937: Guernica is bombed by the _Luftwaffe_. The Basque town was a Republican communications hub during the Spanish Civil War, and Nationalist leader Francisco Franco requested Nazi Germany provide aerial support to destroy it. The bombing killed at least 400 and as many as 1,600 civilians, and was one of the deadliest aerial attacks on a civilian target at the time. It later became the subject of a famous anti-war painting by Pablo Picasso.

1956: The first container ship is launched. The SS _Ideal X_ was originally a World War II oil tanker that was converted to carry commercial shipping containers. On its first voyage it carried 58 containers from Newark, New Jersey, to Houston, Texas.

1962: Ranger 4 crashes into the Moon. The lunar probe was designed to transmit pictures of the Moon's surface to Earth for ten minutes before a planned crash, but its onboard computers failed, and it was unable to send any data. It was, however, the first U.S. spacecraft to reach another celestial body.

1986: The Chernobyl disaster. A nuclear reactor in the Chernobyl Power Plant exploded, leading to a serious fire raging through the building. The fire burned for nine days, creating updrafts that expelled large amounts of radioactive particles into the atmosphere. The particles resulted in nuclear fallout affecting Europe and the USSR in what was the worst nuclear disaster to date. The site is surrounded by a 19-mile exclusion zone that has largely reverted to a natural state. It is estimated that it will be unsafe for human habitation for at least 20,000 years.

1991: Seventy tornadoes hit the central United States. The tornadoes touched down in ten states from Texas to Wisconsin over a period of 19 hours, killing 21 people and injuring hundreds. Property damage was estimated to be above $250 million.

April 27

1667: John Milton sells *Paradise Lost* for £10. The English poet was destitute, and sold the publication rights to the publisher Samuel Simmons. Milton had dictated the poem to his family, because he had gone blind a few years earlier. The piece, which tells the Biblical story of Adam's fall, went on to be considered one of the greatest epic poems ever written in the English language, and Milton followed it in 1671 with *Paradise Regained*.

1861: President Abraham Lincoln suspends *habeas corpus*. The law prohibits unlawful imprisonment without charge: a writ of *habeas corpus* is an order to bring the prisoner before the court to determine whether they are being lawfully held. Lincoln suspended it to allow the Union to imprison suspected Confederate agents and sympathizers without sufficient evidence.

1945: Benito Mussolini is captured by Italian partisans. The fascist dictator was trying to escape to Switzerland with his partner Clara Petacci when they were caught. Two days later they were summarily executed.

1953: Operation Moolah begins. During the Korean War the United States offered North Korean pilots $50,000 if they would defect to the South with a MiG-15 jet fighter. At the time the MiG was outperforming the United States' aircraft, and the U.S. was desperate to acquire and study one. While no pilot defected during the war, No Kum-Suk did so in 1953, although he was unaware of the program. He was paid $100,000 and relocated to the United States.

1967: Expo 67 opens in Montreal. It was the most successful World's Fair of the 20th century, with 62 nations participating and more than 50 million visitors attending. Among other exhibits, Buckminster Fuller's geodesic dome attracted much attention, as did Moshe Safdie's futuristic Habitat 67 housing complex.

1981: The computer mouse is introduced. Before its invention, the trackball was the pointing device most widely used to interface with computer displays. Douglas Engelbart, an electrical engineer working at the Stanford Research Institute, was looking for new ways to innovate computer systems. His team had already conceived of several new user interfaces, including bitmapped screens and hypertext, when he designed a prototype of a mouse. Engelbert's first mouse was made of a simple track wheel housed in a wooden case, and a single button for point-and-clicking. Engelbart patented the devices in the 1970s, but the first one was not sold for several years.

1994: Nelson Mandela is elected President of South Africa. The elections were the first in South Africa that extended universal suffrage to all citizens, and nearly 20 million South Africans lined up over three days to vote. Mandela's African National Congress, once a banned organization, won 63% of the popular vote and a solid majority in the National Assembly.

2006: Construction begins on the Freedom Tower in New York. The structure, later renamed One World Trade Center, was the main building in the World Trade complex that replaced the World Trade Towers destroyed in the September 11, 2001, terrorist attacks. It is 1,776 feet tall and has 104 floors.

April 28

1253: Nichiren Buddhism is founded. The Japanese priest Nichiren espoused a doctrine of achieving enlightenment by focusing on the Lotus Sutra with ritual chanting. The religion spread slowly at first, but caught on outside of Japan in the 20th century, eventually attracting a relatively large following in North America, with more than 12 million practitioners worldwide.

1503: The Spanish win the Battle of Cerignola. The battle is notable because it was the first time in European history that a force armed with gunpowder weapons was victorious. When a combined force of Swiss pike-men and French cavalry charged a line of Spanish musketeers, they opened fire, decimating the attackers. The French lost 2,000 men, and the Spanish 500.

1789: The crew of the HMS *Bounty* mutinies. The ship's captain, William Bligh, was a strict disciplinarian who remained aloof from the crew, as was the tradition in the Royal Navy. The ship sailed from England to gather breadfruit from Tahiti and take it to the Caribbean. After spending five months in Tahiti, however, discipline had deteriorated and Bligh attempted to regain control with harsh punishments. When the ship left Tahiti, morale broke down, and the crew mutinied. They placed Bligh and eighteen crewmen loyal to him in a large open row-boat. Bligh and his men journeyed for more than 4,000 miles to a British settlement in Timor, surviving on daily rations of an ounce of bread and a half a cup of water per man. They made the journey in six weeks, and Bligh returned to England in April 1790. The mutineers settled on Pitcairn Island and were found in 1808. Only one, John Adams, was still alive.

1881: Billy the Kid escapes from jail in Lincoln County. Billy had been an outlaw ever since he killed a sheriff during the Lincoln County War of 1878. He was captured by Pat Garrett and found guilty of the sheriff's murder, but escaped from jail, killing two deputies. He was found by Garrett a few months later and killed in a shoot-out.

1930: The first night game of organized baseball is played. The Independence Producers, a minor league team from Independence, Kansas, bought lights from the Giant Manufacturing Company and installed them on towers around the field. They played the Muskogee Chiefs in the first night game, and night baseball quickly caught on in the minor leagues.

1947: Thor Heyerdahl sets out on the vessel *Kon-Tiki* from Peru. The Norwegian explorer and writer was attempting to prove his theory that Polynesia could have been settled by seafaring explorers from South America. He used only materials and technologies available to pre-Columbian peoples to construct his raft, which he sailed more than 4,300 miles over 100 days to the Tuamotus, an archipelago in the South Pacific. Heyerdahl wrote a book about the expedition that was a best seller.

1967: Muhammad Ali refuses induction into the U.S. Army.
A Muslim and follower of Elijah Muhammad's Nation of Islam, Ali refused on religious grounds. For doing so, he was stripped of his boxing titles and found guilty of draft evasion. He appealed the decision all the way to the United States Supreme Court, which overturned the conviction in 1971.

April 29

1429: Joan of Arc arrives at the Siege of Orléans.
She had been granted permission by the French Dauphin to travel to Orléans, and many believed she was the fulfillment of prophecies that had been circulating France and foretold an armored maiden who would lead them to victory. At Orléans, the besieged French attacked the English, driving them from their defenses in the first major victory since their defeat at Agincourt fifteen years earlier.

1862: David Farragut captures New Orleans.
The Union naval commander had already won the Battle of Mobile Bay and outfought Confederate forts guarding the Mississippi river. This left virtually no defenses between his fleet and the city. Rather than destroy New Orleans, he established martial law and made Benjamin Butler military commander of the city.

1910: Parliament passes the People's Budget.
The Budget was backed by David Lloyd George and Winston Churchill, and set high taxes on the holdings of wealth Britons to collect funds for welfare programs established by the Liberal government. It was the first program in British history that expressly attempted to redistribute wealth from the rich to the poor.

1944: Nancy Wake parachutes into France.
Wake was a British special operations agent and leader of the French Resistance. She had been living in Marseille, France, when the

Nazis invaded, and when Paris was captured she became a courier for the Resistance. She was highly successful in her missions, managing to evade capture even as she became the Gestapo's most wanted target, with a price on her head that rose to five million francs. In 1943 she escaped France by crossing the Pyrenees, reaching England later that year. She joined the British Special Operations Executive and was sent back to France to coordinate with French guerillas. Wake turned the guerillas into a serious force and led attacks on Gestapo stations. Her network caused over 1,400 German casualties during the war.

1945: Hitler gets married. While in an underground bunker in Berlin, Adolf Hitler married his longtime partner Eva Braun, and designated Admiral Karl Dönitz as his official successor. He had realized that Berlin was being captured by the Soviet Red Army earlier that week, and knew the end of the German Reich was coming very soon. They would commit suicide together in the bunker the following day.

1946: Hideki Tojo and 28 other Japanese leaders are tried for war crimes. The International Military Tribunal of the Far East was convened by the Allies following World War II. Charges ranged from prisoner abuse to conspiracy to start and wage war. The trials lasted two years, and at their conclusion Tojo and six other defendants were sentenced to death.

1968: *Hair* opens on Broadway. The musical celebrated the hippie counterculture and sexual revolution, and it was highly controversial when it opened. It followed a "tribe" of hippies living in New York and attempting to avoid the Vietnam draft. The production invented the genre of rock musical theater, and several of its songs became anthems of the antiwar movement.

2015: Zero fans attend a baseball game in Baltimore. Following the killing of Baltimore resident Freddy Gray by police on April 19, riots erupted. The Baltimore Orioles played the Chicago White Sox in an empty stadium as a result, setting the record for the lowest attendance at a major league game.

April 30

1492: Columbus is granted a commission by Spain. The Spanish King realized that if Columbus was correct that he could reach Asia by sailing westward, it would provide Spain with a new route in the spice trade. Columbus reached the Americas, instead, ushering in a new era of European exploration.

1598: Henry IV issues the Edict of Nantes. The declaration extended basic rights to the Calvinist Protestant minority in France called the Huguenots. Protestants in the largely Catholic country were treated as heretics before the Edict, with few civil or legal rights. The Edict was revoked in 1685 by Louis XIV, and Protestant rights were not reinstated again until 1787.

1789: George Washington takes the oath of office. Washington had been unanimously elected by the Electoral College, and John Adams was elected his Vice President. He took the oath of office in New York City on the balcony of Federal Hall. He was reelected in 1792, and declined to run for a third term.

1897: J.J. Thomson announces the discovery of the electron. Thomson, an English physicist, determined that cathode rays could travel much farther through air than they should be able to if their smallest constituent parts were atom-sized. He deduced that this meant smaller particles were present and concluded atoms were made up of electrons.

1938: *Porky's Hare Hunt* opens in movie theaters. The short film shows the hunter Porky Pig pursuing an unnamed rabbit. The rabbit, which was shown chewing on a carrot, would eventually become Bugs Bunny, one of the most iconic cartoon characters in animated history.

1973: President Nixon's top aides are fired or resign. As the Watergate scandal wore on, it became increasingly clear that top members of the Nixon administration were implicated in the conspiracy. In an attempt to insulate himself, Nixon had John Ehrlichman and H.R. Haldeman resign, and fired White House Counsel John Dean. Dean later testified that tapes of Oval Office conversations existed. The discovery of the tapes would lead to Nixon's resignation.

1975: The fall of Saigon. The People's Army of Vietnam and the Viet Cong occupied strategic positions in the city, including the presidential palace. Nearly all American civilian and military personnel were evacuated from the city, as were thousands of South Vietnamese who had cooperated with the South Vietnamese government. South Vietnamese general Duong Van Minh surrendered and the government capitulated soon after.

1997: Ellen DeGeneres comes out of the closet. DeGeneres, a comedian and actor, was the star of the highly-rated sitcom *Ellen.* In the show's fourth season, her character Ellen Morgan realizes that she is lesbian and comes out to her therapist, played by Oprah Winfrey. DeGeneres came out publicly on the *Oprah Winfrey Show* as well. The episode of Ellen was one of its highest-rated episodes, but media coverage was very antagonistic toward her decision. ABC cut back on promoting the show as a result, and it was cancelled after the fifth season. ABC's parent company Walt Disney had decided it was uncomfortable with the show's subject matter now that the lead character was openly lesbian. DeGeneres returned to performing stand-up comedy and later became one of the most successful talk show hosts in the United States with *The Ellen DeGeneres Show*, which premiered in 2003. It won four Daytime Emmy Awards in its first season. She was awarded the Presidential Medal of Freedom in 2016 by Barack Obama.

Chapter 5

May

May 1

1707: The Kingdom of Great Britain is formed. Although England and Scotland had been ruled by a single monarch for more than a century, they had remained separate kingdoms, with the ruler literally wearing two crowns. The Acts of Union, passed by the Parliaments of the two nations, united them as Great Britain.

1753: Carl Linnaeus publishes the *Species Plantarum*. The two-volume work described more than 7,000 species of plants. Linnaeus classified each one by species and genus, giving them two-part Latin names. It was the first time that plants were classified using the system of binomial nomenclature.

1776: The Illuminati are founded in Ingolstadt. The Bavarian secret society was founded to oppose the spread of superstition and diminish the influence of religious institutions and the state over public life. The group, which was banned along with all other secret societies, declined in the 1780s, although modern-day conspiracy theorists credit them with a persistent worldwide influence.

1840: The first adhesive postage stamp is issued. The British Penny Black postage stamp featured the image of Queen Victoria in profile. Prior to its issue, the cost of postage was typically charged by distance the mail traveled and paid by the recipient on delivery. The Penny Black allowed senders to post up to a ounce of mail any distance.

1863: The Battle of Chancellorsville begins. The week-long engagement between the Army of Northern Virginia and the Army of the Potomac has been called General Robert E. Lee's "perfect battle." Outnumbered two to one, Lee launched multiple attacks on the Union Army and forced it to withdraw.

1884: Major unions begin agitating for the eight-hour workday in the U.S. At a national convention, the Federation of Organized Trades and Labor Unions passed a resolution that all labor organizations under its umbrella demand an eight-hour workday. The convention set May 1, 1886, as the date the eight-hour workday would begin to be observed by its members.

1930: Pluto is officially named. Clyde Tombaugh had discovered the dwarf planet in February at the Lowell Observatory in Flagstaff, Arizona. The observatory accepted suggestions from around the world; Pluto, the god of the underworld, was submitted by an English schoolgirl named Venetia Burney.

1960: Gary Powers' U-2 spy plane is shot down over the USSR. The plane was shot down while performing aerial reconnaissance and Powers was captured. Although the United States first denied that the plane was on a military mission, the Soviets produced the captured pilot along with spy cameras from the plane's wreckage that included photos of military targets. A summit between Soviet and Western leaders was overshadowed by the incident, which sharply increased tensions between the Cold War adversaries. Powers was tried and found guilty of espionage, and sentenced to ten years in prison. He was exchanged after two years for Rudolf Abel, a Soviet spy who had operated in New York City.

2003: George W. Bush declares "Mission Accomplished." Bush landed on the aircraft carrier USS *Abraham Lincoln* on an S-3 Viking jet to deliver a speech declaring that "major combat operations" were completed in the invasion of Iraq. The speech was criticized after a major guerrilla insurgency developed that would plague U.S. forces in Iraq for several years.

May 2

1611: The King James Bible is published in London. It was the third translation of the Bible into English. The translation took seven years to complete, and was printed by Robert Barker, the King's Printer. The King James version is one of the best-selling books in history.

1863: Stonewall Jackson is wounded by friendly fire. Thomas J. Jackson was a Confederate general who led his Second Corps in numerous victorious engagements for Lee's Army of Northern Virginia. Jackson was wounded during the Battle of Chancellorsville. He was returning to camp with his staff at night, and Confederate soldiers mistook them for Union cavalry and opened fire. Jackson died a week later from complications due to his wounds. Lee was reported to have said "I have lost my right arm" when he heard the news.

1885: King Léopold of Belgium establishes the Congo Free State. Léopold convinced other European leaders that Belgium intended to engage in humanitarian efforts in the central African state, but his intentions were anything but altruistic. Belgium forcibly extracted natural resources from the region for the next twenty years, committing brutal atrocities that resulted in the deaths of up to 13 million Congolese.

1920: The Negro Baseball League plays its first game. The League was organized by Rube Foster, the owner of the Chicago American Giants, and seven other teams. The Negro League operated throughout the Midwest until 1931, and included teams such as the Kansas City Monarchs and the St. Louis Stars.

1945: The Red Army captures Berlin. After an intense two-week offensive, an army of more than two million Soviet troops captured the city with brutal house-to-house fighting. Red Army soldiers were photographed raising the Soviet flag at the top of the Reichstag, the seat of German political power, in an image that became an iconic symbol of the defeat of the Third Reich.

1955: Tennessee Williams wins the Pulitzer Prize. His play *Cat on a Hot Tin Roof* was a Broadway hit that depicted the family dynamics of a Southern cotton planter's family and explored deceit, sexuality, and death. It was made into a film that starred Elizabeth Taylor and Paul Newman in 1958.

2000: GPS access is made public. Until President Bill Clinton signed an order abandoning the practice, Global Positioning Systems incorporated a feature that added intentional errors at all levels except military ones. The feature was intended to prevent enemies, including state actors and terrorists, from using civilian GPS to pinpoint guided weapons targets.

2011: Osama bin Laden is killed by U.S. special forces. Bin Laden, the mastermind of the September 11, 2001 terrorist attacks on the World Trade Center and Pentagon, had been a fugitive since he escaped a U.S. encirclement in the mountains of Tora Bora during the War in Afghanistan. He was located living in Abbottabad, Pakistan, and a CIA-led force of Navy SEALs assassinated him there during a nighttime raid on his compound.

May 3

1715: A total solar eclipse occurs in northern Europe and Asia. Edmond Halley, who also discovered the comet that bears his name, accurately predicted the eclipse to within four minutes. The eclipse passed directly over London, which experienced three and a half minutes of totality.

1848: The Benty Grange helmet is discovered in England. The Anglo-Saxon helmet was found in an earthen mound that was being excavated by Thomas Bateman, an amateur archaeologist. The helmet has an iron boar with bronze and garnet eyes on the crest, similar to those described in the poem *Beowulf*.

1913: The first Indian feature film is released. The film, *Raja Harishchandra*, was based on the legend of the same name from the ancient epic poem *Ramayana*. The film initiated the Indian film industry known variously as Bollywood and Indywood. The industry eventually grew to become the largest in the world with more than 1,500 films produced each year.

1921: Ireland is partitioned. The partition of Ireland divided the island into the Free State of Ireland in the south and the smaller region of Northern Ireland, both of which initially remained a part of the United Kingdom. The partition sparked a civil war between former comrades who either supported or opposed the partition. Pro-Free State forces led by Michael Collins won the conflict. The North remained in the U.K.

1937: Margaret Mitchell wins the Pulitzer Prize. Mitchell's novel *Gone with the Wind* told the story of Scarlett O'Hara, the scion of a Southern plantation owner who was reduced to poverty following the Civil War. The novel was adapted into a film in 1939 starring Clark Gable and Vivien Leigh. The film won eight Academy Awards, a record at the time.

1952: Two Americans land a plane at the North Pole. Pilots Joseph O. Fletcher and William P. Benedict flew to the North Pole with Albert P. Crary, a polar geophysicist. The three men were the first Americans to set foot on the North Pole. Crary later became the first person to walk on both the North and South Poles when he visited Antarctica in 1962.

1963: Bull Connor turns fire hoses against civil rights marchers in Birmingham. The Southern Christian Leadership Conference was engaged in a campaign of nonviolent direct action to highlight discrimination in the city. The campaign included sit-ins and demonstrations that resulted in mass arrests of adults and children, filling the city jails. Connor, in an effort to drive protesters from the city center, ordered police to attack them with dogs and firefighters to turn high-pressure fire hoses on them. The water pressure was set to a level so high that it could peel bark from trees, and the hoses were aimed at nonviolent teenagers and children. Images of the demonstrators being blasted with water were printed in *Life* magazine and broadcast on national news, and a photo of a police dog biting a teenager was printed in *The New York Times*. By May 10, the SCLC had secured an agreement from the city that public places would be desegregated.

1978: The first spam email is sent. The message, which advertised a new computer model produced by Digital Equipment Corporation, was sent by Gary Thuerk. It was received by nearly four hundred recipients on the ARPANET, a precursor of the modern internet. The community was incensed that the network had been used for a mass email, although it also resulted in several sales.

May 4

1776: Rhode Island renounces allegiance to Great Britain. The colony was the first in North America to do so; the other twelve colonies disavowed the British Crown soon after. The state also fielded the first unit of African-American troops during the war, the First Rhode Island Regiment.

1886: The Haymarket riot begins. Thousands of workers had begun a general strike on May 1 to demand the eight-hour workday be enforced, with 80,000 workers marching in Chicago. Strikers at the McCormick Plant on the city's South Side had been attacked by police, and several workers were killed. Labor and anarchist organizers called for a rally at the city's Haymarket, and a few thousand workers assembled there. As the rally was winding down, police arrived in force and began marching towards the stage to clear the crowd. An unknown person threw a bomb at the police, killing one and fatally wounding six others. The police opened fire on the crowd. Six anarchist organizers were later tried and sentenced to death. Four of them were executed. They were posthumously pardoned in 1893 by Governor John Altgeld.

1904: Construction begins on the Panama Canal. Following several failed attempts by the French to build the Canal, the

United States took over the project in 1904. The 48-mile canal took another ten years to complete.

1932: Al Capone goes to prison. Capone, a notorious gangster who controlled bootlegging and organized crime in Chicago from his stronghold in nearby Cicero, had evaded prosecution for years. He was convicted of tax evasion and sentenced to eleven years in prison. He was paroled after seven years.

1942: The Battle of the Coral Sea begins. The Pacific naval battle between elements of the Allied and Japanese navies was the first in history to see aircraft carriers fighting one another. Neither sides' ships ever came into view of or fired upon each other, which was also a first. The Japanese succeeded in sinking more Allied ships, but the Allies stopped the Japanese advance.

1959: The first Grammy Awards are held. The awards for musical accomplishments were presented at two events held simultaneously in Los Angeles and New York. Henry Mancini won Album of the Year, Ella Fitzgerald won Best Female Vocal Performance, and Perry Como won Best Male Vocal Performance.

1970: The National Guard opens fire on students at Kent State. The students were taking part in a mass demonstration against U.S. involvement in the Vietnam War. The guardsmen first attempted to disperse the crowd with tear gas and by advancing with fixed bayonets. They then opened fire without warning, killing four students and wounding nine others. A nationwide student strike that involved 4 million students was called in response.

1978: Margaret Thatcher is elected Prime Minister of the U.K. Thatcher was the first woman elected to the position, and had the longest administration of any British Prime Minister of the 20th century. She implemented a slate of economic and social policies that dismantled social programs, promoted deregulation, and advocated an aggressive stance internationally.

May 5

1809: Mary Kies is the first woman to receive a U.S. patent. Kies developed a new method of weaving straw, silk, and thread together to make inexpensive work bonnets. The invention improved the local cottage economy of hat-making in New England. In 2006, the National Inventors Hall of Fame inducted her for her contributions.

1860: Garibaldi leads an expedition to conquer Sicily. He set sail from Genoa with a force of only a few hundred volunteers, and added to its ranks with local Sicilians. Garibaldi is considered one of the Fathers of Italy for his efforts to unite the disparate Sicilian and Neapolitan kingdoms as one nation.

1864: The Battle of the Wilderness begins. A Union Army led by General Ulysses S. Grant invaded Virginia with the objective of wiping out Robert E. Lee's Army of Northern Virginia. Lee attacked Grant in the wilderness near Spotsylvania, and after two days and nearly 30,000 casualties suffered by both sides, Grant withdrew. The bloody battle was a harbinger of the coming engagements between the two armies, which would prove very costly for both.

1868: The first Memorial Day is declared. General John A. Logan, the commander of the Grand Army of the Republic, an organization of Civil War Veterans, issued a proclamation declaring May 5 to be "Decoration Day." The holiday spread through the Northern states and was adopted as a national holiday in 1968.

1877: Sitting Bull escapes to Canada. One year after he and other Lakota Sioux warriors wiped out General George Custer's 7th Cavalry at the Battle of Little Big Horn, Sitting Bull led them and their families to Canada to escape harassment by the U.S. Army led by Nelson Miles. While there, James Morrow Walsh, the regional commander of the Canadian Mounted Police, negotiated a truce with Sitting Bull that allowed them to safely remain in Canada provided they obeyed British laws. Sitting Bull and Walsh eventually became good friends and remained so for the rest of their lives. The Lakota band remained in Canada for four years, but found life there difficult, with the buffalo that they relied on for food much scarcer. They returned to the United States in 1881, and Sitting Bull surrendered at Fort Buford, North Dakota.

1904: Cy Young pitches the first perfect game in modern baseball. The right-handed pitcher faced 27 batters over nine innings, and not a single batter made it on base. Young's team, the Boston Americans, defeated the Philadelphia Athletics 3–0. Major League Baseball's annual pitching award is named for Cy Young.

1945: A Japanese balloon bomb explodes in Oregon. The balloons had been released from the Japanese home islands several months before, and were carried across the Pacific by prevailing winds. The bomb that landed in Oregon killed a pregnant woman and five children, and was the only fatal attack on U.S. soil during the war.

1973: Secretariat wins the Kentucky Derby. The Thoroughbred racehorse went on to become the first winner of the Triple Crown—a combination of the Derby, the Preakness Stakes, and the Belmont Stakes—in 25 years. Secretariat is considered one of the best racehorses of the 20th century, second only to Man o' War. He lived until 1989.

May 6

1536: The siege of Cuzco begins. Incan forces commanded by Yupanqui attacked a garrison of Spanish soldiers led by Hernando Pizarro. The Spaniards, numbering less than 200, held out for ten months of fierce fighting against an army of thousands of Incan warriors. The siege was unsuccessful, and the Inca eventually withdrew.

1835: The first issue of the *New York Herald* is published.
James Gordon Bennett Sr. published the paper, which would go on to become one of the most popular American newspapers of the 19th century. Bennet believed that the purpose of a newspaper was "not to instruct but to startle and amuse," and the *Herald* reflected this. In 1874 it published the New York Zoo Hoax, in which it reported that animals had escaped and were attacking New Yorkers.

1876: The painting *Duchess of Devonshire* by Thomas Gainsborough is sold for 10,000 guineas. The painting had been lost to obscurity for more than a century when it was recovered and auctioned at Christies. The price was the highest ever paid for a piece of art at the time. A few weeks later the painting was stolen, and later recovered when it was found in the possession of a master thief named Adam Worth. After being imprisoned for other crimes for four years, Worth was released and later negotiated a sale of the painting back to its original owners for $25,000 with the help of American detective William Pinkerton. Worth was nicknamed "the Napoleon of the criminal world" by Scotland Yard Detective Robert Anderson, and became the inspiration for Moriarty, Sherlock Holmes' nemesis.

1915: Babe Ruth hits his first major league home run. Ruth, a pitcher who was then playing for the Boston Red Sox, would go on to be one of the greatest hitters in baseball. Over his career, he hit 714 home runs and over two thousand runs batted in.

1925: The first transoceanic fax is sent. The fax was sent by radio and telephone from Honolulu, Hawaii, to New York City.

1937: The Hindenburg explodes. It was the largest airship ever built, and exploded while touching a mooring mast in Lakehurst, New Jersey. The airship was filled with highly flammable hydrogen, and the explosion was probably caused by a buildup of static electricity. Thirty-six passengers and crew were killed in the disaster.

1949: EDSAC performs its first operation. The Electronic Delay Storage Automatic Calculator was constructed by a team of British computer scientists to solve complex math problems. Sandy Douglas later programmed it to play tic-tac-toe in what is considered the world's first computer game.

1954: Roger Bannister breaks the four-minute mile. He ran the distance on a running track in Oxford, England, in three minutes and 59.4 seconds. Bannister's record was broken in under two months.

1960: Princess Margaret marries Anthony Armstrong-Jones. It was the first royal wedding to be televised, and 20 million viewers tuned in around the world to watch the ceremony, which was held in Westminster Abbey.

May 7

1763: Pontiac's War begins. A confederation of Native American tribes under Pontiac, the leader of the Ottawa, attacked British forts and settlements along the colonial frontier. They destroyed eight forts and killed or captured hundreds of colonists. The war continued for over a year until the Treaty of Fort Niagara was signed.

1840: The Great Natchez Tornado hits Mississippi. It was the second deadliest tornado in United States history, killing 317 people and injuring another 109. Of the deaths, 269 occurred when boats on the Mississippi River were sunk.

1915: A U-boat sinks the RMS _Lusitania_. During World War I, the German navy was fighting an ongoing naval battle with maritime shipping carrying arms to Great Britain, which was engaged in a naval blockade of Germany. The _Lusitania_ was sunk off the coast of Ireland when it was attacked with torpedoes by the German _U-20_ submarine. Although the ship was carrying arms for the wartime effort, it was officially classified as a civilian ship, and had over 1,200 civilian passengers and 696 crew members.

The sinking of the passenger ship killed 1,198 civilians, including more than 100 American citizens. It caused an international outcry against Germany, and is widely credited with accelerating the American entry into World War I.

1945: Germany formally surrenders in World War II. General Alfred Jodl, representing the German High Command, signed terms of unconditional surrender in Rheims, France. He was later tried and executed for war crimes.

1946: The Tokyo Telecommunications Engineering company is founded. It would later be renamed Sony and become one of the largest electronics and entertainment corporations in the world.

1952: Geoffrey Dummer publishes his idea for an integrated circuit. Dummer's concept, for integrating multiple electronic circuits into a single device, revolutionized the electronics industry, making a wide range of products possible.

1992: The Space Shuttle *Endeavor* is launched. It was the fifth and last operational shuttle built during NASA's Space Shuttle program. It replaced the *Challenger*, which exploded in 1987 on liftoff. Endeavor flew 25 missions between its initial launch and final one in May 2011.

2007: The tomb of Herod I is discovered. The ruler of Judea was referenced in the Christian Scriptures of the Bible following the birth of Jesus. He ruled until his death in 1 CE. His tomb was discovered south of Jerusalem by a team of Israeli archaeologists, who searched the area based on writings by Titus Josephus, a first-century Romano-Jewish historian.

May 8

1541: Hernando de Soto discovers the Mississippi River. He named it the "Río de Espíritu Santo," or River of the Holy Spirit, near Sunflower Landing, Mississippi. It took his expedition nearly a month to cross the river; they had to construct flatboats to do so. They crossed the river at night to avoid being seen by Native Americans in the area.

1846: American forces win the Battle of Palo Alto. Led by Zachary Taylor, a detachment of U.S. soldiers attacked and defeated a smaller Mexican force in the first major battle of the Mexican War. The Mexican forces were forced to retreat back across the Rio Grande. It was the first time that American troops had defended the newly annexed state of Texas.

1861: Richmond is named the capital of the Confederacy. The Confederate capital had been provisionally located in Montgomery, Alabama. Relocating to Richmond was tactically flawed because it was so far north and at the very end of a long supply line. The move required the Army of Northern Virginia to spend a great deal of time and effort defending it from Union armies in 1862 and again in 1864–65. Richmond fell to the Union in April 1865.

1877: The first Westminster Kennel Club Dog Show opens. The show is the longest-running sporting event held continuously in the United States, second only to the Kentucky Derby. It was started by a group of hunters and dog breeders in New York City. The first show was so popular, with over  1200 dogs in attendance, that it was extended for an additional day. Dogs were judged according to breed standards, which include the ability of the dogs to perform a certain set of tasks related to hunting, as well as aesthetic considerations.

1914: Paramount Pictures is founded. The film magnate W. W. Hodkinson created the company, and it grew to become one of the most powerful film studios in America by acquiring competing production houses and movie theaters across the country. In 1994, Paramount was acquired by Viacom.

1945: The Sétif massacre is committed by French soldiers. Muslim Algerians who were celebrating the Allied victory in World War II marched through Sétif to demonstrate against colonial rule. French police opened fire on the crowd, sparking reprisal attacks on French colonists in the surrounding country-side. Subsequent repression of Algerians was a major factor in the Algerian War, which began nine years later.

1972: American Indian Movement activists surrender at Wounded Knee. The activists had taken over the site of the 1890 massacre to protest federal policies on Native reservations. They were surrounded by a heavily armed force of U.S. Marshals and FBI agents. After a 71-day standoff, the activists surrendered. Two of their leaders, Dennis Banks and Russell Means, were later acquitted of conspiracy charges.

1980: Smallpox is officially eradicated. The World Health Organization (WHO) led a decades-long international effort to stop smallpox outbreaks and vaccinate everyone living near them. The effort succeeded, with the last naturally occurring case detected in 1975.

May 9

1662: *Punch* first appears in London. The puppet show, which would later become famous as *Punch and Judy*, traditionally features Punch in a jester's costume and speaking in a squawking voice. Martin Powell popularized the performance at his "Punch's Theatre," and many imitations sprang up around the United Kingdom. Punch and Judy remain popular in England.

1671: Thomas Blood attempts to steal the Crown Jewels. He first entered the Tower of London where they were kept in April, disguised as a parson and accompanied by a woman who claimed to be his wife. They befriended Talbot Edwards, the Master of the Jewel House, and Blood continued to visit over the next few weeks. On the day of the theft, Blood came to the Tower with three other men. They attacked Edwards and made off with the Jewels, but were captured before they could escape the Tower grounds. Blood was brought before King Charles, who was amused by his daring and pardoned him and granted him land in Ireland. Blood then became a celebrity in England.

1877: An earthquake strikes Iquique, Peru. The magnitude 8.5 earthquake killed more than 2,500 people and destroyed buildings all along the Peruvian coast. It also triggered a tsunami that made landfall on coastal regions around the Pacific Ocean.

1914: Woodrow Wilson officially proclaims Mother's Day. The holiday was already being observed by all U.S. states, largely due to the efforts of Ann Jarvis. Her mother had been a nurse during the American Civil War and was a pacifist. Ann Jarvis began campaigning for Mother's Day in 1905 following her mother's death. She later complained about the commercialization of Mother's Day by greeting card companies.

1936: Italy annexes Ethiopia. Italian dictator Benito Mussolini gave a speech on the balcony of the Palazzo Venezia to proclaim Italy's conquest of Ethiopia. He declared that Italy was a fascist empire in what he said was the tradition of the Roman Empire.

1941: The Allies capture the Enigma machine. The British Royal Navy captured the German submarine *U-110* in the North Atlantic. A boarding party discovered a codebook and the Enigma machine in the radio room, and took them back to the British destroyer. The machine and codebook helped the code-breaking efforts at Britain's Bletchley Park, led by Alan Turing, which proved vital to the war effort.

1958: Alfred Hitchcock's film *Vertigo* premieres. It was the first film to use the "dolly zoom" effect, in which the camera zooms in on an object while physically moving away from it. The dolly zoom creates a perspective distortion that added to the sense of unease in audiences of the psychological thriller.

1974: Impeachment hearings against President Richard Nixon begin. The hearings, by the House Judiciary Committee, were publicly televised, and resulted in a vote for impeachment. Nixon would resign from office rather than face impeachment three months later.

1995: The United States returns refugees to Cuba. Under President Bill Clinton, the United States changed its policy of sending Cuban refugees to detention camps abroad, instead moving to return any found at sea (those who made it to the U.S. were allowed to stay).

May 10

1775: The Green Mountain Boys capture Fort Ticonderoga. The Green Mountain Boys were a militia led by Ethan Allen and Benedict Arnold that mustered in Vermont, fought in New York and invaded Canada during the American Revolutionary War. After they captured the fort, Allen sent its cannons and other arms to Boston with Henry Knox, where they were used to break the Siege of Boston.

1865: Jefferson Davis is captured. The former President of the Confederate States of America was attempting to flee with his family when Union forces apprehended them in Georgia. Because he was wearing his wife's shawl when he was captured, it was reported that Davis was trying to escape in women's clothes.

1872: Victoria Claflin Woodhull is nominated for President. Woodhull was the first woman to be nominated to run for President of the United States. She was selected by the National Woman Suffrage Association in New York City to run as the

Equal Rights Party candidate. Her running mate was famed orator Frederick Douglass.

1924: J. Edgar Hoover is appointed the first Director of the FBI. Hoover would remain the Director of the investigative Bureau until his death in 1972. During his tenure, the FBI tracked and ambushed Depression-era gangsters such as John Dillinger, investigated the American Mafia, and infamously engaged in the COINTELPRO program, which targeted political dissidents in the 1960s and '70s, including Dr. Martin Luther King, Malcolm X, and the Black Panther Party for Self-Defense.

1933: Nazi book-burnings begin. The German Student Union led the campaign, and burned more than 25,000 books declared to be "un-German" in bonfires at the State Opera in Berlin. Joseph Goebbels addressed a crowd of more than 40,000 at the burning.

1954: "Rock Around the Clock" is released. Written by Max Freedman and James Myers and performed by Bill Haley and the Comets, it was the first rock and roll record to reach number one on the Billboard charts.

1962: The Incredible Hulk debuts. The Marvel Comics superhero first appeared in *Tales to Astonish*, and received a standalone series written by Marvel icons Stan Lee and Jack Kirby.

1970: Bobby Orr scores "The Goal." In the 1970 Stanley Cup Finals, Orr's Boston Bruins were playing the St. Louis Blues. In Game 4, the score was tied 3–3 with the Bruins leading the series three games to none. Forty seconds into overtime, Orr scored a series-winning goal. A picture of him flying through the air, arms raised, became the iconic image of his career.

2005: Vladimir Arutyunian attempts to kill George W. Bush. While the American President was delivering a speech in Tbilisi, Georgia, Arutyunian threw a hand grenade at the podium. The grenade failed to detonate, and Arutyunian was later captured with the assistance of the FBI and sentenced to life in prison.

May 11

868: The Diamond Sutra is printed in China. The book is the earliest known printed book in existence. The archaeologist Aurel Stein found the Sutra among 40,000 other texts when he was exploring the Mogao Caves, a system of nearly 500 temples in central China.

1857: The Indian Rebellion begins. Indian soldiers working for the British East India Company mutinied at their garrison north of Delhi. The rebellion quickly spread to other cities and garrisons in central India and continued for over two years. It led to the disbanding of the British East India Company, and India was thereafter ruled directly by the British government with the establishment of the British Raj.

1894: The Pullman strike begins. The nationwide railroad strike shut down almost all freight and passenger train traffic in the American Midwest and profoundly affected U.S. labor law. The Pullman Company manufactured railroad cars on the South Side of Chicago, where its workers lived in a community owned by the company. Workers were not allowed to buy their own houses, and instead had to rent housing from the company. In 1893, Eugene Debs founded the American Railway Union, which began organizing Pullman workers. The company refused to recognize the union and cut wages, and the ARU declared a boycott of all trains using Pullman cars. Over 125,000 workers walked off the job around the country. President Grover Cleveland ordered U.S. Army troops to break the strike, and 30 strikers were killed. Following the strike, Cleveland established Labor Day as a holiday in a concession to the union.

1910: Glacier National Park is established. The park, located in northern Montana, includes two mountain ranges, over 100 lakes, and 35 glaciers. The glaciers began rapidly melting in the 20th century, and the National Park Service estimate that they will all be gone by 2030.

1945: Kamikaze planes attack the USS *Bunker Hill*. Two Japanese kamikaze pilots crashed into the aircraft carrier near Okinawa, killing 346 crewmen and wounding 300 more. The carrier was damaged in the attack but not sunk, and made its way back to the United States mainland, where it remained until the end of the war.

1960: Adolf Eichmann is captured. The Nazi war criminal managed the logistics of the Holocaust, including the mass deportation of Jews to ghettos and concentration camps. Agents of the Mossad, Israel's intelligence agency, located him living under an assumed name in Argentina. They brought him to Israel, where he was tried and hanged for war crimes in 1962.

1997: Deep Blue defeats Kasparov. In the sixth and final game of a rematch between the IBM supercomputer and the chess grandmaster, Deep Blue beat Kasparov. It was the first time in history that a computer defeated a reigning chess champion.

1998: India tests the bomb. India tested three atomic bombs in underground tests, becoming the first nation in nearly 25 years to do so. It was the second nuclear test conducted by India, the first having been in 1974.

May 12

1784: The Treaty of Paris takes effect. The treaty formally ended the American Revolutionary War and established the boundaries between the United States and the British Empire. Benjamin Franklin, John Jay, Henry Laurens and John Adams represented the United States during the negotiations.

1846: The Donner Party begins its journey west. The party would undergo incredible hardships during its year-long journey when it became stranded in the Sierra Madre. It would be rescued the following February after enduring starvation and resorting to cannibalism to survive.

1864: Fighting rages at the Bloody Angle. During the Battle of Spotsylvania Courthouse, Union and Confederate troops fought one another in a vicious engagement at a series of wooden breastworks that became known as the "Bloody Angle." Fighting in ankle-deep mud during a torrential rainstorm, the Union Army sent brigade after brigade of men against the Confederate defenses. The Confederates held them off for 20 hours of continual fighting, but each side suffered horrific casualties. After nearly 20,000 casualties on both sides, the fighting ended in a stalemate, with the lines exactly where they had been when it began.

1903: President Theodore Roosevelt is filmed in San Francisco. The footage was recorded by cameraman H.J. Miles. It was later shown in nickelodeon theaters and was one of the first motion picture documentaries ever shown to the public.

1908: Nathan Stubblefield patents the wireless telephone. He demonstrated his invention in 1902 in Philadelphia, giving New Year's greetings to crowds around the city. It initially worked using magnetic induction, but he later switched to using ground currents. He was unable to commercialize the invention, and lived in seclusion until his death in 1928.

1926: The *Norge* flies over the North Pole. The Italian airship was the first verified expedition to travel over the North Pole. It was piloted by Umberto Nobile and included Norwegian explorer Roald Amundsen, who became the first person to reach both the North and South poles.

1937: King George VI and Queen Elizabeth are crowned. George had not been expected to inherit the throne, but did so after his brother Edward abdicated. They ruled until George's death in 1952, when he was succeeded by Queen Elizabeth II.

1943: Axis forces surrender in North Africa. After a long campaign fought across North Africa, the last Axis forces surrendered in Tunisia. The 1st Italian Army, comprised of more than 275,000 men, surrendered to the British 8th Army, ending the

North Africa campaign and resulting in all of the Italian colonies in Africa being captured by the Allies.

1982: Juan María Fernández y Krohn attacks Pope John Paul II. A former Catholic priest, Krohn attempted to stab the Pope with a bayonet when he was overpowered by guards. He claimed his motive was due to his belief that the Pope was a Soviet agent. He served three years in prison for the attack.

May 13

1568: Mary, Queen of Scots, is defeated in the Battle of Langside. A confederation of Scottish Protestants under the command of James Stewart, the Earl of Moray and half-brother of Mary, defeated her Army in the battle, near Glasgow. Mary fled to Cumberland after the battle to seek the protection of Queen Elizabeth I.

1787: The First Fleet leaves England. Captain Arthur Phillip led the fleet of eleven ships filled with convicts to Botany Bay, Australia, where they would establish the first penal colony the following January. That colony became Sydney, Australia.

1880: Thomas Edison sues the American Mutoscope and Biograph Pictures company. Edison claimed that the studios were infringing on his patent for the Kinetograph movie camera. Edison later became a shareholder in the Biograph and Mutoscope company.

1888: Brazil abolishes slavery. With the passage of the Golden Law, signed by Isabel, the Princess Imperial of Brazil, all slaves in Brazil were unconditionally freed. The law caused widespread dissent among slaveowners and the Brazilian upper class, and ultimately resulted in the overthrow of the monarchy and establishment of the republic the following year.

1917: Three children claim to see the Virgin Mary appear at Fatima, Portugal. The children were herding sheep in the countryside when they reported apparitions of Mary, who asked them to recite the Rosary every day to bring about an end to World War I. The report drew thousands of pilgrims to the site, which became a holy shrine in the Catholic faith. The children were canonized as saints in 2017.

1940: Churchill addresses the Nazi threat in Parliament. Germany had just invaded France, and Churchill sought to rally Parliament behind his newly formed government and England behind the war effort. He famously declared "I have nothing to offer but blood, toil, tears and sweat." The speech electrified Parliament, and the phrase has become synonymous with his administration.

1982: The European Economic Community calls for rubber bullets to be banned from use in the conflict in Northern Ireland. The weapons were invented by the British for use in crowd control in Northern Ireland in 1970, and were intended to be non-lethal. However, they resulted in dozens of deaths until they were finally banned in 2005.

1985: The Philadelphia police drop a bomb on a neighborhood. Following years of demonstrations by the radical liberation group MOVE, the Philadelphia police became engaged in a standoff with the group at their home in West Philadelphia. The mayor declared MOVE to be a terrorist organization based on their public speeches advocating the overthrow of the government, and the police obtained arrest warrants for four members of the group. The MOVE members refused to come out of their rowhouse even after police lobbed tear gas canisters at the building and fired on it with automatic weapons. Police commissioner Gregore Sambor then ordered police to drop an explosive incendiary device from a helicopter on the building in an attempt to burn the MOVE members out. Eleven people, including five children, were killed in the fire, which spread to adjoining houses and eventually destroyed much of the neighborhood. The only surviving MOVE member, Ramona Africa, was jailed for seven years; the city later settled a civil suit with her for $1.5 million.

May 14

1264: Henry III is defeated in the Battle of Lewes. An army led by Simon de Montfort won the field and negotiated terms of surrender from the king. Henry was forced to turn over his son, Prince Edward, to be held as a captive to guarantee the king's good behavior towards noblemen. This made Montfort the effective ruler of the country, and was a major step towards a more democratic England.

1643: Louis XIV is crowned. He ascended the French throne after the death of his father, Louis XIII, when he was just four years old. Known as the Sun King, he reigned until his death in 1715, becoming the longest-ruling monarch of a sovereign country in recorded European history.

1787: The Constitutional Convention convenes in Philadelphia. The Convention was called to write the United States Constitution after it became clear that the Articles of Confederacy gave too little power to the Federal government to be effective. Because of the slowness of travel at the time, not enough delegates arrived on the 14th for there to be a voting quorum. As a result negotiations did not begin until May 25, when delegates from seven states were present.

1804: Lewis and Clark set out from St. Louis. The two explorers led a party to map the American Northwest, search for a nautical Northwest Passage to the Pacific, and reach the Pacific Ocean. The explorers' party, called the Corps of Discovery, included 28 men and a Native American guide named Sacajawea. The Corps traveled up the Missouri River to the Great Plains, and followed other rivers Northwest, meeting various Native American tribes and French trappers along the way. They reached the Yellowstone River the following April, and crossed the Rocky Mountains that August. They reached the Pacific Ocean at the mouth of the Columbia River in November of 1805. It took almost another year for them to return to St. Louis.

1897: "The Stars and Stripes Forever" debuts. John Phillip Sousa's stirring march was played for a crowd that included President William McKinley at the unveiling of a statue of George Washington in Philadelphia.

1913: The Rockefeller Foundation is established. American industrialist John D. Rockefeller gave $100 million to establish the Foundation, and New York Governor William Sulzer approved the organization's charter.

1948: The State of Israel is declared. David Ben Gurion, the chairman of the Jewish Agency, proclaimed the founding of the State of Israel in Tel Aviv. The declaration ended British rule of Mandated Palestine and established the first Jewish state in over 2,000 years. Israel was immediately attacked by its neighbors, triggering the 1948 Arab-Israeli War.

1955: The Warsaw Pact is signed. Eight communist countries including the Soviet Union signed the treaty, which established a mutual defense organization that could counter the military might of the North Atlantic Treaty Organization. The Warsaw Pact was not an alliance of equals, however: it placed the militaries of seven member states under the control of the Soviets.

May 15

1536: The trial of Anne Boleyn begins. She was the second wife of Henry VIII, and his marriage to her led to the establishment of the Church of England. Boleyn had three miscarriages, and Henry, desiring a son and wishing to leave her for Jane Seymour, was looking for a reason to end their marriage. He had her tried on charges of treason, adultery, and incest. The trial lasted one day, and she was found guilty and sentenced to death. She was beheaded four days later.

1648: The Treaty of Westphalia is signed. The Treaty ended the Thirty Years War between the Habsburgs and Anti-Habsburg Alliance as well as the Eighty Years' War between

Spain and the Dutch. The Treaty established the modern concept of the sovereign nation-state.

1718: James Puckle invents the first machine gun. The weapon was a manually-operated revolver that could fire nine rounds per minute. It was first called the "Puckle Gun" and renamed the "machine gun" in 1722.

1756: England declares war on France. The war, which would become a global conflict involving both of the world powers' navies and fighting in their colonies, became known as the Seven Years' War in Britain and the French and Indian War in the American colonies. The British initially lost numerous battles with the French and their allies, but in 1758 William Pitt began leading them to victory. The French were unable to resupply their colonial forces because they would not risk sending large naval colonies across the Atlantic, where they could be attacked by the British Navy. Outnumbered in the colonies, the French eventually lost the war and were forced to cede territories east of the Mississippi River to Great Britain.

1905: Las Vegas is founded. An area of 110 acres that later became downtown Las Vegas was auctioned off to buyers. It would eventually become the most populated city in Nevada and the self-titled Entertainment Capital of the World.

1911: Standard Oil is broken up. In the Supreme Court Decision *Standard Oil of New Jersey vs. United States*, the Court ruled that the company was an "unreasonable" monopoly that violated the Sherman Antitrust Act. It was broken up into companies that later became Exxon and Mobil.

1928: Mickey Mouse is introduced. The iconic Disney cartoon character made his debut in the silent short animation *Plane Crazy*. The film was screened for test audiences but was not picked up by a distributor.

1941: Joe DiMaggio begins a 56-game hitting streak. It was and is the longest hitting streak in Major League Baseball.

1970: The first female generals in the United States Army are appointed. Anna Mae Hays and Elizabeth Hoisington were appointed the rank of Brigadier General by President Richard Nixon.

May 16

1532: Sir Thomas More resigns. More, the Lord Chancellor of England, resigned in protest of Henry VIII's separation from the Catholic Church. He refused to recognize Henry as the Head of the Church of England or the annulment of Henry's marriage to Catherine of Aragon. He was convicted of treason and beheaded for his decision.

1843: The first wagon train sets out on the Oregon Trail. The wagon train included over one thousand pioneers and left Elm Grove, Missouri, headed for the Pacific Northwest. The wagon train was led by John Gantt, who charged a dollar per person to guide the train west. The wagon train faced many obstacles on its journey, including having to cut a path through forests covering the Blue Mountains in Oregon. By early October, nearly all of the settlers who had left Elm Grove made it to the Willamette Valley, where they divided land into parcels of more than 300 acres. Once the trail had been established by Gantt's wagon train, more settlers followed, with nearly half a million traveling west over the next twenty years.

1868: President Andrew Johnson is acquitted. Johnson, the first U.S. president to be impeached and face a trial in the U.S. Senate, was charged with eleven high crimes and misdemeanors, including the violation of the Tenure of Office Act. He was acquitted in the Senate by a single vote.

1918: Congress passes the Sedition Act. The Act made it illegal to criticize the U.S. government during wartime. Few people were prosecuted under the Act; Eugene Debs, the labor organizer and socialist, was convicted under the Act and served two years in prison. The Act was repealed in 1921.

1925: The first air-to-ground telephone call is made. Arthur Atwater Kent was flying a blimp named *The Los Angeles*, when he called his wife, Mabel Lucas Kent, who was riding in an automobile in Philadelphia.

1951: Regular transatlantic flight begins. Flights between Idlewild (now JFK) Airport in New York City and Heathrow Airport in London were operated by El Al Israel Airlines. Transatlantic flight grew rapidly, with modern traffic including between two and three thousand flights daily.

1974: Josip Tito is elected President-for-Life in Yugoslavia. He ruled as president until his death in 1980, during which time he remained a popular figure at home and abroad, widely considered a benevolent, if politically repressive, dictator. Along with Nehru and Nasser, Tito led the Non-Aligned Movement during the Cold War.

1988: Surgeon General C. Everett Koop declares nicotine to be highly addictive. Koop declared in an official Report that the addictive properties of nicotine were similar to those of heroin or cocaine. While he was Surgeon General, Congress mandated warning labels on cigarette packs and advertisements, and smoking rates in the U.S. declined by over ten percent.

2011: Final flight of *Endeavor*. The Space Shuttle carried payloads to the International Space Station on its flight. The *Endeavor* returned to earth on June 1 of that year.

May 17

1673: Marquette and Jolliet explore the Mississippi River. The two explorers were the first Europeans to travel the northern portion of the River, departing in two canoes with five other voyageurs. They journeyed from the site of present-day Portage, Wisconsin, to the mouth of the Arkansas River, about 400 miles from the Gulf of Mexico. They were later the first Europeans to spend the winter in what would become present-day Chicago.

1809: Napoleon I annexes the Papal States. The territories, ruled by the Catholic Pope, became part of the French Empire. They remained French possessions until Napoleon's death in 1814, when they were restored to the Church.

1875: The first Kentucky Derby is held. The event was organized by Meriwether Clark Jr., the grandson of William Clark, the explorer (with Meriwether Lewis) of the Louisiana Territory. Clark had recently witnessed the English Derby in Epsom and organized a Jockey Club when he returned to Kentucky. The club raised money to build a track that would become Churchill Downs after the owners of the land it was built on. The first Kentucky Derby was 1.5 miles long, and fifteen horses competed before an audience of 10,000 spectators. It was won by Aristides, ridden by the jockey Oliver Lewis. The race was later shortened to its current distance of 1.25 miles. The Derby has been held every year since 1875 and is the longest-running sporting event in the United States.

1902: The Antikythera mechanism is discovered. It was an ancient Greek analogue computer that was used to predict astronomical events such as eclipses and the position of stars and planets. It was a clockwork mechanism made of bronze gears. Archaeologist Valerios Stais found the mechanism in a shipwreck. It was dated to around 100 BC.

1939: The first televised sports event in the U.S. is held. Columbia University's baseball team, the Lions, played the Princeton Tigers. The first professional baseball game would be televised later that summer.

1954: The Supreme Court rules that "separate but equal" is unconstitutional. In the landmark case *Brown vs. Board of Education*, the Court ruled unanimously that separate facilities for education were "inherently unequal." The ruling overturned the 1896 precedent set by *Plessy vs. Ferguson* that had established separate-but-equal standards.

1977: The first Chuck E. Cheese restaurant opens. Originally called "Chuck E. Cheese's Pizza Time Theatre," it was opened by Nolan Bushnell, the co-founder of Atari and creator of video games such as Pong. It featured arcade games, animatronic and costumed shows, and amusement rides. It later merged with competitor ShowBiz Pizza.

1995: Shawn Nelson steals a tank in California. Nelson, a U.S. Army veteran, stole the M60 tank from a National Guard Armory in San Diego. He went on a rampage with the tank, destroying cars, fire hydrants, and damaging a bridge before he was shot and killed by police.

2004: The first same-sex marriages are performed in the United States. Massachusetts became the first state to legalize same-sex marriage. Other states soon followed, and in a landmark Supreme Court case in 2013, it was effectively made federally legal.

May 18

1804: Napoleon Bonaparte is proclaimed Emperor of France. The military leader gained prominence during the French Revolution. Following several foiled assassination plots, he convinced the French Senate to adopt an imperial system similar to that of ancient Rome. He ruled as Emperor until 1814, when he was defeated in the War of the Sixth Coalition and exiled to the island of Elba.

1860: Abraham Lincoln is nominated for President. At the Republican National Convention, held in Chicago, the Kentucky native and Illinois Representative won the nomination on the third ballot. His campaign cultivated an army of youth volunteers called the Wide Awakes. They organized voter registration drives among other young people. The Republican Party was new at the time, and Lincoln correctly surmised that the young party would need the youth vote to win the White House. He was considered moderate on the issue of slavery at the time, but his abolitionist sentiments grew as he campaigned, and with them so did his base. When he was elected President, several Southern states immediately seceded, triggering the Civil War. Lincoln guided the nation through the War, and is widely regarded as one of the greatest American presidents in history.

1917: The Selective Service Act is passed. The Act gave the United States President the power to draft soldiers into the Armed Forces for service in World War I. It was cancelled at the end of the war.

1933: The Tennessee Valley Authority is created. The federally-owned corporation was founded as part of Franklin D. Roosevelt's New Deal to provide economic relief to the Tennessee Valley region during the Great Depression. It built hydroelectric facilities, dams, fossil fuel plants, and nuclear plants over the next 80 years.

1953: Jackie Cochran breaks the sound barrier. Cochran was a former Air Force pilot who had distinguished herself during World War II. She became the first woman to break the sound barrier, flying a borrowed Canadian jet at 652 mph over the Rogers Dry Lake in California. Cochran went on to set numerous other flying records for distance, altitude, and speed during her career.

1969: Apollo 10 is launched. It was the fourth manned mission to the Moon, although it stopped short of actually landing on the lunar surface. It was crewed by Thomas Stafford, John Young, and Eugene Cernan. Its NASA call-signs were "Charlie Brown" and "Snoopy."

1980: Mount St. Helens erupts. The explosion destroyed over 200 square miles of wilderness and killed 57 people. The volcano was last active between 1831 and 1857. The eruption was preceded by steam venting from the mountain, and the northern flank of the mountain collapsed in the explosion, which was followed by massive lava and mudflows. The volcano's ash plume reached 16 miles above sea level.

2005: Two new moons of Pluto are verified. The moons had been photographed by the Hubble telescope three days before, and a second photograph confirmed their existence.

May 19

1652: Rhode Island outlaws slavery. It was the first law in the English-speaking colonies in North America that made the practice illegal.

1743: The Centigrade temperature scale is invented. Jean-Pierre Christin developed the scale, where zero was the freezing point of water, and 100 degrees was its boiling point. He created a mercury thermometer that measured the scale, and published his design. Working independently of Christin, the astronomer Anders Celsius developed an identical scale at the same time, with the numbers in reverse order—zero was the boiling point, and 100 the freezing point. In 1948 the Centigrade scale was renamed Celsius in the astronomer's honor.

1836: Cynthia Ann Parker is kidnapped by Native Americans. Members of the Comanche, Kiowa, and Caddo tribes attacked Fort Parker in Texas. They killed most of her family and kidnapped Cynthia, who was ten years old. She was taken to a Comanche settlement, where she was treated well and raised as one of their own. She lived with them for nearly 25 years, adopting their language and customs and completely forgetting her earlier life. She married a Comanche chief named Peta Nocona and had three children with him. Their village was attacked by Texas Rangers in 1860 who were searching for

Cynthia Ann. They killed her husband and captured her, taking her with them back to Texas. Her story attracted attention around the country, and the Texas legislature granted her a large piece of land and annual pension of $100. She never adjusted to life among white people, and died unhappily seven years later.

1928: The first frog-jumping contest is held. Inspired by the writings of Mark Twain, the event was held in Calaveras County, California. The winning frog jumped three feet, four inches, to beat 49 competitors and win the contest.

1957: Marilyn Monroe sings "Happy Birthday" to President Kennedy. At a celebration of JFK's 40th birthday, the actress sang the traditional tune with the words "Mr. President" inserted for his name. She wore a skin-tight dress designed by Jean Louis that had 2,500 rhinestones sewn into it. It was one of Monroe's last public appearances before her death later that year.

1963: The "Letter from Birmingham Jail" is published. Dr. Martin Luther King Jr. wrote the letter while he was jailed for organizing protests against segregation in Birmingham. It was published in *The New York Post Sunday Magazine*. In the letter, he responded to criticism from moderate clergy and white liberals who had denounced his nonviolent direct-action tactics. The letter became one of the central texts of the Civil Rights Movement.

2015: The Refugio oil spill occurs. More than 100,000 gallons of crude oil spilled from a corroded pipeline, damaging one of the most biologically diverse areas of the American West Coast and killing thousands of animals.

May 20

325: The First Council of Nicaea opens. The Council of Christian bishops was called by Roman Emperor Constantine I to set the official consensus of the doctrines of the Christian Church. Among the tenets they established were the divine nature of the Son of God and the Holy Trinity, the date of Easter, and the Nicene Creed, which is still observed by Catholics to this day.

1498: Vasco da Gama discovers a sea route to India. Da Gama sailed from Lisbon, Portugal, around the southern tip of Africa and across the Indian Ocean to Calcutta, India. The discovery established the first maritime passage to India, which Europeans had sought for centuries to make the import of spices easier.

1609: Shakespeare's sonnets are first published. The collection of 154 sonnets discussed themes of love, mortality, and beauty. The first 126 are addressed to a young man, and the rest to a woman. They established a new, modern style of love poetry, and with the Romantic movement in the 19th century, became considered some of the greatest poetry ever written in the English language.

1873: Levi Strauss and Jacob Davis patent blue jeans. The pair invented the use of copper rivets in denim blue jeans. The copper rivets greatly improved the seam strength of jeans, making them ideal work pants for miners in the California Gold Rush. The style has remained popular for more than a century, and the name "Levi's" has become synonymous with blue jeans.

1940: The first prisoners arrive in Auschwitz. The infamous concentration camp, in annexed areas of Poland, first held Polish political prisoners. The first were exterminated in September 1941. The camp then became a major site in the Holocaust, as thousands of Jews deported from German-occupied Europe were brought there to be killed in gas chambers. It is estimated that 1.3 million people were sent to Auschwitz, and 1.1 million killed there. Ninety percent of the prisoners killed in the camp were Jewish. The camp was liberated by Soviet troops in January 1945, and in 1947 it became a museum of the atrocities committed by the Nazis during the Holocaust.

1964: Cosmic microwave background radiation is discovered. Robert Woodrow Wilson and Arno Penzias accidentally discovered cosmic radiation while testing a highly sensitive radio antenna system. The background radiation is what's left over from the Big Bang, when the universe rapidly expanded from a state of very high density and temperature. Its discovery offered significant evidence for the Big Bang theory. Penzias and Wilson shared the 1978 Nobel Prize in Physics for their discovery.

1978: Mavis Hutchinson runs across America. She became the first woman to do so, jogging 2,871 miles from Los Angeles to New York City in 69 days, 2 hours and 40 minutes. She was 53 years old.

May 21

1471: Henry VI is murdered in the Tower of London. The deposed king had been kept alive by his successor Edward IV to prevent an uprising on his behalf. But when Henry's son was killed at the Battle of Tewkesbury earlier that month, Edward decided there was no longer any reason to keep him alive. Richard III, Edward's brother, is widely credited with the murder.

1792: Mount Unzen erupts in Japan. The eruption caused the southern flank of the mountain to slide into the Ariake Bay outside of Nagasaki. This created a mega-tsunami that inundated the city and killed more than 14,500 people. It was the worst volcanic disaster in recorded Japanese history.

1871: French troops attack the Paris Commune. The Commune was a revolutionary socialist government that ruled

Paris for two months after the collapse of the French Second Empire. The French Third Republic sent troops to besiege Paris, and after four months they invaded the city. During a week of fighting later called "The Bloody Week," more than 20,000 members of the Commune were killed, many in summary executions after they had surrendered. Another 43,000 were arrested. The leaders of the Commune were tried and several were sentenced to death. The Paris Commune would be a major influence on the works of Karl Marx, who considered it a successful example of his idea of the "dictatorship of the proletariat."

1881: The U.S. Lawn Tennis Association is founded. The organization is the main governing body for the sport in the United States. It held the first U.S. Open in August of its first year at the Newport Casino in Rhode Island.

1924: Bobbie Franks is kidnapped and murdered. His killers, Richard Loeb and Nathan Leopold, Jr., were well-to-do students at the University of Chicago who murdered him for thrills. They believed that they were smart enough to plan and get away with the "perfect crime." They could not: Leopold's glasses were found near the boy's body, and Loeb later confessed that they had murdered him. They were defended by Clarence Darrow in what was called the "trial of the century" and sentenced to life imprisonment. Loeb was murdered in prison, while Leopold was released after 33 years.

1945: Humphrey Bogart and Lauren Bacall marry. They began their romance when they were co-stars in the movie *To Have and Have Not*. The couple continued to perform on-screen together, starring in several other films before Bogart's death in 1954.

1968: Contact with the submarine USS *Scorpion* is lost. The submarine disappeared in the North Atlantic with all 99 crew members. Its wreckage was later discovered on the sea floor 400 miles southwest of the Azores islands.

1991: Indian Prime Minister Rajiv Gandhi is assassinated.
A female suicide bomber named Thenmozhi Rajaratnam detonated a suicide belt hidden under her dress while Gandhi was
at a reelection campaign event. She was acting on behalf of the
Tamil Tigers, a separatist guerilla army in Sri Lanka.

May 22

1455: The Wars of the Roses begin. Richard, the Duke of York,
led his army against King Henry VI's forces at the First Battle
of St. Albans, 20 miles northwest of London. Richard defeated
the Royal army and captured his cousin Henry in the battle. He
was then appointed Lord Protector of England and Henry was
briefly imprisoned. The battle was later depicted in the second
part of Shakespeare's trilogy *Henry VI*. The Wars of the Roses,
so named because the opposing forces' symbols were white and
red roses, respectively.

1761: The first American life insurance policy is sold. The
policy, issued in Philadelphia, Pennsylvania, was sold by the
Corporation for Relief of Poor and Distressed Presbyterian
Ministers and of the Poor Distressed Widows and Children of
Presbyterian Ministers.

**1819: The first steam-powered ship to cross the Atlantic
departs.** The *Savannah* left its port in Georgia and arrived in
Liverpool, England, on June 20.

1856: Preston Brooks attacks Charles Sumner in Congress.
Brooks, a Representative from South Carolina, beat Sumner
nearly to death in retaliation for an anti-slavery speech delivered
by Sumner. Brooks was lauded as a hero in the South.

1906: The Wright Brothers receive their first patent. The pair
had successfully achieved heavier-than-air flight three years prior
to being granted a patent. The patent was granted not specifically
for a flying machine, but for the system of aerodynamic control
that manipulated its wings.

1939: Germany and Italy sign the Pact of Steel. The agreement was the formal beginning of the Axis powers in Europe. The two fascist nations agreed that they would give one another military and political support in the coming years. Germany attacked Poland three months later, beginning World War II.

1947: The Truman Doctrine goes into effect. The Doctrine provided for military and economic assistance to Greece and Turkey following World War II. It was designed to contain Soviet influences in the region, and became the foundation of American Cold War policy. It eventually led to the formation of the North Atlantic Treaty Organization.

1964: President Lyndon B. Johnson launches the Great Society. It was a series of domestic programs designed to eliminate poverty and racial disparities in the United States. Major spending programs in education, medical care, and rural initiatives were announced. The Great Society increased federal aid to the poor from $9.9 billion to $30 billion, and reduced the number of African Americans living in poverty by eighteen percent.

1972: Richard M. Nixon arrives in Moscow. It was the first visit to the Soviet Union by an American president. During the visit, he negotiated a number of agreements with the Soviets, including a joint space flight initiative and the groundwork of the Strategic Arms Limitation Treaty.

May 23

1430: Joan of Arc is captured. The French military leader was commanding an army that was traveling to the city of Compiegne to relieve the siege by a combined force English and Burgundians when her party was ambushed and captured by Burgundians. She was imprisoned and made several attempts to escape, but was unsuccessful. The Burgundians turned her over to the English after they agreed to pay a large price for her capture. She was taken to Rouen, the English headquarters in France, where she was tried for heresy and cross-dressing.

During her trial, her forces attacked Rouen several times in an effort to free her, but could not break the English defenses. She was found guilty and burned at the stake. Her aggressive military tactics, which had included the use of frontal assaults and artillery bombardments, changed French warfare for the remainder of the Hundred Years' War.

1844: The Declaration of the Báb. Sayyeed 'Ali Muhammad Shirazi declared himself a prophet and sparked a religious movement in Persia. He preached the religion for six years before being imprisoned and executed. The religion was the forerunner of the Baha'i faith.

1911: The New York Public Library is dedicated. The building, with its iconic lion statues at the entrance, has become a symbol of New York City. The library has appeared in dozens of movies and television shows and has been the setting of several novels and children's books.

1934: The Battle of Toledo begins. During a strike of the Electric Auto-Lite plant in Toledo, Ohio, a force of 1,300 National Guard troops attacked about 10,000 picketers. The strikers fought back in a pitched battle that lasted five days. Two strikers were killed and hundreds more wounded in the battle.

1945: Heinrich Himmler commits suicide. The former chief of the Nazi SS and assistant chief of the Gestapo had been the chief architect of the Holocaust. He swallowed a cyanide capsule one day after he was captured, while held in a prison in Luneberg, Germany, by Allied forces.

1949: Stalin lifts the Berlin Blockade. Stalin had blockaded Berlin in order to wrest economic control of the city from the Allies, but a ten-month airlift of food and supplies kept the city from starving to death. The airlift cost the lives of 78 U.S. airmen in accidents, but kept West Berlin free from Soviet control.

1992: Italian judge Giovanni Falcone is assassinated. The judge was known for his efforts to prosecute the Sicilian mafia. He was killed by the Corleonesi Mafia after he sent several of their members to prison. They killed him using an improvised explosive that was so powerful it registered on earthquake monitors. The killing sparked a major crackdown on the Italian mafia.

2013: The Interstate 5 bridge collapses in Washington state. The bridge collapsed after an oversize load struck the support beams of the bridge. Three people fell into the river as the bridge collapsed, but were rescued. The bridge was replaced the following September.

May 24

1738: John Wesley is born again. Wesley's conversion effectively sparked the Methodist movement, as he would preach his newfound ministry across England. The event is celebrated by Methodists as Aldersgate Day.

1856: John Brown launches an attack on Pottawatomie Creek, Kansas. The attack was one of many bloody confrontations that occurred in Kansas in the run-up to the Civil War. The conflict in Kansas over whether it would enter the Union as a slave state or free state was vicious enough that it earned the period the name "Bleeding Kansas." Brown traveled to Kansas specifically to take up arms against pro-slavery factions in the state. He was particularly angry over an attack on abolitionist newspaper offices that had occurred in Lawrence, Kansas, and the tepid abolitionist response. Brown and his men went to the houses of five known pro-slavery settlers, abducted them and killed them with broadswords. The murders sparked one of the most severe outbreaks of violence during Bleeding Kansas, with a month of retaliatory attacks and counterattacks that left 29 more people dead.

1859: "Ave Maria" is performed for the first time. The piece, written by French composer Charles Gounod, was performed by Madame Caroline Miolan-Carvalho. It became an incredibly popular devotional song.

1883: The Brooklyn Bridge opens. It took 14 years to complete the bridge, which connected Brooklyn with the island of Manhattan across the East River. It was the first steel-wire suspension bridge ever built. It went on to become an iconic symbol of New York City, and was named a National Historic Landmark in 1964.

1903: Michel Renault dies in a severe crash during an auto race. Renault's death, along with that of his riding mechanic, while engaging in a race from Paris to Madrid, was the second during an auto race between cities in Europe that year. It brought an end to the practice, and auto racing was switched to closed tracks.

1935: The first Major League Baseball night game is played. The game took place at Crosley Field in Cincinnati. The Cincinnati Reds beat the Philadelphia Phillies, 2–1.

1941: The *Bismarck* sinks the HMS *Hood*. The *Bismarck* was Germany's largest battleship, and the *Hood* was the biggest battle cruiser in the British Royal Navy. They engaged in battle off the coast of Iceland during the Battle of the Atlantic. The British, incensed at losing the pride of their Navy, relentlessly pursued the *Bismarck* and sank it while it was heading for occupied France for repairs.

1956: The first Eurovision Song Contest is held. The contest was held in Lugano, Switzerland, and seven countries participated. The host nation won the initial contest, which continues to be held every year and now involves every country in Europe as well as Australia, Israel, and others.

1994: Convictions are reached for the World Trade Center bombing. Four terrorists are convicted for the bombing and sentenced to 240 years in prison each.

May 25

1521: Charles V issues the Edict of Worms. It declared Martin Luther to be an outlaw and officially ended the Diet of Worms, which had convened to address the question of whether Luther had committed heresy.

1659: Richard Cromwell abdicates his position of Lord Protector. Cromwell had been appointed Lord Protector of England and served in the position until Parliament was restored. His resignation marked the beginning of a short period of democratic rule known as the Commonwealth of England. The Commonwealth was overthrown by Charles II exactly one year later.

1878: The *H.M.S. Pinafore* opens in London. The comic opera by Gilbert and Sullivan had an opening run of 571 performances, which was the second-longest run of any musical at the time. It became an international sensation, and is still performed around the world today.

1925: John T. Scopes is indicted. The Tennessee schoolteacher was indicted for teaching Charles Darwin's *Theory of Evolution* in a public school. In the highly sensational trial, he was defended by Clarence Darrow and found guilty. The law was later overturned.

1961: President John F. Kennedy announces the Apollo program. Stating that Americans would go to the Moon because the effort would "measure the best of our energies and skills," Kennedy proclaimed the mission of landing a human on the Moon by the end of the decade. Although Kennedy would not live to see it, the Apollo program successfully landed astronauts on the Moon in just eight years.

1977: *Star Wars* is released in theaters. It was the first film in the *Star Wars* trilogy, which would spark an additional series of prequels, re-releases and sequels, as well as substantial product marketing, an entire literary genre, and animated series. The film was written and directed by George Lucas, and starred Carrie Fisher, Mark Hamill, Harrison Ford, and Alec Guinness. With the exception of Guinness, all of the film's stars were relative newcomers to the big screen. *Star Wars* earned a total of $775 million, making it the highest-grossing film ever released until *E.T. the Extra Terrestrial* was released in 1982. Because it cost only $11 million to produce, it remains one of the most profitable films ever made. *Star Wars* was nominated for ten Academy Awards, of which it won seven. It completely revolutionized science fiction films, inventing the concept of the "space opera" and becoming an iconic pop culture reference.

2000: Israeli Defense Forces withdraw from Lebanon. The IDF had begun occupying territory in southern Lebanon after its invasion in 1978. The United Nations passed a resolution demanding the IDF withdraw five days after the invasion; they complied after 22 years.

2011: The final episode of *The Oprah Winfrey Show* airs. The syndicated talk show had run for 25 seasons, broadcast from Oprah's studio in Chicago, Illinois. It was the highest-rated daytime show of all time.

May 26

1637: The Pequot War begins. English captain John Mason led a force of Puritan settlers and Mohegan warriors against a Pequot village in Connecticut. The attackers killed 500 Pequot men, women, and children in the raid. The Pequot war lasted another two years and practically annihilated the tribe.

1736: The Battle of Ackia is fought near Tupelo, Mississippi. A combined force of British and Chickasaw fighters repelled an attack by French and Choctaw soldiers. The battle was a major turning point in the Chickasaw Campaign, an early prelude to the French and Indian War.

1830: President Andrew Jackson signs the Indian Removal Act. The Act gave him the authorization to negotiate the removal of Native American tribes in the states of Georgia, Florida, Alabama, and Mississippi to lands west of the Mississippi River. Although white settlers in these states overwhelmingly supported the Act, it was resisted by whites in the Northeast and the Natives living in the affected areas. The Cherokee nation in particular worked to try to oppose their relocation, but were unsuccessful. As a result of it, thousands of Native Americans were forced to move west in an event called the "Trail of Tears." Of the more than 16,500 Cherokees who were forcibly removed, between 2,000 and 6,000 died along the way. The Seminole of Florida refused to be relocated and disappeared into the vast swamps in the Everglades, from where they launched guerilla attacks for the next twelve years.

1865: General Edmund Kirby Smith surrenders at Galveston, Texas. Smith was the commander of the Trans-Mississippi Division of the Confederate Army. He was the last Confederate general to surrender to the Union at the close of the Civil War.

1896: Nicholas II is crowned Tsar of Russia. He would be the last Tsar to rule Imperial Russia, and was overthrown in the February Revolution of 1917.

1897: Bram Stoker publishes *Dracula*. The gothic horror novel became an instant best-seller. In the 20th century it was adapted into the film *Nosferatu*, and spawned an entire genre of vampire novels, films, and television shows.

1940: Operation Dynamo begins. The massive Allied evacuation of troops from the beaches of Dunkirk, northern France, took ten days to complete. Over 330,000 Allied soldiers were rescued when they became trapped on Dunkirk by seven German divisions. British civilian ships, including fishing vessels, yachts, and lifeboats were used in the evacuation. It became a rallying point for Great Britain during the war.

1960: The United States announces the discovery of bugs in its embassy. U.S. Ambassador to the United Nations Henry Cabot Lodge announced the discovery of listening devices planted in its embassy in Moscow. Although the U.S. had discovered the devices years before, it did not make its knowledge public until the Soviet Union captured U-2 spy plane pilot Gary Powers.

2004: Terry Nichols is found guilty. Nichols had helped Timothy McVeigh carry out the Oklahoma City bombing, which killed 168 people when it destroyed the federal building. Nichols was sentenced to life imprisonment without the possibility of parole for his role in the bombing.

May 27

1703: St. Petersburg is founded. Tsar Peter the Great established the city as the new capital of Imperial Russia. During his rule, he instituted sweeping economic, political, and cultural reforms that moved Russia out of the medieval era and established it as a major world power.

1831: Jedediah Smith is killed. While the colorful mountain man was riding alone on the Santa Fe Trail, he was ambushed and killed by a Comanche war party. He had been an instrumental figure in opening up the West, and was regularly hired by companies on the East Coast to establish trapping and trading opportunities in North America.

1907: Bubonic plague strikes San Francisco. Following the massive destruction of the 1906 San Francisco earthquake, the epidemic sprang up randomly around the city. Some cases were reported across the San Francisco Bay in Oakland. To control the disease, city health officials killed millions of rats and California ground squirrels. The plague was contained after killing 78 people.

1927: The last Model T is manufactured. The Ford Motor Company switched from production of the iconic, inexpensive automobile in favor of the Model A, a larger and more modern car. The Model A was also a huge success, selling a million models in the first year of production.

1933: Disney releases the cartoon *Three Little Pigs*. The animated short film was based on the folktale of the same name, and was part of Disney's Silly Symphony series. It was produced by Walt Disney and directed by Burt Gillet, who went on to produce a number of other films for the Silly Symphony franchise. *Three Little Pigs* notably featured a musical score that included the song "Who's Afraid of the Big Bad Wolf," which became a hit almost immediately and remains one of Disney's best-known and most iconic songs. The eight-minute film won that year's Academy Award for Best Animated Short Film.

1937: The Golden Gate Bridge opens. The bridge was initially only open to pedestrian traffic. It linked San Francisco to Marin County, on the far side of the Golden Gate Strait, which connects San Francisco Bay to the Pacific Ocean.

1940: British troops are massacred by the SS. Near the port of Dunkirk, where British troops were being evacuated from France, a force of 99 British Expeditionary troops became cut off near Le Paradis. They agreed to surrender only after they ran out of ammunition. The Waffen-SS lined the soldiers up against a wall and machine-gunned them to death. The commander of the SS unit, Fritz Knöchlen, was convicted for the war crime after the war and executed.

1971: Sweden admits to aiding the Viet Cong. Although Sweden insisted that it did not provide military aid to the communist Vietnamese guerilla force, only $500,000 in medical supplies, its allies in the West widely condemned the act.

May 28

1588: The Spanish Armada departs from Lisbon, Portugal.
The supposedly invincible Armada included 130 ships and more
than 25,000 crewmen and soldiers. Another 30,000 Spanish
soldiers were waiting for the Armada in the Netherlands. The
Armada intended to seize control of the English Channel to
allow the combined force of soldiers to invade Great Britain.
They were delayed by storms off the coast of Portugal, and by the
time they made it to England, the British were ready for them
with a much larger fleet. The two fleets engaged one another in
battle over several days, and the Spanish were unable to link up
with their troops in the Netherlands. They sailed around the
northern coast of England and were ultimately smashed against
the coast of Ireland by a series of severe storms. Only 67 ships
made it back to Spain.

1754: The first battle of the French and Indian War is fought.
A 22-year-old Lieutenant Colonel George Washington led the
Virginia militia in an engagement with a French reconnaissance
patrol in the Battle of Jumonville Glen. Washington's militia was
victorious in the skirmish; they killed ten French soldiers while
only losing one of their own.

1802: The Guadeloupe slave rebellion ends. Led by Louis
Delgrés, an army of 400 former slaves held off the French army
at Fort Saint Charles for several days. Realizing they could not
escape the French encirclement, the rebels ignited the fort's gunpowder stores, killing themselves rather than being recaptured
and enslaved.

1892: John Muir founds the Sierra Club. Muir, a naturalist
and early conservationist, founded the Club to lobby for the
preservation of nature. Muir had previously convinced President
Theodore Roosevelt to preserve part of the Sierra Madre as
Yosemite National Park, and he founded the Club to continue
this mission. The Sierra Club still operates today, advocating

for clean energy, environmental stewardship, and protecting the planet from the excesses of human activity.

1934: The Dionne quintuplets are born. The children, born to Oliva and Elzire Dionne in Ontario, Canada, were the first set of quintuplets to survive infancy. Born two months premature, all five survived to adulthood. While still children, they starred in three Hollywood films about their lives.

1948: Daniel Malan is elected Prime Minister of South Africa. Malan was the creator of the South African apartheid system, which institutionalized racial discrimination in the country. He died in 1959.

1987: Mathias Rust lands a private plane in Moscow's Red Square. The 18-year-old amateur pilot flew his plane from West Germany to Moscow, evading Soviet air defenses. His flight was meant to reduce tensions during the Cold War, but he was detained, imprisoned, and later expelled from the USSR. However, the incident also allowed Mikhail Gorbachev to dismiss many military officials who opposed his policies of liberalization.

2002: Cleanup of Ground Zero officially ends. The last steel girder is removed from the wreckage of the World Trade Center, and cleanup duties are ended with an official closing ceremony honoring those who were killed there in the September 11 attacks.

May 29

1453: Constantinople is captured by the Ottomans. Following a 53-day siege by an Ottoman army led by Sultan Mehmed II, the capital's garrison, led by Byzantine Emperor Constantine XI, surrendered. The capture of Constantinople spelled the end of the Byzantine Empire, which had lasted for more than fourteen centuries. The fall of the Byzantine Empire was a high point in the history of the Ottoman Empire, which could now menace Europe without any significant adversaries in the Middle East to threaten its power. Mehmed II adopted the title Mehmed the Conqueror, and ruled the Ottoman Empire until his death in 1481. Under his reign, the Ottomans expanded their territory with conquests of the Black Sea coastal regions, Bosnia, Moldavia, and Albania.

1886: The first Coca-Cola advertisement runs. John Pemberton, the founder of Coca-Cola, placed an ad for the beverage in *The Atlanta Journal*. He claimed the soda was a "brain tonic" that could cure headaches, calm the nerves, and invigorate whoever drank it.

1919: Arthur Eddington tests Einstein's theory of general relativity. Eddington confirmed the theory by observing stars during a solar eclipse. The starlight that passed near the Sun appeared to be slightly shifted. This was because the light had been curved by the Sun's gravitational field.

1953: Edmund Hillary and Sherpa Tenzing Norgay reach the summit of Mount Everest. In doing so they became the first humans to summit the mountain at 29,029 feet. Following the achievement, Hillary climbed other mountain peaks in the Himalayas and visited the South Pole. Norgay became the first Director of the Himalayan Mountaineering Institute in 1954.

1973: Tom Bradley is elected mayor of Los Angeles. Bradley was the first African-American mayor of L.A. He served until 1993, and upon his retirement he was the city's longest-serving mayor. He left office after the Rodney King verdict and subsequent L.A. riots caused his poll numbers to drop precipitously.

1988: President Ronald Reagan visits the Soviet Union. While in Moscow, he took part in a summit with Soviet leader Michael Gorbachev.

1999: The Space Shuttle *Discovery* docks with the International Space Station. In doing so, *Discovery* became the first Shuttle to dock with the ISS. It continued to fly missions to the ISS, delivering supplies and rotating crew members, until it was retired from service in 2011.

May 30

1536: King Henry VIII marries Jane Seymour. Henry's former wife Anne Boleyn had been executed only weeks before for treason. Seymour, a lady-in-waiting to Henry's first two wives, died from complications from childbirth weeks after giving birth to Edward VI. She was the only wife of Henry to be given a queen's funeral and be buried beside him at Windsor Castle.

1631: The first newspaper in France is published. The *Gazette de France* was published by Théophraste Renaudot and initially was the newspaper of a Royalist political faction. It continued to be published for nearly three centuries, and was discontinued in 1915.

1806: Andrew Jackson kills Charles Dickinson in a duel. The two men had a disagreement that arose from a disputed horse race. Dickinson published an article in the newspaper calling Jackson a coward, and Jackson demanded a duel. The two men traveled to Kentucky for the duel, as it was illegal in Tennessee. Each man shot the other in the chest, but Jackson survived Dickinson's shot, carrying the bullet in his chest, inches from his heart, for the rest of his life.

1899: Pearl Hart robs a stagecoach in Arizona. Her robbery was one of the last recorded stagecoach robberies in the Old West. She was captured a week later and imprisoned for five years. Hart gained fame for the crime because of her gender and briefly worked for Buffalo Bill's Wild West Show.

1911: The first Indy 500 is held. The 500-mile auto race was held at the Indianapolis Motor Speedway, a 2.5-mile long oval racetrack. Racers were competing for a prize purse of $27,550, the largest sum ever offered for automobile racing at the time. The high prize offer drew 46 contestants from the United States and Europe. Cars were lined up in rows of five, with the racing positions determined by the date contestants had completed their official entry forms. The 200-lap race was won by Ray Harroun, the defending AAA-level racing champion. Following his victory, Harroun never raced again. The Indy 500 continued to be held every year, and remains one of the premiere auto racing events in the United States.

1922: The Lincoln Memorial is dedicated. The memorial, located on the western end of the National Mall in Washington, D.C., faces the Washington Monument.

May 31

1578: Construction of the Pont Neuf begins in Paris. King Henry III laid the first stone of the bridge, which was completed in 1604. The Pont Neuf is the oldest surviving bridge in Paris, and connects the Île de la Cité in the middle of the River Seine with the mainland.

1790: The first U.S. Copyright law is passed. The Copyright Act of 1790 provided the sole right of reprinting material to authors for a period of 14 years. Following the first 14 years, the author was permitted to reapply if they were still alive.

1864: The Battle of Cold Harbor begins. One of the last battles of General Ulysses S. Grant's Overland Campaign to invade Virginia, it was also one of the bloodiest battles of the war. Over nearly two weeks, Union forces attempted to breach Confederate defenses with frontal assaults on entrenched positions. Thousands of Union soldiers died in the fighting. Grant later said that he always regretted ordering the final, doomed assault on Confederate lines.

1879: Gilmore's Garden is renamed Madison Square Garden. The New York City arena had been the site of entertainment presented by Patrick Gilmore since 1876. Renamed by its new owner William Vanderbilt, the Garden continued to be the site of sports entertainment, including tennis, boxing, bicycle racing, and the Westminster Kennel Dog show until 1890, when it was demolished. It was rebuilt in 1925.

1889: The Johnstown Flood kills thousands. The South Fork Dam, 14 miles upstream of Johnstown, suffered a catastrophic failure. More than 14 million cubic meters of water cascaded down the river, inundating the Pennsylvania town and killing 2,209 residents.

1921: The Tulsa race riot. In one of the worst incidents of racist rioting in United States history, a white mob attacked the African-American community in Tulsa for two days. More than 35 city blocks were destroyed and 300 civilians were killed by the mob. Many of the survivors left Tulsa, and the community never fully recovered from the attack.

1985: Forty-one tornadoes hit North America. Tornados in Pennsylvania, New York, and Ontario touched down over a period of eight hours. Twelve people were killed and the tornadoes caused close to $1 billion in damage across the region.

2005: *Vanity Fair* magazine names Mark Felt as "Deep Throat." Felt was the former Associate Director of the FBI. He was given the pseudonym by Bob Woodward and Carl Bernstein, reporters for *The Washington Post* who broke the Watergate scandal in 1973. Felt's identity remained a closely guarded secret for decades in spite of intense speculation.

Chapter 6

June

June 1

1495: Scotch whisky is invented. A monk named John Cor recorded the first written reference to Scotch whisky being distilled. Cor was a monk at the Lindores Abbey in Fife, Scotland, and a servant to King James IV.

1660: Mary Dyer is hanged. She was a Puritan settler in the Massachusetts Bay Colony who converted to the Quaker religion. The Quaker religion was forbidden by the Puritans, and Quakers were not allowed to set foot in the Colony. Several had been punished by having their ears cut off and being subjected to whippings simply for being Quakers. Dyer defied the order repeatedly, returning to New England several times to preach the Quaker faith. She was arrested along with two other Quakers, William Robinson and Marmaduke Stephenson. Robinson and Stephenson were hanged, but Dyer was granted a reprieve while she stood on the gallows. She refused to accept it, demanding that she join the others in martyrdom and refusing to leave the Colony. She was hanged in Boston. Edward Wanton witnessed her execution, and was so moved by it that he converted to Quakerism. He went on to become a leader in Rhode Island, where the faith blossomed in several communities and continues to be practiced to this day.

1813: The USS *Chesapeake* is captured. The *Chesapeake* and the British ship HMS *Shannon* engaged in a brief but violent encounter in Boston Harbor during the War of 1812. 252 men were killed or wounded during the battle, which lasted only fifteen minutes. The Chesapeake was commanded by James Lawrence, who famously shouted "don't give up the ship!" even as he was mortally wounded.

1916: Lewis Brandeis is appointed to the U.S. Supreme Court. He was the first Jewish Supreme Court Justice appointed to the bench. Brandeis served until 1939, and ruled on landmark cases that defended the right to free speech and the right of privacy. He died in 1941.

1946: Ion Antonsescu is executed. The Romanian dictator collaborated with the Nazis during World War II and was responsible for sending over 350,000 of his countrymen to their deaths in pogroms and massacres. After the war, he was found guilty of war crimes and shot.

1974: The Heimlich maneuver is published. Named for its creator, Dr. Henry Heimlich, the first aid procedure is a fast means of clearing obstructions from the upper airway. Heimlich also attempted to promote the use of the maneuver for drowning and asthma attacks, but the American Heart Association and Red Cross challenged this claim, and he abandoned it.

1980: The Cable News Network begins broadcasting. CNN was launched by Ted Turner as one of the first satellite-distributed television channels. The network was broadcast from Atlanta, and aired an interview with President Jimmy Carter as one of its first segments. It rose to national prominence during its live, on-the-ground coverage of the first Gulf War in 1990. CNN pioneered the genre of cable news television, and in so doing fundamentally altered journalism.

2001: The Nepalese Royal massacre. Nine members of the Nepalese Royal Family, including the king and queen, were killed in the palace by Crown Prince Dipendra. The prince then mortally wounded himself.

June 2

1692: Bridget Bishop is tried for witchcraft. Bishop was the first person to be tried during the Salem Witch Trials. Bishop owned a tavern in her home, spurned the modest dress adopted by her devout neighbors, preferring to wear bright red, and was outspoken. She was found guilty and hanged. After her trial, another 72 residents were tried for witchcraft, and 19 executed.

1774: The Quartering Act is passed. It was one of the so-called Intolerable Acts that were cited among the grievances of American revolutionaries in the Declaration of Independence. The Quartering Act allowed colonial governors to order British troops to be housed in private residences if no other quarters were available. It later inspired the Third Amendment to the Constitution.

1835: P.T. Barnum launches his first tour of the United States. Barnum began his career as a showman by displaying a woman named Joice Heth, an enslaved African American. He falsely claimed Heth had been the nursemaid of George Washington and that she was 161 years old.

1896: Guglielmo Marconi patents the wireless telegraph. Marconi was an Italian inventor who also invented the radio and shared the 1909 Nobel Prize in Physics for his developments in wireless technology.

1919: Anarchists bomb eight U.S. cities. Followers of Luigi Galleani, who advocated what he called the "propaganda of the deed," set off large bombs in cities around the country almost simultaneously. They targeted government officials who had passed anti-sedition laws, deported immigrants, and imprisoned anarchists. The bombings prompted the Palmer raids, in which 10,000 people were arrested and about 550 anarchists deported.

1967: *Sgt. Pepper's Lonely Hearts Club Band* is released in the United States. The Beatles' eighth studio album was a huge commercial and critical success that spent 15 weeks as the number one record in the U.S. The album included several experimental compositions, including "Lucy in the Sky with Diamonds," "A Day in the Life," and its titular song. The record's producer George Martin worked with sound engineer Geoff Emerick to electronically modify the sounds of these and other songs on the album. They also incorporated a 40-piece orchestra in many of the album's compositions, which helped make it a crossover between pop music and legitimate art in the eyes of many critics. It is considered one of the first art rock albums and helped spark the progressive rock genre. The iconic cover was designed by Jann Hanworth and Peter Blake, and featured the members of the Beatles surrounded by multiple celebrities and figures from history. It has sold more than 30 million copies, and it was added to the Library of Congress National Recording Registry in 2003.

1979: The Pope visits Poland. The visit by Pope John Paul II was the first to Communist country by the pontiff. He was greeted by enthusiastic crowds, and the visit helped spark the formation of the Solidarity labor movement the following year.

2003: The *Mars Express* is launched. It was the first probe launched by the European Space Agency to voyage to another planet. The Express included an orbiter and a lander named the *Beagle 2*.

June 3

1140: The French monk Abelard is found guilty of heresy. Abelard is best known for his epistolary love affair with the abbess Heloise. His advocacy of a rational interpretation of the dogma of the Holy Trinity led to his trial for heresy. He was excommunicated by Pope Innocent II and ordered to perpetual silence. Abelard died in 1142.

1539: Hernando de Soto claims Florida for Spain. De Soto, a conquistador who led the first expedition of Europeans into

North America, landed in Tampa Bay with nine ships and 620 men. He was later also the first European known to have crossed the Mississippi.

1621: The Dutch West India Company receives a royal charter. The charter was granted to the company to establish colonies in New Netherlands, or present-day New York and Delaware. The company was paramilitary in nature, and in 1628 its ships captured the Spanish silver fleet on its way back to Europe.

1861: Union forces are victorious in the Battle of Philippi. The battle was the first engagement of the Civil War fought on land. Roughly 3,000 Union troops attacked a smaller force of Confederates and routed them. The battle instantly made General George McClellan famous in the North. It also encouraged the citizens in the area loyal to the Union to break away from Virginia and form their own state, which is now West Virginia.

1937: The Duke of Windsor marries Wallis Simpson. The Duke, formerly the British king Edward VIII, abdicated his throne in order to be able to marry her. Edward had initially requested a morganatic marriage, which would have allowed him to retain his throne and excluded his wife and potential heirs from the line of royal succession. When his request was denied, he chose to abdicate instead. Simpson had not yet finalized her divorce from her previous husband when she wed Edward.

1943: The Zoot Suit Riots begin. So-named because of the style of suits worn by Mexican-American youths on the West Coast at the time, they involved hundreds of white servicemen attacking Mexican Americans, African Americans, and Filipino Americans in Los Angeles. The zoot suit included flamboyant jackets that came down almost to the knee and baggy pants that were pegged at the ankle. The youths who wore them called themselves "pachucos" and were primarily Mexican American. When World War II broke out, rationing regulations banned the production of clothing that used large amounts of fabric. White government officials publicly criticized the wearing of zoot suits as unpatriotic and against wartime laws. When a fight broke out between a white sailor and a pachuco in L.A. on May 30, it was followed by large-scale confrontations in which white servicemen and civilians randomly attacked Latinos and other ethnic minorities. More than 150 youths were severely beaten in the riots, and over 500 arrested.

1965: *Gemini 4* is launched. The spacecraft was the first to carry a crew of NASA astronauts for multiple days in orbit. During its flight, Ed White stepped outside the craft to complete the first spacewalk by an American.

June 4

1783: The Montgolfier brothers show off their hot-air balloon. The brothers, a pair of French paper manufacturers, began experimenting with the idea for a hot-air balloon in 1782. They built a balloon from burlap and paper, held together with 1,800 buttons. Their demonstration, before a group of French dignitaries, saw the unpiloted balloon travel a little over a mile before it landed. It quickly made them famous in France, and they began testing piloted flights later that fall.

1855: Major Henry Wayne sets out to procure camels for the U.S. Army. The Camel Corps was originally the brainchild of Major George Crosman, who believed camels would be useful as pack animals in the American southwest. Wayne traveled to Greece, Turkey, and Egypt on the USS *Supply*, purchasing 33 camels in all. Development of the Camel Corps was interrupted by the Civil War and eventually abandoned.

1876: The Transcontinental Express train arrives in San Francisco. The train had departed New York City only 83 hours and 39 minutes before its arrival. It traveled via the First Transcontinental Railway, and set a record at the time for travel across the United States.

1913: Emily Wilding Davison is trampled by King George V's horse. Davison was a militant agitator for the right of women to vote in the United Kingdom. She had been arrested nine times for various offenses that included breaking the windows of a building where a men-only political meeting was being held and throwing stones at a cabinet minister's car. She repeatedly went on hunger strikes in prison and was force-fed 49 times; she later described the experience as something that would haunt her the rest of her life. At the Epsom Derby, she brought two flags bearing the suffragists' colors of violet, green and white, and waited at the track's final corner before the home stretch. As the King's horse passed, she ran onto the track and attempted to grab the horse's reins, but was knocked down and trampled. Davison died eight days later from a skull fracture.

1917: The first Pulitzer Prizes are awarded. The newspaper magnate Joseph Pulitzer left provisions in his will to Columbia University to establish a journalism school and award the Prize. Herbert Swope won the first Pulitzer for journalism, and Laura Richards, Maude Elliot, and Florence Hall received the first Pulitzer for a biography.

1939: The *St Louis* is denied entry to the United States. The ship, carrying nearly 1,000 Jewish refugees from Germany, had already been turned away from Cuba. It was later denied entry to Canada and forced to return to European countries. Roughly 200 of the ship's passengers were later murdered in Nazi concentration camps.

1944: The Allies capture Rome. It was the first capital of an Axis country to fall to the Allied advance; Berlin and Tokyo would be captured the following year.

1961: The Berlin Crisis begins. Soviet premier Nikita Khrushchev threatened to sign a separate treaty with the government of East Germany that would have prevented access to Berlin by American, British, and French troops. The crisis culminated in the partitioning of the city and the construction of the Berlin Wall.

June 5

1817: The *Frontenac* is launched. It was the first steamboat launched on the Great Lakes, making regular runs across Lake Ontario between Toronto (then Kingston) and Niagara on the Lake. A round-trip fare was $18, but the ship rarely broke even, as the population at the time could not support it.

1862: The Treaty of Saigon is signed. The treaty gave France control of Vietnamese territory following a French invasion in 1848. A guerilla resistance to the French broke out almost immediately, led by Tru'o'ng Dinh It lasted two years before being crushed by the French. Dinh committed suicide rather than be captured.

1916: The Arab Revolt against the Ottoman Empire begins. Led by the Sharif of Mecca, it was one of the first manifestations of Pan-Arab nationalism. The rebels fought for independence and to establish a unified Arab state from Syria to Yemen. The British sent Army officers to advise assist Arab forces, and among them was a young Captain T.E. Lawrence, who would gain fame for his exploits as Lawrence of Arabia.

1947: The Marshall Plan is announced. George Marshall, the U.S. Secretary of State, announced that a massive aid initiative for Western Europe was to be undertaken by the United States. The plan spent over $13 billion to help rebuild war-torn cities, modernize industry, and jump-start the economies of Western European nations. The Marshall Plan had the additional goal of preventing the spread of communism in Western Europe.

1956: Elvis Presley introduces "Hound Dog." Elvis performed the song on *The Milton Berle Show*, and famously scandalized the audience with suggestive, hip-gyrating dance moves.

1967: The Six-Day War begins. Tensions between Israel and the neighboring states of Egypt, Jordan, and Syria were extremely high in the weeks leading up to the war. Pan-Arab nationalists were very hostile to the existence of Israel, and aggrieved by the issue of Palestinian refugees and Israel's invasion of Egypt during the 1957 Suez Crisis. Egypt blockaded the Straits of Tiran, preventing Israel from accessing the Red Sea, in May of 1967. Israel responded by launching a surprise attack on Egyptian airfields, annihilating their air force. The Israeli Defense Force simultaneously invaded the Gaza Strip and Sinai Peninsula. They pursued retreating Egyptian troops across the Sinai, inflicting heavy casualties. Egypt asked Jordan and Syria to intervene, and in the resulting battles Israel captured the West Bank and East Jerusalem from Jordan. Over 20,000 Arab troops were killed, while the Israelis lost several hundred soldiers. Many civilians became permanent refugees in the war: 300,000 Palestinian refugees fled the West Bank, and 100,000 Syrians were driven from the Golan Heights. A ceasefire was signed on June 11.

1968: Sirhan Sirhan assassinates Robert F. Kennedy. As the U.S. Senator and presidential candidate was leaving a campaign event at the Ambassador Hotel in Los Angeles, Sirhan shot him three times. One of the bullets struck Kennedy in the back of the head, and he died the next day. Sirhan was initially sentenced to death, but his sentence was commuted to life in prison after California abolished the death penalty in 1972. Kennedy was buried near his brother John at Arlington National Cemetery.

1981: The Centers for Disease Control unknowingly announces the discovery of AIDS. The CDC reported that five men in Los Angeles were diagnosed with a particular kind of pneumonia. It was later discovered that these were the first cases of Acquired Immune Deficiency Syndrome (AIDS) in the U.S.

June 6

1844: The Young Men's Christian Association is founded.
George Williams, an English philanthropist, founded the organization in London to promote young men's healthy development of mind, body, and spirit. It eventually became an international organization based in Geneva, Switzerland.

1844: The first artificially frozen ice rink opens. The Glaciarium opened in Chelsea, London, and featured an ice rink that was frozen by a series of cooling pipes beneath the ice. The rink featured an orchestra and walls decorated with murals of the Swiss Alps.

1882: A cyclone devastates Mumbai. The massive cyclone touched down in the Arabian Sea, off the shore of the city. It caused huge waves to sweep through Mumbai's harbor and into the city, inundating it. More than 100,000 residents of the city were killed when the waves came ashore.

1892: The Chicago "L" begins operating. The elevated train— or "L" for short—began service as a steam locomotive pulling four wooden coaches. The elevated tracks it traveled along are still in use, although they have been modernized. The train service was extended through the 20th century to eventually cover 224 miles of track with 145 stations.

1945: The Allies invade Normandy on D-Day. The invasion, code-named Operation Overlord, involved 150,000 Allied troops simultaneously attacking five beaches in Normandy. It was the largest amphibious invasion in human history. The beaches were code-named Gold, Juno, Sword, Omaha, and Utah, and objectives were divided among American, British, and Canadian assault troops. The day before the landings, the Allies pounded the German defenses along the Normandy coast with an aerial and naval bombardment. Just after midnight, 24,000 Allied airborne troops were dropped behind the coastal defenses. Infantry and armored troops began landing on the beaches at 6:30 in the morning. The invaders had to cross open beaches under withering machine-gun fire from entrenched German positions. Allied casualties were heaviest at Omaha beach, which was under constant fire from the high cliffs above. The Allies suffered 2,000 casualties on Omaha and more than 10,000 in total. The Normandy landings secured a foothold on the beaches that permitted them to invade France and begin turning the tide of the war in Europe.

1946: The Basketball Association of America is founded. The BAA would merge with the National Basketball League in 1949 to form the National Basketball Association. The Philadelphia Warriors won the first BAA championship in 1947.

1982: Israel invades Lebanon. The Palestinian Liberation Organization (PLO) had been using southern Lebanon as a base from which to launch attacks on Israel, prompting the invasion. The Israeli Defense Forces (IDF), fighting alongside Maronite Christians, attacked PLO, Lebanese, and Syrian targets. The resulting war lasted until June 1985 and culminated in the expulsion of the PLO from Lebanon and an Israeli occupation of southern Lebanon that lasted 15 years.

2002: An asteroid explodes in the upper atmosphere over the Mediterranean. The asteroid, only ten meters in diameter, entered the Earth's atmosphere between Greece and Libya over the Mediterranean Sea. It caused an explosion that had an estimated force of 26 kilotons, comparable to the atomic bomb dropped on Nagasaki during World War II.

June 7

1892: Homer Plessy is arrested for refusing to give up his railway seat. The resulting court case went all the way to the Supreme Court and resulted in the landmark decision *Plessy vs. Ferguson*. Plessy, a mixed-race citizen of New Orleans, agreed to help a group of early Civil Rights activists pursue a test case challenging a separate-but-equal law that mandated segregated rail cars in Louisiana. Plessy lost, and "separate but equal" remained legal until it was eventually overturned by *Brown vs. Board of Education* in 1954.

1899: Carrie Nation destroys all the liquor in a saloon in Kiowa, Kansas. Nation, a temperance crusader, would become famous for smashing up bars and liquor stores with a hatchet. She was often accompanied by a choir of women singing hymns while she destroyed the establishments. Nation was arrested 30 times over the next ten years for her attacks, and paid her fines with speaking fees and sales of souvenir hatchets.

1917: The British blow up German trenches during World War I. Before the Battle of Messines, British engineers dug tunnels to the German trenches and filled them with mines. At the beginning of the battle, the British detonated the mines, creating 19 huge craters and killing about 10,000 German troops. The explosion is one of the largest non-nuclear blasts in history.

1940: The Norwegian government goes into exile. Following Hitler's invasion of Denmark and Norway the previous April, King Haakon VII and his son Prince Olav initially led the military resistance against the Nazis. They escaped to London for the remainder of the war, where they led a government in exile.

1977: Queen Elizabeth II celebrates her Silver Jubilee. The 25th anniversary of the Queen's coronation was viewed by 500 million people in Commonwealth nations worldwide, making it one of the most-watched TV events in history.

1981: Israel attacks Iraq's main nuclear reactor. An Israeli F-16 bombed the reactor after Israel claimed it was being used to develop nuclear weapons. The United Nations condemned the attack. The reactor was later completely destroyed by the United States in the Persian Gulf War.

1982: Graceland opens to the public. The former estate of Elvis Presley, located in Memphis, Tennessee, was converted into a museum. The mansion had extraordinary amenities, including a swimming pool in one of the eight bedrooms, jukeboxes, stained glass windows, and marble floors. The estate famously features gates shaped like a book of sheet music, with a silhouette of Elvis surrounded by musical notes. When Elvis died in 1977, he was laid in state in a copper coffin in the mansion's foyer, and more than 3,500 fans paid their respects. Elvis' family decided to turn the estate into a museum when they could no longer afford the upkeep costs. It became a major tourist destination for fans.

June 8

632: Muhammad, the founder of Islam, dies in Medina. Muhammad was born in Mecca around the year 570 and became the prophet of Allah in the third branch of Abrahamic religions. He conveyed the scriptures that were later collected in the Koran, Islam's holy book. The religion eventually grew to be one of the largest in the world, with over 1.8 billion followers.

793: The first Vikings raid Britain. A Viking raiding party attacked the abbey of Lindisfarne in Northumbria. The Vikings continued to maraud the British Isles for the next 300 years.

1789: James Madison introduces the Bill of Rights. Madison originally proposed nine amendments to the U.S. Constitution. Congress ultimately approved 12 amendments, and ten of these were ratified by the states, becoming Amendments One through Ten of the Constitution.

1794: The Cult of the Supreme Being is inaugurated.
Maximilien Robespierre, one of the leaders of the French
Revolution, established the deist religion as the state's official
religion as a means of opposing the political power of the
Catholic Church.

1906: Theodore Roosevelt signs the Antiquities Act. The law
gave the President the power to reserve parcels of public land
deemed to have historical or environmental value. Under the Act,
National Monuments and pristine land across the country have
been converted to National Parks.

**1929: Margaret Bondfield is appointed to the British
Cabinet.** As the Minister of Labour, Bondfield, a former
Member of Parliament, became the first woman to join the
Cabinet. She served until 1931.

1959: The U.S. Navy attempts to deliver mail by missile.
The submarine USS *Barbero* helped the U.S. Postal Service in
its attempts to find a rapid means of delivering mail. The Navy
retrofitted a Regulus cruise missile that was initially designed to
deliver nuclear warheads with two Postal Service mail containers. The Barbero was designated as an official Post Office branch
and carried 3,000 pieces of mail addressed to President Dwight
Eisenhower, the Postmaster General, and other government officials. The mail was loaded into the Regulus' containers, and the
submarine fired the missile at the Naval Auxiliary Air Station
in Mayport, Florida. Just over twenty minutes later, the missile
landed at the Naval Station, its descent slowed by onboard parachutes. While the demonstration was a success, the high cost of
sending mail by missile made its adoption unlikely. The delivery
remains the only use of a missile to send mail in U.S. history.

1972: Phan Thi Kim Phúc is burned by napalm. The nine-
year-old girl and her family were caught in a napalm attack
during the Vietnam War. A photograph of the girl running
down the road after the attack won Nick Ut the Pulitzer Prize
for Photography.

2002: Serena Williams wins the French Open. She beat her sister Venus Williams to win the Grand Slam tournament title. The Williams finished the year ranked first and second in the tennis world, becoming the first pair of sisters to do so.

June 9

1534: Jacques Cartier discovers the Saint Lawrence River. Cartier, a French explorer of what is now Canada, was the first European to encounter the river. He sailed up the river, meeting members of the Iroquois tribe along the way. French fur traders and whalers later settled on the shores of the river.

1856: Edmund Ellsworth departs Iowa City with 500 Mormons. Ellsworth led a train of Mormon "handcart pioneers," so named because they carried their belongings in handcarts they pulled themselves, rather than using horses or oxen. The pioneers made the 1,300-mile journey along the Mormon Trail to Salt Lake City in just three months, arriving on September 26.

1862: Stonewall Jackson's Valley Campaign ends. Jackson led an army of 17,000 men over 640 miles in just 48 days, winning multiple battles along the way. Jackson moved swiftly and unpredictably through the Shenandoah Valley, confounding Union commanders and preventing them from reinforcing the attack on Richmond.

1930: *Chicago Tribune* reporter Jake Lingle is murdered by mobsters. The reporter was gunned down by assassin Leo Brothers during rush hour as he waited for an Illinois Central commuter train. Brothers was a bouncer and manager at the Green Mill, a jazz club on Chicago's North Side owned by associates of the ruthless gangster Al Capone. The murder initially sparked outrage at the killing of a journalist, and Lingle was hailed as a martyr. The *Tribune* offered a $25,000 reward for information leading to the arrest of his killers. The subsequent investigation revealed that Lingle had gangland connections and was involved in gambling and bootlegging. Ultimately it came to light that the *Tribune* had been aware of his criminal activity but had turned a blind eye to it.

1959: The first submarine carrying nuclear weapons is launched.

The USS *George Washington* carried two Polaris missiles designed for retaliatory strikes in the event of nuclear war. The submarine engaged in successful test-firing of its missiles in July 1960. It patrolled the Pacific, armed with nuclear missiles, until 1985.

1967: Israel captures the Golan Heights.

The 410-square-mile plateau north of Jordan was part of Syrian territory when the Six-Day War broke out. The Israeli army captured the Heights during the war. Between 80,000 to 130,000 Syrian refugees fled the territory, which Israel later annexed.

1973: Secretariat wins the Triple Crown.

The three-year-old thoroughbred racehorse won the Belmont Stakes to complete the victory trifecta needed to win the Crown. He had previously won the Kentucky Derby and the Preakness Stakes. Secretariat set a record for the fastest 1 miles and the fastest 1 miles run by a racehorse.

1978: The Mormon Church opens its priesthood to "all worthy men."

The Church of Jesus Christ of Latter-Day Saints, which then had a membership of 4.2 million, had excluded black men from the priesthood for 148 years.

1993: Japanese crown prince Naruhito weds Masako Owada.

The Imperial Household Agency, which oversees the Japanese state's affairs regarding the Imperial Family, disapproved of the pairing because Owada was not of royal descent. The prince proposed to her anyway, and they were married in formal ceremonies whose traditions are over a thousand years old.

June 10

1719: James Francis Edward Stuart's forces win the Battle of Glen Shiel. The Royal Navy had destroyed the Jacobite's supply base during the rebellion, and poor weather prevented the Spanish fleet from relieving them. The "Old Pretender," as Stuart was called, ascended the Scottish throne following the battle.

1791: The British Parliament passes the Canada Act. The legislation divided Canada between Upper and Lower Canada at the Ottawa River. Upper Canada was chiefly populated by British Canadians, while Lower Canada was settled by French Canadians. Those of French descent were the largest segment of the population at the time, numbering roughly 140,000 compared to the British population of about 100,000.

1829: The First Oxford-Cambridge Boat Race is held. The race, which became an annual sporting event in 1856, was won by the Oxford rowing crew in 14 minutes. The race was run from the Hambledon lock to the Henley Bridge on the Thames.

1898: U.S. Marines land on Cuba. The Spanish-American War was fought in two theaters, the Caribbean and the Pacific. In the Caribbean, Cuba and Puerto Rico were the main sites of battle, and U.S. Marines were aided in their invasion of Cuba by local resistance fighters. The last Spanish soldiers left the island by July 3.

1935: Bob Smith and Bill Wilson found Alcoholics Anonymous. Smith, a physician, had been an alcoholic for over thirty years, continuing to drink during Prohibition because of an exemption for medicinal alcohol. Smith met Bill Wilson, a recovering alcoholic, in May of 1935, and Wilson convinced him to stop drinking. The two developed the twelve-step program of spiritual and character development as well as the organization's anonymous member model. In 1939 they wrote a book about their project that outlines the program's reliance on taking a "moral inventory" and a higher power for guidance. The success of the program is debated, as many participants ultimately relapse in less than a year, but AA claims that it has helped tens of thousands of alcoholics recover in its tenure. Bob Smith remained sober until his death in 1950.

1940: Italy declares war on France and the United Kingdom. The majority of Italian military action would occur in North Africa and the Balkans until the Allied invasion of Italy in 1943. Italy surrendered in 1945, and its leader, Benito Mussolini, was executed by Italian partisans in April.

1944: Joe Nuxhall becomes the youngest player in a Major League Baseball game. Nuxhall, just fifteen years old, pitched $2/3$ of an inning for the Cincinnati Reds. Nuxhall was asked to play because of player shortages due to World War II. He yielded five walks, two hits, and five runs.

1975: Pelé joins the New York Cosmos. The Brazilian soccer legend joined the North American Soccer League team for a record salary of $1.4 million per year. Pelé played his last game in October 1977. The Cosmos played Pelé's former team, the Brazilian club Santos. Pelé played one half for each team.

2003: The *Spirit* rover is launched. The robotic rover was active on Mars from 2004–2010. It became stuck in soft soil in May 2009, but continued to run experiments and gather data. Its final communication was on March 22, 2010.

June 11

1184 BC: Troy is sacked and burned. According to the Greek historian Eratosthenes, the city was conquered by the Achaeans on this date during the Trojan War. The War was started when Paris, the Prince of Troy, eloped with Helen, the Queen of Sparta. Paris was killed later in the war.

1509: Henry VIII marries Catherine of Aragon. Catherine was the first wife of Henry. They were married until 1533, when he had their marriage annulled in order to marry Anne Boleyn.

Banished from the royal court, Catherine lived out her days at Kimbolton Castle east of London, where she died in 1536.

1770: Captain James Cook runs aground on the Great Barrier Reef. Cook was engaged in a voyage of discovery in the Pacific at the time. His ship was badly damaged by the reef, and it took his crew seven weeks to repair it while they stayed on the shore of Botany Bay, Australia.

1919: The racehorse Sir Barton wins the Belmont Stakes. In winning, he became the first horse to win the Kentucky Derby, Preakness Stakes, and Belmont Stakes in a single year. The achievement would later be known as the Triple Crown.

1920: The phrase "smoke-filled room" is coined. The Associated Press invented the phrase, which indicates a back-room political deal is being hatched. It used the phrase to describe the gathering of a group of Republican Party leaders at the party's national convention at the Blackstone Hotel in Chicago to select a candidate for U.S. President.

1963: A Buddhist monk burns himself to death in Saigon. During the Crisis of 1963, the South Vietnamese government was persecuting Buddhist monks. Ngo Dinh Diem, the President of South Vietnam, was a member of the country's Catholic minority, and his repressive policies towards Buddhists sparked widespread protests by monks in the capital. After police and soldiers attacked a demonstration of praying monks on June 3, pouring chemicals on their heads, the monks escalated their demonstrations. Thich Quang Duc made plans with the Buddhist leadership to commit suicide by self-immolation as a response. Duc sat in the roadway, poured gasoline over himself, said a prayer and struck a match. Reporters were informed ahead of the demonstration that it would be important, and several were in attendance; Malcolm Browne won a Pulitzer Prize for his photograph of the event. Diem's regime was criticized internationally as a result of the protest. In November, Diem was toppled in a military coup and executed.

1968: Lloyd J. Old discovers cell surface antigens. The discovery of the molecules that allow cells to identify different cell types revolutionized the field of immunology and cancer research.

1982: *E.T. the Extra-Terrestrial* opens in theatres. The film, about a young boy who befriends a space alien that is stranded on Earth, was a major blockbuster. It surpassed *Star Wars* to become the highest-grossing film of all time. *E.T.* held that record until *Jurassic Park* surpassed it in 1993.

2008: The Fermi Gamma-ray Space Telescope is launched. The Fermi Telescope mission was a joint venture by NASA and agencies in France, Germany, Japan, Italy, and Sweden. The telescope is used to observe gamma rays, investigate dark matter, and perform surveys of the heavens.

June 12

1817: Karl von Drais rides the first bicycle. The early form of the machine lacked pedals, and the rider powered it by kicking his feet along the ground. Von Drais rode the *Laufmaschine* ("running machine") about four miles in a little over an hour.

1942: Anne Frank receives a diary for her birthday. The young girl's diary would later become an important document of the history of World War II and the Holocaust. Frank, whose family was Jewish, was forced to hide from Nazi occupiers of the Netherlands during the war. Her family had moved there from Frankfurt, Germany, when the Nazis came to power there. They hid in concealed rooms behind a bookcase in the building where Anne's father had worked. Anne's family stayed there for two years until they were discovered by the Gestapo in August 1944. They were sent to concentration camps; her father was the only member of the family who survived. He had the diary published. In it, Anne discussed the conditions they lived under as well as her various family members and others who lived in hiding with them. Her final entry was dated August 1, 1944.

1964: Nelson Mandela is sentenced to life in prison. The South African anti-apartheid activist and leader of the African National Congress was convicted of sabotage and conspiracy to overthrow the government. He remained in prison until 1990, and was later elected President of South Africa.

1967: Anti-miscegenation laws are ruled unconstitutional. The Supreme Court of the United States made the landmark ruling in the case *Loving vs. Virginia*. The plaintiffs, Mildred and Richard Loving, had been sentenced to a year in prison by the state of Virginia for marrying one another. They appealed, and the Supreme Court sided with them. June 12 became an informal holiday for multiracial couples, called "Loving Day."

1987: Ronald Reagan calls on Mikhail Gorbachev to "tear down this wall." Reagan was delivering a speech at the Brandenburg Gate in (then) West Berlin, Germany when he challenged the Soviet leader to dismantle the Berlin War. His speech was protested by tens of thousands of West Germans who opposed his escalation of Cold War tensions with the Soviet Union.

1991: Boris Yeltsin is elected President of Russia. He was a former member of the Politburo and the first person to resign from it, which he did so in protest of Mikhail Gorbachev's program of *perestroika*. He served as president until 1999. During his tenure, Yeltsin shifted Russia's socialist economy to a capitalist market economy and implemented nationwide privatization. After his resignation, Prime Minister Vladimir Putin took over as acting president.

1994: Nicole Brown Simpson and Ron Goldman are murdered. The investigation of their killings soon focused on O.J. Simpson, Nicole's estranged ex-husband. After a high-profile trial that lasted nine months, he was acquitted in October 1995. In a 1997 civil trial, however, he was found liable for the wrongful death of Goldman and ordered to pay $19.5 million in damages.

2016: A gunman kills 49 people in Orlando. The gunman sprayed a crowd in the Pulse nightclub with gunfire before later being killed in a shootout with police. It was the deadliest act of violence targeting LGBT people in United States history.

June 13

1525: Martin Luther marries Katharina von Bora. The marriage was against the celibacy rule for Catholic priests and nuns. By marrying, they set the precedent in the Protestant faith that allows members of the clergy to marry.

1777: The Marquis de Lafayette arrives in America. He was the liaison for the French government to the American revolutionaries. Lafayette became a member of George Washington's staff and led Continental Army troops in several battles. He was wounded at the Battle of Brandywine, lived with the Army at Valley Forge during the winter of 1777–78, and was made a major general.

1804: Lewis and Clark sight the Great Falls of the Missouri River. The massive waterfalls forced the expedition to carry their boats and supplies overland for 18 miles. The detour around the Great Falls took the party a full month.

1917: German warplanes bomb London. The air raid was the deadliest carried out over London during World War I, killing 162 people, including 46 children and injuring more than 400.

1948: Babe Ruth visits Yankee Stadium for the last time. Appearing at the 25th anniversary celebrations of "The House That Ruth Built," Ruth was near death from throat cancer. He was underweight and had difficulty walking, using a baseball bat as a cane. He bid farewell to fans, calling baseball "the only real game, I think, in the world." Ruth died three months later.

1967: Thurgood Marshall is nominated to the U.S. Supreme Court. Nominated by President Lyndon Johnson, Marshall was the first African American to serve on the Court. He had previously served on the Second Circuit U.S. Court of Appeals, appointed to that position by President John F. Kennedy, before becoming the United States Solicitor General. Early in his career as a lawyer, Marshall had founded the NAACP Legal Defense and Education Fund. In his capacity as the Fund's director, he argued several landmark cases before the Supreme Court, including *Brown vs. Board of Education*, which held that racial segregation in public education was unconstitutional. Marshall was confirmed to the Supreme Court by the Senate with an overwhelming majority and served on the Court for 24 years. He wrote majority opinions and dissents on a number of landmark cases during his tenure. Marshall died in 1993 and was buried in Arlington National Cemetery.

1971: The Pentagon Papers are published. The Papers, which were released by Daniel Ellsberg, a military analyst who worked for the RAND corporation, were initially published by *The New York Times*. They showed that the United States had secretly widened the scope of its involvement in the Vietnam War with bombings of Laos and Cambodia, and that the Johnson Administration had lied to cover up this activity.

1983: *Pioneer 10* leaves the central Solar System. As the space probe passed the orbit of Neptune, it became the first human-made object to travel that far. Radio communications with *Pioneer* were lost on January 23, 2003, as its transmitter lost power. At the time it was 12 billion kilometers from Earth. It is traveling in the direction of the constellation Taurus.

2000: The first inter-Korea summit begins. Kim Dae-jung, the President of South Korea, met with Kim Jong-il, the leader of North Korea, in Pyongyang. It was the first meeting of the leaders of the two nations since the peninsula was divided in 1945.

June 14

1645: Parliamentarian forces are victorious in the Battle of Naseby. The Royalist Army of King Charles I was defeated by the Parliamentarian New Model Army, led by Thomas Fairfax and Oliver Cromwell. The Royalists suffered about 6,000 casualties out of 7,400 soldiers, while the New Model Army lost only about 400 men. The battle effectively defeated the Royalist cause, and the Parliamentarians won the Civil War later that year.

1777: The Continental Congress adopts the Stars and Stripes. It became the national flag of the young United States, replacing the British Flag of the Grand Union. The date later became a national holiday, Flag Day, celebrating the adoption.

1846: The Bear Flag Revolt begins. A group of white settlers in Sonoma, California, proclaimed the State of California. They hoisted a flag depicting a brown bear that symbolized the fierce independence of the settlers as well as their willingness to stand their ground. California would not officially become a state for four more years. When it was, the brown bear was retained on the state flag, which also features the Sierra Nevada.

1881: John Tammany patents the player piano. The device featured a system that could play the piano by reading a paper roll and striking keys with spring-loaded mechanical "fingers."

1909: Anna Shaw calls for a suffragette alliance. Shaw, a physician and ordained Methodist minister, became the president of the National American Woman Suffrage Association. She opposed the kind of militant tactics that were being used by their British counterparts in the movement for the women's vote. Other American suffragists favored these tactics, and the resulting split caused organizational difficulties. In 1909, Shaw called for an alliance between the militant wing and the politically-minded wing to overcome these difficulties. This led to the formation of a powerful organizing coalition and ultimately in the ratification of the 19th Amendment in 1920.

1937: Congress passes the Marihuana Tax Act. The Act placed a tax on the sale of cannabis sativa, which was sold by physicians and pharmacists at the time. The American Medical Association lobbied for cannabis to instead be regulated by the Narcotics Tax Act, but was unsuccessful. In 1969, the Act was found to be unconstitutional, as it required anyone reporting the sale to incriminate themselves in trafficking in cannabis.

1954: The words "under God" are added to the Pledge of Allegiance. President Dwight D. Eisenhower signed a bill that added the words after extensive lobbying by the Catholic group the Knights of Columbus. The addition was approved in part to differentiate the United States from the Communist Soviet Union, which was officially an atheist country.

1959: The Disneyland Monorail System opens. The train was the first monorail system that operated daily to be opened in the Western Hemisphere. It carries visitors to the theme park from a station in the Downtown Disney district to the Tomorrowland exhibit.

1982: The Falklands War ends. Argentinian forces in Stanley, the capital of the Falklands Islands, surrendered to British forces following two months of conflict. Great Britain and Argentina severed diplomatic relations until 1989 as a result of the war. Argentina continued to dispute the British rule of the Islands, and added its claim to them to its constitution in 1994.

June 15

1667: The first human blood transfusion is performed. Jean-Baptiste Denys, a French physician, transfused sheep's blood into a 15-year-old boy who had been bled with leeches. The boy survived, and Denys continued his experiments, but two of his other patients died from the transfusions. He was tried for murder, acquitted, and quit practicing medicine.

1752: Benjamin Franklin proves that lightning is electricity.
Franklin performed his famous experiment in Philadelphia. He flew a kite in an electrical storm, using a wet string made of hemp to conduct the electricity to a house key attached to its end. The hemp string was connected to a dry silk string that Franklin's son William held. The silk string was intended to insulate him from any electricity that the kite might pick up in the storm. While popular culture often depicts the experiment with the kite being struck by lightning, this did not in fact happen. The string picked up electricity from the charged air of the storm, and the loose threads of the hemp string stood up. Franklin moved his hand near the key and saw an electric spark, proving the lightning storm was conducting electricity. He wrote an article describing the experiment in the *Pennsylvania Gazette* that October. His experiments led him to invent the lightning rod, which are still used to protect buildings from lightning storms.

1775: George Washington is appointed the commander of the Continental Army. Washington's military experience in the French and Indian War, prestige among Virginians, and credentials as a dedicated Patriot made him an ideal candidate in the eyes of the Continental Congress. Washington recruited Baron von Steuben to train the new Army and established an officer corps that included Nathaniel Greene, Daniel Morgan, and Henry Knox. After eight years of fighting, the colonists won their independence with the signing of the Treaty of Paris in 1783.

1846: The Oregon Treaty is signed. The Treaty established the 49th parallel as the border between the United States and Canada from the Rocky Mountains to the Pacific Ocean.

1864: Arlington National Cemetery is established. The cemetery was created on land around Arlington Mansion, which had formerly belonged to Confederate General Robert E. Lee that was confiscated during the Civil War. Originally built for Civil War dead, it became the final resting place of many of the soldiers who fought in America's wars.

1888: Wilhelm II becomes the last Emperor of Germany. He established an aggressive foreign policy posture that resulted in Germany supporting Austria-Hungary in the crisis of July 1914 that brought about World War I. Wilhelm was not an effective wartime leader, and by 1918 he had lost the support of his generals. He abdicated the throne and fled Germany for exile in the Netherlands, where he lived until his death in 1941.

1919: The first nonstop transatlantic flight is completed. John Alcock and Arthur Whitten-Brown landed in Ireland sixteen hours and 27 minutes after taking off from Newfoundland.

1934: Great Smoky Mountain National Park is dedicated. The Park consists of 800 square miles of public land, and has a greater diversity of trees than all of the forests of Europe combined. The U.S. government acquired much of the land from logging and pulpwood companies.

June 16

1779: The Great Siege of Gibraltar begins. Spain declared war on Great Britain, joining France as an ally and beginning the longest siege the British military has ever endured. A combined fleet of Spanish and French ships blockaded Gibraltar, while the Spanish army dug in around the island fortress. The siege lasted three years and seven months, during which time over 300 British soldiers were killed. Another 500 died from disease.

1858: Abraham Lincoln delivers his "House Divided" speech. Upon accepting the Republican Party's nomination for U.S. Senate, Lincoln delivered the speech in which he declared that a "house divided against itself cannot stand." He said that the United States could not continue to exist with half of the Union as slave-owning states and half as free states. The speech was designed to distinguish Lincoln's position from that of Stephen A. Douglas, his Democrat opponent. The two debated one another in seven debates around the state of Illinois, which primarily discussed the issue of slavery. Lincoln lost the election and published edited texts of the debates in a book, which raised his national profile and established him as a moderate regarding the issue of slavery.

1884: The Switchback Railway opens at New York's Coney Island. It was the first roller coaster designed as an amusement park ride in America. Riders paid five cents to board a rail car with a single long bench that was then released on a track from a 600-foot tower to coast over rolling tracks at about 6 mph.

1904: James Joyce meets Nora Barnacle. The Irish author and his future wife met while she was working as a chambermaid at a hotel in Dublin. The two went on a long walk through the city, which he later used as the inspiration for his novel *Ulysses*. In the novel, Leopold Bloom goes on a day-long stroll through Dublin, intended to parallel the odyssey told by Homer.

1909: Glenn Curtiss sells his first airplane. Curtiss was the main competitor of the Wright Brothers in developing new aviation technology. He founded the first commercial aviation company in the United States in 1907, and sold his first airplane for $5,000.

1933: The Federal Deposit Insurance Corporation is created. The FDIC was created by the U.S. Congress following the stock market crash of 1929 and ensuing bank runs that led to the Great Depression. The Corporation guarantees Americans' principal deposits in banks against market crashes up to $100,000.

1976: The Soweto Uprising begins. A peaceful march by 15,000 students protesting Apartheid in South Africa was attacked by police, who fired into the crowd. The march then turned into riots that lasted for several days. Police killed 176 students during the Uprising.

1977: Oracle Corporation is founded. The company was originally called Software Development Laboratories, and was incorporated in Redwood Shores, California. It went on to become one of the largest software development companies in the world.

June 17

1631: Mumtaz Mahal dies in childbirth. Her husband, Shah Jahan I, the ruler of the Mughal Empire, spent the next 17 years overseeing the construction of her mausoleum. The Taj Mahal became the centerpiece of a large complex that includes a mosque and a guest house. It was designated as a UNESCO World Heritage Site in 1983, and attracts more than 7 million visitors per year.

1767: Samuel Wallis sights Tahiti. The English sea captain and explorer was the first European to make contact with the Pacific island.

1775: Don't fire until you can see the whites of their eyes! This famous command was shouted to American Patriots fighting for control of Boston at the Battle of Bunker Hill. The colonial troops were besieging the city, which was occupied by a large force of British troops. They learned that the British were planning to seize several hills on the outskirts of the city, which would allow them to control the Harbor. The colonists occupied Bunker Hill and Breed's Hill and prepared to defend them from British assaults. The British attacked the colonists' positions twice, being driven back with heavy losses both times. The colonists held their fire until the British troops were so close that they could "see the whites of their eyes." By waiting to fire, they inflicted the maximum possible damage on the British troops. On a third and final assault, they were able to drive the colonists from their positions only after the defenders had run out of ammunition. The colonists' resolve made the British more cautious when engaging them in later battles.

1876: The Battle of Rosebud is fought. Crazy Horse led 1,500 Sioux and Cheyenne warriors against General George Crook's forces at the Rosebud Creek in Montana. Crook was driven back by Crazy Horse and unable to continue his campaign until August.

1885: The Statue of Liberty arrives in New York Harbor. The Statue was sent from France aboard the ship *Isere* as a gift to the people of America. Its head is ten feet wide, and the elevated arm, which carries the torch, is 42 feet long. The Statue was the tallest structure in New York City until 1899.

1901: The first standardized test is introduced. The College Board administered the test, which was named the Scholastic Aptitude Test, or SAT, in 1926.

1932: The Bonus Army demonstrates at the U.S. Capitol. The Bonus Army was a group of 17,000 World War I veterans who massed in Washington, D.C. to demand the government pay the bonuses it owed them for their service in the War. Many of them had been out of work since the beginning of the Great Depression, and while the bonuses were not due until 1945, they demanded the payments be made early to provide some relief. The government refused their demands and used the U.S. Army, with cavalry and tanks, to disperse them. At least two were killed and over 1,000 wounded in the fighting.

1944: Iceland declares its independence. Iceland had been a territory of Denmark since 1918, and the island became an independent republic following a referendum in which 98% of voters chose independence.

1972: Burglars are arrested at the Watergate Hotel. The five men were caught breaking into the offices of the Democratic National Committee, and had wiretap equipment with them. The arrests were later connected to the White House and Nixon Administration.

June 18

1778: British troops abandon Philadelphia. After occupying the city for nearly nine months, General Henry Clinton's troops evacuated Philadelphia. The city, which was the seat of the Continental Congress, was occupied while the Continental

Army froze at Valley Forge. When France entered the war, its Navy made holding the city impossible, and the British retreated to New York City.

1858: Charles Darwin receives a letter from Alfred Russell Wallace. In the letter, Wallace described a hypothesis about the mechanisms driving evolution that were nearly identical to Darwin's own theory of natural selection. Darwin had come up with his ideas years before, but was reluctant to publish them. Wallace's letter prompted him to do so.

1873: Susan B. Anthony is fined for attempting to vote. The suffragist organizer had been agitating for the right of women to vote since her partnership with Elizabeth Cady Stanton began in 1851. She was forbidden from attempting to vote in the 1872 U.S. presidential election; still, she persisted, and was arrested and fined $100 following a widely publicized trial. She refused to pay the fine.

1923: The first Checker Taxi goes into service. The cab company began service in Chicago, Illinois, where multiple railroads had terminals. Its iconic Checker A series sedans that operated between 1959–1982 became the most widely-recognized taxis in America.

1940: De Gaulle delivers a defiant speech. After the German Army conquered France, capturing Paris and sweeping through the country in a lightning attack, General Charles de Gaulle became the leader of the Free French Forces that were evacuated to London. Herni Pétain established the Vichy government of France, which capitulated to and collaborated with the Nazis. De Gaulle realized the importance of communicating to the French people that the Vichy government was illegitimate and that a resistance movement was forming. He received permission from Winston Churchill to broadcast a speech to the French people using the BBC. In the speech, he admitted the long odds facing the French resistance, but declared that the fight was far from over, and famously said, "Is the defeat final? No!" The speech is considered one of the most important ever delivered in French history, in spite of the fact that it only reached a small minority of the French people.

1960: Arnold Palmer wins the U.S. Open golf tournament. In one of the greatest comebacks in the history of golf, Palmer was down by seven strokes entering the final round of the Open. He overcame the deficit and won the tournament with a score of 65, two strokes ahead of the second-place finisher.

1983: Sally Ride lifts off in the *Challenger*. Ride was the first American woman to go into space, and the third woman ever to do so. She was also the youngest American astronaut to travel in space, being only 32 at the time.

1984: Miners clash with police at the Battle of Orgreave. During the U.K. Miners' Strike of 1984–85, some 5,000 miners gathered at a mill in the North of England, intending to shut it down. The government of Margaret Thatcher responded by deploying 6,000 riot police. The police attacked the miners while television crews filmed the violence, and broadcast it on the evening news.

June 19

1586: English colonists leave Roanoke, Virginia. The colonists had suffered two years of deprivation, hardship, and fear in the isolated colony on Roanoke Island. They sailed back to England rather than face another harsh winter alone.

1865: Enslaved people in Galveston, Texas, are freed. The Emancipation Proclamation had declared that all enslaved persons held in Confederate States were to be freed over two years before the news reached Galveston. Texas was more isolated than other Confederate States, and very few battles were fought there during the Civil War. Many slaveholders had moved to Texas during the Civil War to escape the battles, and they brought their slaves with them. This greatly increased the number of people in bondage in Texas by the end of the war. About 1,000 people were enslaved in Galveston and Houston at the beginning of the war, but by its end that number was more than 250,000. On June 18, the Union Army arrived at Galveston under the command of General Gordon Granger. The following day Granger read the "General Order No. 3," which emancipated all enslaved people in Texas. The anniversary of the day is still celebrated as Juneteenth in Texas and other states.

1846: The first recorded baseball game is played.
The Knickerbocker Club of New York City played the New York Nine at a baseball diamond at Elysian Fields in Hoboken, New Jersey. The Knickerbockers won the game 23–1. The journalist Henry Chadwick, who covered cricket for *The New York Times*, later began writing about baseball at Elysian fields. He was the first to note that it could one day become America's National Pastime.

1910: Father's Day is celebrated for the first time in Spokane, Washington.
The celebration was held at a YMCA and organized by Sonora Smart Dodd. Dodd wanted to honor her father, a Civil War veteran and single parent who had raised her and five other children. The holiday later spread across the United States and around the world.

1949: The first NASCAR race is held.
It was the first race to permit only stock cars to enter; this means that the cars cannot have been modified from their original factory configuration in any way. It was held at the Charlotte Speedway in North Carolina. Jim Roper won the race after it was revealed that Glenn Dunaway, who captured the checkered flag, had used altered rear springs on his car.

1953: Julius and Ethel Rosenberg are executed.
The husband and wife had been convicted of passing atomic weapons secrets to the Soviet Union. Although an international campaign sought to have them spared from the death penalty, it was unsuccessful.

1978: *Garfield* debuts. The comic strip, which features the lazy, sarcastic, lasagna-loving cat Garfield, his owner Jon, and the less-than-brilliant dog Odie, went on to become the most widely-syndicated in the world.

1987: The Basque separatist group ETA bombs a supermarket. The bombing killed 21 people and injured 45 and was one of ETA's most violent attacks. The group claimed it had telephoned warnings to the police and the department store before the attack. They were almost universally condemned for it, and 750,000 people marched in Barcelona against the organization.

June 20

1756: British troops are imprisoned in the "Black Hole of Calcutta." The British garrison that was protecting the East India Company's trade in Calcutta was attacked by the Bengali Army and overwhelmed. After the fort was captured, more than 140 British prisoners were crammed into a cell only 14 feet wide and 18 feet long. Only 23 survived the night in the suffocating conditions.

1782: The Great Seal of the United States is incorporated. The seal features an American bald eagle clutching an olive branch in the talons of one foot and 13 arrows in the other. The phrase *E Pluribus Unum*—"out of many, one"—is on a banner held in the eagle's beak.

1791: Louis XVI attempts to flee France. The French King and his family attempted to flee the country in the midst of the French Revolution. They disguised themselves and sought the protection of Loyalist troops. But they were recognized soon after they left Paris and returned to the city. In September, Louis was forced to swear an oath of allegiance to the Constitution of 1791, which abolished the aristocracy and gave the Assembly the right to pass laws. Louis was executed at the *Place de la Révolution* in 1793.

1837: Queen Victoria ascends the throne of England.
Victoria was the daughter of Prince Edward and the grand-
daughter of King George III. Both her father and grandfather
died in 1820, when she was just one year old. She inherited the
throne when she turned eighteen, after all of her father's brothers
died. Her reign, which lasted 63 years, was the longest in English
history at the time, and is known as the Victorian era. The period
saw a significant expansion of the British Empire's holdings
around the world. She famously sparred with the Prime Minister
Benjamin Disraeli regarding foreign policy. Victorian England
was associated with strict moral codes, and saw the birth of the
Industrial Era and the accompanying rapid political, scientific,
and military changes. Queen Victoria died in 1901.

1863: West Virginia joins the Union as the 35th State.
The area of Virginia voted to remain loyal to the Union at
the Wheeling Conventions of 1861 when the rest of the state
seceded. The new state constitution made provisions for slavery
to be abolished gradually. Over 20,000 West Virginians joined
the Union Army during the Civil War.

1942: Kazimierz Piechowski escapes from Auschwitz.
Piechowski was attempting to leave Poland to join the free Polish
Army when he was captured in 1939 and sent to Auschwitz by
the Gestapo. He and three other inmates disguised themselves in
SS uniforms and stole a car, which they drove out the front gate
of the camp. Piechowski joined the resistance after his escape.

1979: *Jaws* is released in the United States. The film was the
highest-grossing ever upon its release, and it started the trend of
summer "blockbuster" movies. The movie, about a massive shark
terrorizing a quiet beachfront community, spawned three sequels.
None of them were as successful as the original.

1990: The Manjil-Rudbar earthquake strikes northern Iran.
The earthquake, which had a magnitude of 7.4, killed more than
35,000 people and injured as many as 100,000. The capital city
of Tehran was heavily damaged in the quake.

June 21

1768: James Otis Jr. infuriates the colonial governor. In a speech to the Massachusetts provincial assembly, the American Patriot referred to the British Parliament as a group of "button-makers, horse jockey gamesters, pensioners, pimps, and whore masters." He was attacking Parliament because they had levied more and more taxes on the colonies. Otis coined the phrase "Taxation without representation is tyranny."

1798: The British defeat Irish rebels at the Battle of Vinegar Hill. The battle was one of the most pivotal ones in the Irish Rebellion of 1798. The British Army attacked a large encampment of rebels in County Wexford that also served as the headquarters of the United Irishmen. Most of the Irish were only armed with pikes, and the British routed them. The battle was the beginning of the end of the rebellion.

1919: The German fleet is scuttled. Admiral Ludwig von Reuter, the commander of the fleet, claimed he had orders from the beginning of World War I that prohibited him from surrendering a single ship in the fleet. Nine sailors were killed as a result, and they were the last casualties of the war.

1940: Henry Larsen leaves Vancouver on the *St. Roch*. Larsen made the first west-to-east voyage of the Northwest Passage over the next two years. The journey took so long because the ship became trapped in ice the over the winters of both 1940–41 and 1941–42. In 1944 he sailed the much speedier return journey with the *St. Roch* outfitted with a more powerful engine.

1964: Civil Rights workers are murdered in Mississippi.
Andrew Goodman, James Chaney, and Michael Schwerner were activists working with the Freedom Summer campaign to register African Americans to vote in Mississippi. They were arrested outside of Philadelphia, Mississippi, and held for several hours. Upon being released, they were followed out of town, pulled over, abducted, and shot. The FBI became involved in the investigation when it became clear that local law enforcement was at best indifferent to the case. Their bodies were located after a two-month search. The FBI determined that the Ku Klux Klan, Sheriff's Department, and Philadelphia, Mississippi, Police were involved in the murders. The killings, and involvement of law enforcement and the Klan, sparked national outrage. It helped lead to the passage of the Voting Rights Act of 1965. The federal investigation into the murders, code-named *Mississippi Burning*, later inspired a film of the same name.

1989: The Supreme Court rules that flag-burning is legal.
In the case *Texas vs. Johnson*, the Court ruled that burning the American flag was a protected form of political protest under the First Amendment to the Constitution.

2004: *SpaceShipOne* completes the first manned private spaceflight. The spacecraft, which used a hybrid rocket motor and was launched from the air, achieved suborbital space flight. It was the first privately funded spacecraft to do so.

June 22

1633: Galileo is forced to recant. The Italian astronomer published his research of the nature of the heavens in 1632 as *Dialogue Concerning the Two Chief World Systems*. In it, he presented evidence that the Sun, and not the Earth, is the center of the Universe. The work was also seen to attack Pope Urban VIII and criticize the Jesuit order of the Catholic Church. Both had been political supporters of Galileo before it was published. They rescinded their support of the astronomer as a result of the book, and Galileo was tried by the Inquisition. The Copernican model of the universe that Galileo defended, so named because it was initially conceived by Copernicus, was heresy in the eyes of the Catholic Church. Galileo was ordered to "abjure, curse, and detest" his heliocentric views by the Holy Office in Rome, and his writings were banned. He spent the rest of his life confined to house arrest. He wrote *Two New Sciences* during that time. He died in 1642. In 1741, Pope Benedict XIV authorized the publication of Galileo's works, including the *Dialogue*.

1807: The Chesapeake-Leopard Affair. The British warship HMS *Leopard* attacked the USS *Chesapeake*, searching for deserters from the Royal Navy. Four American sailors were killed and 17 wounded, and four crew members were tried by the British for desertion. The incident sparked an international incident and led to the Embargo of 1807.

1839: Cherokee leaders are assassinated. Major Ridge, his son John Ridge, and Elias Boudinot had signed the treaty that resulted in the Cherokee nation being forcibly relocated to Oklahoma in the Trail of Tears. The three were killed by a faction loyal to John Ross and the Cherokee National Council, who blamed them for the removal.

1918: The Hammond Circus Train wreck kills 86 people. Another train's locomotive engineer fell asleep and crashed into the rear of the Circus Train. The following train crashed through the caboose and four sleeping cars of the Circus Train, killing most of the victims on impact. The rest died in the resulting fire. It was one of the worst train disasters in U.S. history.

1945: The Battle of Okinawa concludes. American troops captured the island 300 miles south of the Japanese mainland in one of the bloodiest battles of World War II in the Pacific. The battle lasted for three months. The Americans lost over 20,000 soldiers, and the Japanese lost about 100,000.

1969: The Cuyahoga River catches fire. The river in Cleveland, Ohio, was one of the nation's most polluted. It had caught fire at least a dozen times since 1868, but the 1969 fire caught national attention after a *TIME* magazine article. The fire sparked the passage of the Clean Water Act and the creation of the Environmental Protection Agency.

1986: The Hand of God scores a goal. In the quarterfinals of the 1986 FIFA World Cup match between Argentina and England, the Argentinian striker Diego Maradona scored a goal six minutes into the second half. Although he likely touched the ball with his hand, Maradona was not penalized, and Argentina went on to win 2–1. In a postgame press conference, when asked about the goal, Maradona replied that it was scored with his head and "a little bit the hand of God," after which it was forever known as the Hand of God goal.

1990: Checkpoint Charlie is dismantled. It was one of the most famous crossing sites of the Berlin Wall that divided East and West Berlin during the Cold War. The checkpoint was established in 1947 and later became a tourist attraction.

June 23

1314: The Battle of Bannockburn begins. After the Scottish King Robert the Bruce captured the English garrison at Stirling, King Edward II invaded Scotland with a massive army. More than 20,000 strong, the English army encountered the Scots, who numbered only about 6,000 infantry and riders, on the road to Stirling. Robert's army had established a strong defensive position in the roadway, and dug small pits to break up charges by the English heavy cavalry. Before the battle began, Robert engaged in single combat with Henry de Bohun, the nephew of one of the English commanders. Bohun and Robert rode towards one another on horseback, and as they passed, Robert killed Bohun with a single blow of an axe. The Scottish army then charged. They used shiltroms, which were compact, tightly arrayed phalanxes of troops protected by shield walls and wielding long pikes to attack the English. As the Scots overwhelmed the English, Edward fled the field with his bodyguards. More than 11,000 English infantrymen were killed in the battle and its aftermath. Fourteen years later, Scotland won its independence.

1868: Christopher Latham Sholes patents the "type-writer." Although various designs of typewriters with keyboards had existed for 150 years, Sholes' design was the first to use the QWERTY keyboard. That keyboard is still used on typewriters and computer keyboards today.

1905: The Wright brothers launch Flyer No. 3. Their third heavier-than-air craft was the first one to travel in a consistent straight line, and the last one the pioneering brothers designed before Wilbur Wright flew a 77-mile flight on New Year's Eve, 1908.

1914: Pancho Villa captures Zacatecas. The battle for the city was the bloodiest of the war, with the Mexican revolutionaries suffering 1,000 casualties and the Mexican Federal Army losing 6,000 men, or about half their strength. Villa's men surrounded the city and bombarded it with artillery before attacking it from all sides at once. Mexican President Victoriano Huerta fled the country following the battle.

1917: Babe Ruth punches an umpire. After Ruth walked the first batter, he began arguing with the umpire, who ejected Ruth from the game. Ruth punched him and was forcibly removed from the game. Ernie Shore was then brought in to pitch. He picked off the runner at first and then retired the next 26 batters, effectively pitching a nearly "perfect" game. The Red Sox beat the Washington Senators 4–0.

1931: Wiley Post and Harold Gatty take off to travel around the world. The American aviator and Australian navigator left Long Island on the first successful attempt to circumnavigate the globe in a single-engine airplane. They flew to Newfoundland, then to England, across Siberia, Alaska, and Canada before landing in New York. They completed the journey in eight days, fifteen hours, and 51 minutes.

1972: The Higher Education Act is signed by President Nixon. The act included Title IX, which bars discrimination from sports and other activities based on gender. On the same day he signed the Act, Nixon discussed using the Central Intelligence Agency to obstruct the FBI's investigation of the Watergate break-ins with H.R. Haldeman. The conversation was recorded.

2016: Brexit! The United Kingdom voted to leave the European Union in a referendum, by a margin of 52% to 48%. Prime Minister David Cameron, who had campaigned against leaving the EU, resigned as a result of the outcome.

June 24

1374: An outbreak of dancing mania strikes Germany. Dancing mania involved large groups of people fitfully dancing together until they collapsed from exhaustion. Groups included men, women, and children, and could be numbered in the thousands. Participants often experienced strange hallucinations while they were dancing. The phenomenon occurred in Europe many times over the next three centuries, but it is still not completely understood. The outbreak that occurred in 1374 began in Aachen and then spread to other cities in Germany and even as far as Italy and Luxembourg. Several hypotheses have been proposed to explain the phenomenon of dancing mania, including a kind of mass hysteria, an unknown religious cult movement, or even spider bites. It was also proposed that the dancers had eaten bread contaminated with molds that had psychoactive properties. Dancing mania died out by the 18th century.

1604: Samuel de Champlain discovers the Saint John River. De Champlain, a French navigator and explorer, was the first European to see the mouth of the river. The site became the location of the first permanent European settlement in North America, the port city of Saint John, New Brunswick.

1717: The Premier Grand Lodge of England is founded. It was the first Grand Lodge of the Freemasons, a semi-secret fraternal organization that many major figures of the 18th and 19th centuries belonged to.

1812: Napoleon invades Russia. The French Emperor led an army of 685,000 troops in the invasion of Russia. Of them, only 120,000 would survive. The French Army captured Moscow in September, and the Russians retreated to the East. Rather than engage the French in pitched battle, they continued retreating until the French were caught deep in Russia with no winter clothes, few rations, and no supply trains. The Russians finally attacked as the French were retreating across the Berezina River and destroyed the French Army, which could only field about 30,000 effective troops. Napoleon fled to Paris, and his Army left Russia six months after invading.

1880: "O Canada" is first performed. The song would become the national anthem of Canada. It was written by Calixa Lavallée for the Saint-Jean-Baptiste Day celebrations, where it was first performed before the French Canadian National Congress.

1939: Siam is renamed Thailand. Plaek Phibunsongkhram, the nation's third Prime Minister, changed the name during a cultural revolution. Phibunsongkhram was an admirer of the fascist Benito Mussolini and believed the cultural changes would bring the country into the modern age.

1995: South Africa wins the Rugby World Cup. It was the first major sports event to be held in South Africa following the dismantling of apartheid. It was also the first World Cup in which South Africa was allowed to participate, as the nation had been banned from competing during the apartheid era. Following the win, President Nelson Mandela appeared on the field and presented the World Cup to the team captain.

June 25

1530: The Augsburg Confession is presented. The document is the primary confession of faith in the Lutheran denomination. It was presented to German rulers at the Diet of Augsburg, which was called by Holy Roman Emperor Charles V to united the Germans against a Turkish invasion.

1678: Elena Cornaro Picopia is awarded a Ph.D. Upon graduating from the University of Padua, Picopia became the first woman to earn a doctorate of philosophy. She pursued the degree after the bishop of Padua refused to allow her to pursue a doctorate of theology because she was a woman. Still, she persisted, and later became a lecturer of mathematics at the University. She went on to become an esteemed member of multiple academies throughout Europe.

1876: Custer is killed at the Battle of the Little Bighorn. The engagement was called the Battle of the Greasy Grass by the Lakota and Plains Indians who won it, and became known in popular culture as Custer's Last Stand. Custer led the 7th Cavalry Regiment into Montana during the Great Sioux War of 1876. The war was an attempt by the U.S. government to force the Lakota Sioux to cease resisting attempts to confine them to ever-shrinking reservations. The Sioux resolved to fight. In the spring of 1876, they gathered at a Sun Dance in Montana. At the religious gathering, the leader Sitting Bull had a vision of soldiers falling "like grasshoppers." Sitting Bull led a force of over 2,000 warriors against Custer's cavalry, who numbered about 650. Custer believed there were only about 800 warriors in the area, so when he attacked a village near the Little Bighorn, he was completely surprised by the size of the force that he encountered. Five of Custer's seven companies were completely destroyed in the battle, and the 7th Cavalry lost more than 260 men.

1943: The Częstochowa Ghetto uprising begins. The Ghetto was established in 1941 by German occupiers of Poland, and some 48,000 Jews were confined there. The uprising began when the Nazis started deporting Jews to concentration camps and executing elders, children, and intellectuals. In the subsequent fighting, 1,500 Jewish residents were killed. The uprising was suppressed on June 30 after heavy shelling of the Ghetto.

1950: The Korean War begins. North Korea invaded South Korea after a series of skirmishes along the border. The United Nations and principally the United States committed troops to defending South Korea. The Chinese People's Liberation Army entered the war on the side of the North. After four years of brutal combat and close to a million killed, the war ended in a stalemate. Neither side gained any significant territory or concessions from the other.

1978: The rainbow flag is first flown at the San Francisco Gay Freedom Parade. The flag was designed by Gilbert Baker, a San Francisco artist and gay rights activist. It became the symbol of the LGBTQ movement for civil rights and recognition in the United States and around the world.

1981: Microsoft becomes an incorporated business. Founded by Bill Gates and Paul Allen in Redmond, Washington, in 1975, the company went on to become one of the dominant software corporations in the world.

June 26

1409: Popes and antipopes abound. During the Western Schism of the Roman Catholic Church, Alexander V was proclaimed the Antipope—someone who claims to be Pope even though there is already another Pope in the Vatican. In Alexander's case, there were already two other Popes, Gregory XII in Rome, and Benedict XII in Avignon. Eventually the Schism was ended by the Council of Constance, which elected Pope Martin V and deposed the other claimants to the title.

1794: Military aircraft first take to the skies. At the Battle of Fleurus during the French Revolutionary Wars, the French Army used a hot-air balloon to conduct reconnaissance of the Austrian Army's movements.

1906: The first Grand Prix is held. The race was held over two days on closed public roads outside of Le Mans, France. The course was 64 miles long and the race consisted of 12 laps. Ferenc Szisz of the Renault team won the race, with Felice Nazzaro of FIAT coming in second.

1927: The Cyclone opens on Coney Island. The wooden roller coaster features a 2,640-foot track with six turns and twelve drops. Its highest point is the top of the lift hill, at 85 feet. In its nearly 100-year history, three people have died while riding it.

1944: Polish resistance fighters are overwhelmed at the Battle of Osuchy. In one of the largest battles between the Polish resistance and the Nazis, 1,200 resistance fighters were attacked by 30,000 Germans. Although the Germans won, killing 400 and capturing many of the other Polish fighters, the resistance soon struck back with a country-wide offensive in July.

1975: Save the whales! The International Whaling Commission formally banned the hunting of finback whales in 1975, after the species became one of five kinds of whales that was on the verge of extinction. Although the United Nations had sanctioned the moratorium of commercial whaling in 1972, the

Soviet Union and Japan, which hunted whales for food, refused to comply. The focus on saving the whales became a major part of the growing environmental movement around the world, and was spurred in part by the increasing rate of species extinction in the 20th century. Non-government organizations such as Greenpeace and Sea Shepherd joined the fight in the 1970s and 80s with militant direct-action tactics designed to disrupt commercial whaling and fishing efforts. Finback whale populations slowly recovered thanks to these and other efforts, but are not expected to approach even 50% of their pre-whaling numbers by the end of the 21st century. The whales are still hunted commercially by several nations, including Iceland and Costa Rica.

2015: The Supreme Court extends marriage rights to same-sex couples. In the landmark case *Obergefell vs. Hodges*, the Court ruled 5–4 that the right to marry is guaranteed to same-sex couples by both the Due Process Clause and the Equal Protection Clause of the Fourteenth Amendment to the Constitution.

June 27

1743: King George II leads British troops at the Battle of Dettingen. He was the last reigning British monarch to lead troops into battle. George's army, in alliance with Hessian troops, defeated a French army near Frankfurt. He ruled England until his death in 1760.

1844: Joseph Smith is murdered. Smith, the founder of the Latter-Day Saint movement and author of the Book of Mormon, was arrested in Carthage, Illinois, along with his brother Hyrum. As the mayor of Nauvoo, Illinois, Smith had suppressed a newspaper that was critical of Mormonism, had its press destroyed, and declared martial law. He was charged with treason against Illinois and was awaiting trial when a mob of 200 men descended on the jail. They shot the Smith brothers, who have been viewed as martyrs by the Mormon Church since then.

1905: Mutiny on the *Potemkin*. During the Russo-Japanese War, the battleship was serving as part of the Russian Empires' Black Sea Fleet. Morale in the Fleet was very low, and several riots and uprisings were breaking out elsewhere in the country. The Social Democratic Organization of the Black Sea Fleet, a revolutionary anti-Tsarist group, began preparing for a mass mutiny involving the entire fleet. On June 27, the crew of the *Potemkin* were served borscht made with rotten meat that was infested with maggots. They refused to eat it, and the ship's second in command, Ippolit Giliarovsky, threatened to shoot them. The crew mutinied, killing Giliarovsky, and took control of the ship, raising a red flag of the revolution and sailing for Odessa. They captured a military transport that was carrying troops to put down the uprising in Odessa. The mutineers then sailed to Romania, where they were granted asylum, and sank the ship. Their mutiny was marked as the beginning of the Russian Revolution, and later immortalized in the silent film *The Battleship Potemkin*.

1941: The Iasi Pogrom begins in Romania. The pogrom was one of the most violent ones in the history of the Jewish people. More than 13,000 Jewish citizens in Iasi, Romania, were murdered after Ion Antonescu ordered the commander of the city's garrison to "cleanse" the city. The ethnic cleansing was carried out in broad daylight, and several Gentiles who attempted to intervene were killed along with the people they were trying to protect. After the war, Antonescu was convicted of war crimes and executed.

1954: The world's first nuclear power plant goes online. The plant, at Obinsk in what was then the Soviet Union, operated until 2002. After 1959, however, it functioned solely as a research and nuclear isotope production plant.

1988: Mike Tyson knocks out Michael Spinks in 90 seconds. Both boxers were undefeated going into the fight, which was for the heavyweight championship. The fight was one of the most anticipated since the 1971 Fight of the Century between Joe Frazier and Muhammad Ali. It solidified Tyson's reputation as one of the best boxers in history.

1994: Aum Shinrikyo attack Matsumoto, Japan, with sarin gas. The attack killed eight people and injured 660. The millenarian cult perpetrated the attack to test the sarin gas it was manufacturing, and Matsumoto was chosen because its citizens had prevented the cult from establishing a base of operations there. Nine months later Aum Shinrikyo committed another sarin attack, in the subways of Tokyo.

2013: NASA launches the Interface Region Imaging Spectrograph. The space probe is a solar observation satellite that began transmitting images of the Sun in July 2017. Its research has led to a much greater and more complex understanding of the Sun.

June 28

1778: American Patriots are victorious in the Battle of Monmouth Courthouse. The Continental Army attacked the rear of a British Army column in Monmouth, New Jersey. The British quickly regrouped and counterattacked, nearly driving the Americans from the field. George Washington arrived and rallied the Americans along a hedgerow, forcing the British to withdraw. It was the first time Washington's army had fought the British to a standstill. It showed the effectiveness of the training Baron von Steuben, General du Motier, and Marquis de Lafayette had provided over the winter of 1777–78 at Valley Forge.

1846: Adolphe Sax patents the saxophone. The Belgian inventor and musician invented several other brass instruments, but the saxophone became his most popular. The valves he designed were cutting-edge at the time, and remain in use today largely in their original form.

1894: Labor Day becomes an official U.S. holiday. The federal holiday takes place on the first Monday of September and celebrates the American labor movement. President Grover Cleveland advocated the September date instead of May 1, which is traditionally the workers' holiday around the world, because he was afraid Mayday would become the anniversary of the Haymarket Riot of 1886, and he preferred a less militant holiday.

1902: Congress passes the Spooner Act. The legislation authorized President Theodore Roosevelt to acquire land rights from Colombia for the Panama Canal. The Canal was first used on August 15, 1914. It remained an American possession until 1999, when control was transferred to Panama.

1914: Archduke Franz Ferdinand is assassinated. Ferdinand, the presumptive heir to the Austro-Hungarian throne, and his wife Sophie, were in Sarajevo when they were shot and killed by Gavrilo Princip. The assassin was one of six who were organized by the Black Hand, a Serbian secret society dedicated to the unification of the Slavic territories in the region. The first two assassins, armed with bombs and pistols, lost their nerve as the Archduke's open car passed them in the streets of Sarajevo. The third threw a bomb, but missed his car. The motorcade sped away, and the Archduke delivered a scheduled speech at the Town Hall. After the speech, the Archduke's motorcade became lost, and stopped near where Princip was waiting for them. He fired two shots at the car, hitting the Archduke with the first and his wife with the second. Austria-Hungary blamed Serbia for the assassination and declared war, with its ally Germany joining. The mobilization caused Russia to join the war, and soon all of the Great Powers of Europe were embroiled in World War I.

1922: The Irish Civil War begins. Forces loyal to the Free State contingent of the Irish Provisional Government began shelling Irish Republican positions in Dublin. The Republicans rejected the British offer of a Free State, demanding a wholly independent republic. The war lasted less than a year, with the British-backed Free State forces easily defeating their former comrades-in-arms.

1926: Mercedes-Benz is formed. The automobile company was created from a merger of two competing car manufacturing companies owned by Gottlieb Daimler and Karl Benz.

1969: The Stonewall Riots begin. The riots, which took place in New York City after police raided the Stonewall Inn, a gay nightclub, mark the beginning of the Gay Rights Movement. The Stonewall Inn's location was designated a National Historic Landmark in 2000.

June 29

1613: The Globe Theatre burns to the ground. A theatrical cannon misfired during a performance of *Henry VIII*. One man was hurt when his breeches caught fire; they were put out with a bottle of ale. The theater was rebuilt the following year.

1644: Charles I defeats a Parliamentarian Army at the Battle of Cropredy Bridge. The Parliamentarians, led by Sir William Waller, attacked a rearguard of Charles' Royalist Army. The attack was repulsed, and the victory raised the morale of the Royalists, allowing Charles to recover from a series of earlier defeats and continue fighting the Civil War.

1767: The British Parliament passes the Townshend Acts. They imposed taxes on processed products imported by the American colonies including tea, paper, glass, and paint. The duties were designed to support the tax collectors who lived in the colonies, and were hugely unpopular. The colonists organized protests and boycotts of British products, and the Acts were repealed, with the exception of the tea tax, in 1770.

1899: Chicago briefly becomes the largest U.S. city. Several townships around the city of Chicago, including Hyde Park on the city's South Side, voted to be annexed into the city. They chose to do so primarily because the city's municipal plumbing system was superior to their own and provided access to fresh water from Lake Michigan. The arrangement made Chicago the largest city in the U.S. in area and the city with the second-largest population at the time.

1956: The Interstate Highway System is created. The Federal Aid Highway Act of 1956 organized the United States system of interstate highways. The formation of the system was championed by President Dwight D. Eisenhower, and many highways are named for him.

1974: Isabel Perón becomes President of Argentina. Perón was the first female president of the country. She was the former vice president, having served during her late husband's third term as president. Juan Perón suffered a series of heart attacks on June 28, and Isabel was sworn in secretly as acting president. He died on July 1, at which point she formally assumed the role of President of Argentina. Upon being sworn in, she became the first woman anywhere in the world to hold the title of "President" of a country. Perón pledged to uphold the economic nationalism her husband had championed, going so far as to enact pro-labor laws, and she quickly enjoyed the support of much of the nation. But she had left-wing politicians purged from government and signed an Anti-Terrorism Law that took away constitutional rights. As a result, she soon became unpopular. In 1976 she was deposed and arrested, and sent to exile in Spain.

1995: Space Shuttle *Atlantis* docks with Mir. It was the first time a NASA Space Shuttle docked with the Russian space station. The shuttle delivered two cosmonauts, Anatoly Solovyev and Nikolai Budarin, to the station, and picked up astronaut Norman Thagard. The mission was the first of seven flown by *Atlantis* to Mir.

2006: The Supreme Court prohibits military tribunals. The George W. Bush administration was attempting to have detainees held at Guantanamo Bay tried in military tribunals. They argued that their status as "illegal non-combatants" meant that due process and other protections granted by the Constitution did not apply to them. In *Hamdan vs. Rumsfeld*, the Supreme Court disagreed.

2007: The first iPhone is released. The touchscreen-operated smartphone, which includes wi-fi and cellular capabilities, completely revolutionized the cell phone market and led to a new era of wireless personal communication devices.

June 30

1859: Charles Blondin crosses Niagara Falls on a tightrope. The French acrobat went on to cross the Falls on a 1,100-foot tightrope a number of other times. Each time he increased the level of difficulty, crossing with a wheelbarrow, with his manager on his back, and on stilts. He once sat on the middle of the tightrope and cooked an omelet over the gorge.

1860: Evolution is debated at Oxford. Charles Darwin had published *On the Origin of Species*, which set out the basis of modern understanding of the mechanisms of evolution, seven months before. The book caused an uproar in England and around the world, and was challenged by those who believed evolution was contrary to the teachings of the Bible. Thomas Henry Huxley, a biologist who was known as "Darwin's Bulldog" for his fierce advocacy of the theory of natural selection, debated Samuel Wilberforce, a bishop of the Church of England. Although several others participated in the debate, it is best remembered for the exchange between Wilberforce and Huxley. Wilberforce asked him whether he claimed his descent from an ape on his grandmother's or grandfather's side of the family, in an attempt to mock him. Huxley responded by saying that he would rather be related to an ape than a man like Wilberforce, who used his gifts to hide the truth. Both said afterward that they thought they had themselves won the debate.

1905: Einstein introduces Special Relativity. In an article "On the Electrodynamics of Moving Bodies," the physicist first presented the idea that the speed of light is the same in a vacuum for all observers, and that the laws of physics are the same in all systems. The theories made Newton's laws of motion consistent with Maxwell's theories of electromagnetism.

1934: The Night of the Long Knives. Hitler purged all of his political rivals in Germany on this night, sending the Gestapo to carry out a series of extrajudicial executions. Many of those killed were members of Hitler's own Brownshirts organization, as well as Nazis who were more left-wing than Hitler and members of government who had suppressed him ten years before.

1937: The first emergency phone system is introduced. Using the number 999, London set up an emergency telephone number with which to summon police, firefighters or ambulances. The system was successful and copied worldwide. The number 999 is still Britain's emergency phone line.

1953: The first Chevrolet Corvette is built. Manufactured by General Motors, the 'Vette would become the iconic American sports car, with a new generation introduced every decade. GM helped popularize the car when a Chevrolet dealer hit upon the idea of giving free cars to NASA astronauts.

1997: The U.K. relinquishes control of Hong Kong. After occupying the island for over 150 years, only briefly abandoning it during World War II, the United Kingdom agreed to formally turn over control of Hong Kong to China. The event is widely regarded as the final act of the British Empire.

2013: Protests begin against Mohamed Morsi in Egypt. The President, who was a member of the Muslim Brotherhood, rose to power after the 2011 Egyptian uprising during the Arab Spring. Morsi enacted a series of laws that were unpopular, and attempted to consolidate power and rule by decree. The protests eventually led to his ouster in August.

Chapter 7

July

July 1

1643: The Westminster Assembly convenes for the first time. The first council, which consisted of theologians and members of the English Parliament, was organized by the Long Parliament during the First English Civil War. It restructured the Church of England, and among other acts, the Assembly wrote a new Confession of Faith that is still widely influential in the Protestant religions.

1690: William of Orange defeats James II at the Battle of the Boyne. The battle, fought near Drogheda, Ireland, spelled the end of James' effort to regain the British crown, and cemented the victory of William's Glorious Revolution. The battle is still commemorated by Loyalist Protestants in Ireland, although the celebrations take place on July 12.

1770: Lexell's Comet records a near miss. The comet, which passed the Earth at a distance of just 0.015 astronomical units (1,400,000 miles), came closer than any other comet in recorded history. It has not been seen since then, and astronomers consider it a "lost comet."

1858: Darwin announces the Theory of Evolution by Natural Selection. At the Linnean Society of London, papers by Charles Darwin and Alfred Russel Wallace describing the Theory were read together. Darwin had first discovered the Theory many years earlier following a voyage around the world on the HMS *Beagle*. Having observed geological formations that obviously took eons to create, and ancient fossils of extinct animals, Darwin became convinced that the Earth was old enough for evolution to have occurred. When he observed numerous species of finches on the Galapagos Islands, he realized they had all descended from a common ancestor. Later, he read an essay by Thomas Malthus that showed populations would quickly outbreed their food supplies. He combined these observations to realize that only the fittest, or best adapted, members of each species survive to pass their traits on. Darwin did not publish his findings for 20 years, as he was afraid of the backlash they would spark. But when Wallace sent him his own very similar findings on natural selection, Darwin's hand was forced. The following year, he published a book detailing the Theory of Evolution.

1863: The Battle of Gettysburg begins. The battle was the bloodiest conflict of the war, with more than 50,000 casualties, and is often considered the turning point in the Civil War. In its aftermath, Lee abandoned his plans to invade the North. The war dragged on for nearly two more years.

1908: Save Our Ship! SOS was adopted as the international distress signal following the second International Radiotelegraphic Convention, held in 1906. The signal, in International Morse Code, consists of three dots, three dashes, and three dots, with no spaces between the letters.

1960: Ghana gains independence. Kwame Nkrumah became the republic's first president, and British Queen Elizabeth ceased to be the country's head of state.

1963: ZIP codes are introduced in United States mail. The acronym stands for *Zone Improvement Plan*, and the system was adopted to make mail sorting and delivery faster. There are about 43,000 ZIP codes in the U.S.

2002: The International Criminal Court is established. The Court at The Hague in the Netherlands prosecutes individuals for war crimes, genocide, and crimes against humanity.

July 2

1698: Thomas Savery invents the steam engine. Savery, an English inventor and engineer, patented a steam pump that was designed for stationary use in mills and factories. He was granted a 21-year patent for his device by Parliament.

1839: African prisoners take over the slave ship *Amistad*. The prisoners, who were Mende people from Sierra Leone, were being transported from Havana, Cuba, to plantations in the Southern United States. Led by Sengbe Pieh (later called Joseph Cinque in the U.S.), the captives held below decks sawed through their manacles with a file and attacked the crew with canes and knives. They commandeered the vessel and demanded the ship return them to Africa. The ship's navigator agreed, but deceived them, sailing the ship to New York, where it was captured by the USS *Washington*. Cinque and his comrades were put on trial in the United States for mutiny. In 1841 the Supreme Court ruled that the captives had been captured illegally and taken the ship in an act of self-defense. Thirty-five of the survivors returned to Africa in 1842 with the help of New York abolitionists.

1881: Charles Guiteau shoots President James Garfield. Guiteau, a loyal Republican, believed himself to be essential to Garfield's victory in the 1880 presidential election. When he heard that Garfield planned to do away with the political jobs patronage system, he became convinced that he had to kill him. He later claimed God told him to assassinate Garfield. He shot the President while he was delivering a speech in Baltimore. Garfield succumbed to his wounds more than two months later.

1897: Guglielmo Marconi patents the first radio. The Italian inventor had previously discovered that he could increase his radio's transmission range by raising the height of its antenna. That breakthrough was critical and led to radios that could transmit at greater and greater distances.

1937: Amelia Earhart disappears while attempting to fly around the world. Earhart and her navigator Frederick Noonan sent their last transmission to the US Coast Guard cutter *Itasca* in a futile attempt to communicate their position.

1962: The first Wal-Mart opens in Rogers, Arkansas. The retail company was founded by Sam Walton. It would go on to become the largest employer in the United States and the world's largest company in terms of revenue, generating $480 billion in 2016.

2000: Vicente Fox Quesada is elected President of Mexico. He was the first candidate from an opposition party to win an election for President in more than 70 years. Fox was the 55th President of Mexico, serving until 2006.

2013: Pluto's moons Kerberos and Styx are named. The fourth and fifth moons of Pluto were discovered in 2011 and 2012, respectively.

July 3

1775: George Washington formally takes command of the Continental Army. In the ceremony, Washington drew his sword at Cambridge Commons in Massachusetts, rode out in front of the assembled American troops, and was declared commander of the Army. He led the American forces through the remainder of the American Revolutionary War, which ended eight years later with the Treaty of Paris.

1789: The first bank opens in the United States. The Bank for Savings opened for business in New York City. Its total deposits on the first day totaled $2,807.

1863: "General Lee, I have no Division." In the previous two days of fighting at the Battle of Gettysburg, General Robert E. Lee's Army of Northern Virginia had attempted unsuccessfully to turn both the left and right flanks of the Union Army's line. Lee chose on the third day to try attacking the Union Army in the center. He ordered General George Pickett to lead three divisions of Confederates straight toward Winfield Scott Hancock's Second Corps, which occupied the high ground in the middle of Cemetery Ridge. The attackers had to cross nearly a mile of open ground before reaching the Union lines, and they were battered by artillery fire the entire way. Upon reaching the Union position, Pickett's division, commanded by General Armistead, withered in the face of unrelenting volleys of musket-fire from the Union troops. The attack failed, and more than half of the Confederate soldiers who set out on the charge were killed or wounded. After, Lee asked Pickett to rally his division for the anticipated Union counterattack. Pickett replied, "General Lee, I have no Division." The battle over and Confederates defeated, they retreated back into Virginia, and would not invade the North again.

1898: Captain Joshua Slocum completes the first solo circumnavigation of the world. Slocum, a Canadian adventurer, sailed west around the globe in a fishing boat named *Spray*. He finished his voyage at Newport, Rhode Island.

1920: William Tilden wins at Wimbledon. Tilden was the first American tennis player to win the men's singles title at the storied club.

1940: The British Navy destroys the French fleet at Merl-el-Kebir. With German forces approaching the city, the British opted to destroy the French ships rather than let them fall into German hands. The Allies feared the Germans would be able to use the ships to mount an invasion of the British Isles.

1962: Jackie Robinson is inducted into the National Baseball Hall of Fame. Robinson, who had broken the color barrier as the first African-American player in the Major Leagues in 1947, was also the first African-American inducted into the Hall.

1976: The Raid at Entebbe. After guerillas acting on behalf of the Popular Front for the Liberation of Palestine hijacked an Air France passenger jet with over 200 passengers, they ordered its crew to land in Entebbe, Uganda. Idi Amin welcomed the hijackers. The Israeli Defense Forces sent commandos to free the hostages, and after a ninety-minute nighttime firefight, the hostages were freed. Several commanders, three hostages, and all of the hijackers were killed in the battle.

July 4

1054: The Crab Nebula lights up. The supernova was first recorded by Chinese and Japanese astronomers as a bright "guest" star in the constellation of Taurus the Bull. The supernova was visible even in the daytime sky for more than three weeks.

1653: Oliver Cromwell dissolves the Parliament. Cromwell had become frustrated with the Parliament's inability to pass strict political or religious tests. He replaced its members with a select group of fanatical Puritans in a body that became called "The Barebones Parliament." The moniker came from the London merchant named Praise-God Barebone, a committed Puritan. That Parliament also failed.

1776: The United States declares its independence. In what has become one of the most famous political moments in human history, representatives from each of the thirteen American colonies passed the Declaration of Independence on this day in Philadelphia. The Declaration famously declared that "all men" had "certain inalienable rights," and that when a tyrannical government seeks to limit these rights, the people can and should endeavor to overthrow it. The document also laid out a litany of abuses of the colonies by King George III. It was drafted by a committee that included John Adams, Benjamin Franklin, and Thomas Jefferson, and signed in a meeting at the Pennsylvania State House by 56 American patriots on July 2. Two days later the Continental Congress approved the wording of the document, making American independence from Great Britain official.

1817: Construction begins on the Erie Canal. The Canal connected Lake Erie to the Hudson River, and contributed considerably to the financial growth of New York City. It also helped boost migration to the Midwest and allowed Midwestern farmers to sell their products in markets back east.

1845: Henry David Thoreau goes to Walden Pond. The writer, naturalist, and philosopher moved to a small shack near the pond, where he wanted to get away from modern life and celebrate his relationship with nature. He kept a journal during his time there, and the journal became the basis for the book *Walden*, one of the most important and enduring books in American literature.

1855: Walt Whitman publishes *Leaves of Grass*. The book, which contained just a dozen poems, would later grow to include more than 400. Whitman's poems explore themes of life, nature, philosophy, and humanity. The book has become one of the most influential poetry volumes in the history of American literature.

1879: British forces win the Battle of Ulundi. The battle, fought between the British and the Zulu army led by Ziwedu kaMpande, was the last one in the Anglo-Zulu War. The war was provoked by the British, who considered an autonomous Zulu kingdom to be a threat to their plan of consolidating their control of South Africa.

1903: The first transpacific telegraph cable is completed. The cable connected the islands of Honolulu, Guam, Midway, and Manila. U.S. President Theodore Roosevelt sent the first telegraph over the cable.

1997: NASA's *Pathfinder* rover lands on Mars. It used parachutes to slow its descent to the surface of Mars and then deployed large air bags to cushion its impact. It was the first spacecraft to touch down on Martian soil in more than two decades. On its first day, it began sending images back, which were immediately put on the Internet by NASA.

July 5

1294: Pietro di Murrone is elected Pope Celestine V. He had been a devout hermit before his election, and he became overwhelmed by the political and social pressures placed on him as Pope. Celestine resigned after just five months in the position, making him the first to do so. His successor, Boniface VIII, had him imprisoned for resigning.

1295: Scotland and France declare an alliance. The "Auld Alliance" against England, designed to thwart their numerous invasions of their neighbors, is one of the oldest mutual defense pacts in history. It was replaced in 1560 by the Treaty of Edinburgh, in which Scotland switched its alliance to England.

1687: Newton's *Principia* is published. One of the most important publications in the history of science, the *Philosophiæ Naturalis Principia Mathematica* (Mathematical Principles of Natural Philosophy) laid out Sir Isaac Newton's laws of motion, which are the foundation of classical mechanics and physics. The work also described Newton's law of universal gravitation and laws on planetary motion. The *Principia* was published by the Royal Society, which printed between 300–400 copies. The publication of Newton's work is widely considered the last great event in the Scientific Revolution, which had begun in 1543 when Copernicus discovered that the Earth was not the center of the Universe. The *Principia*'s findings about the mechanical laws under which the universe apparently operates helped spark the Enlightenment of the 18th century. The work dominated scientists' view of the universe for the next three centuries, until Einstein discovered the Theory of Relativity in the early 20th century.

1811: Simon Bolivar declares Venezuela's independence from Spain. Bolivar, who was from an upper-class family, had been educated in Europe, where he was exposed to the ideas of the Enlightenment that supported autonomous, democratic rule. The Venezuelan Declaration of Independence established a government committed to individual equality and freedom of expression. Bolivar led the Venezuelan cause in a war for independence that lasted until 1823.

1921: Jury selection begins in the "Black Sox" trial. Players for the Chicago White Sox, including "Shoeless" Joe Jackson, Buck Weaver, and Eddie Cicotte, were accused of conspiring to throw the World Series in exchange for payoffs from professional gamblers. Although they were eventually found not guilty by the jury, they were banned from professional baseball.

1942: Ian Fleming graduates from spy school. Fleming was the first graduate from a training program in espionage in Canada that was called "Special 25." Fleming would later go on to invent the famous secret agent James Bond based in part on his experiences at the school.

1950: The Korean War commences in earnest. North Korean forces had crossed into South Korea in June, and the United States sent troops to support the South. Private Kenneth Shadrick, of Skin Fork, West Virginia, became the first American serviceman to die in that conflict on July 5 when he was shot by a machine gunner on a North Korean T-34 tank. Over 30,000 Americans would die in the war over the next three years.

1950: Israel enacts the Right of Return. The Right guarantees all Jewish people around the world free and automatic Israeli citizenship if they decide to immigrate to that country.

July 6

1519: Charles V of Spain becomes Holy Roman Emperor. Charles reportedly bribed the German electors to pick him as Emperor. The German princes limited his authority, and Charles abdicated the throne in favor of his brother in 1556.

1699: Pirate Captain William Kidd is captured. Formerly a merchant from New York, Captain Kidd turned to piracy after the British Navy attempted to impress his crew into service. He sailed the high seas for three years before his capture. He was tried in England and hanged.

1912: "Thanks, King." These two words were Native American athlete Jim Thorpe's simple response to the King of Sweden, who had declared him the world's greatest athlete after Thorpe swept the track and field portion of the 1912 Olympic Games in Stockholm.

1917: Arab forces capture Aqaba. T.E. Lawrence, an English soldier and spy, served as a military advisor and leader of the Arab forces fighting for independence from the Ottoman Empire. Along with Auda ibu Tayi, Lawrence led the forces as they swept into Aqaba, a strategically important Red Sea port. The battle was later depicted in the film *Lawrence of Arabia*.

1942: Anne Frank's family goes into hiding. The Nazis invaded and occupied the Netherlands beginning in 1940, and by 1942 their persecution of Dutch Jews was being committed in earnest. As a result of the persecution, Anne's family had originally planned to go underground on July sixteenth, but when Anne's sister Margot received an order to report to a work camp, they moved their plans up. Anne and her family concealed themselves in a secret apartment behind a bookcase in the building where her father worked for almost two years, during which time Anne kept her now-famous diary. The family was betrayed and arrested in August 1944.

1946: "Bugs" Moran is captured. The gangster had been one of the major players in organized crime during Prohibition, although his influence was severely curtailed after Al Capone had his top henchmen gunned down in the Saint Valentine's Massacre. The FBI never stopped looking for him, though, and he later died in prison.

1957: Paul McCartney meets John Lennon. The two met at a church picnic in Woolton, a village near Liverpool, England. Lennon's band The Quarrymen was playing at the picnic, and he and McCartney hit it off. They would later form the Beatles, one of the most influential pop music bands of the British Invasion.

1986: Davis Phinney wins a road stage on the Tour de France. He was the first American cyclist to do so, winning Stage 3, which is on the island of Corsica.

2003: Calling All Stars: The Yevpatoria Planetary Radar Station sends a METI message to five stars in the Earth's neighborhood. The messages will arrive to these stars in 2036, 2040, 2044, and 2049, respectively.

July 7

1456: Joan of Arc is posthumously acquitted. King Charles II of France acquitted the military leader of crimes including heresy and witchcraft, for which she had been executed.

1520: The Battle of Otumba. About 500 Spanish conquistadores led by Hernan Cortes defeated a much larger force of Aztec warriors. The Aztec force numbered between ten and twenty thousand troops, but they had never encountered mounted cavalry in battle. The conquistadores' victory set the stage for their later invasion and capture of Tenochtitlan, the Aztec capital, and the establishment of New Spain.

1834: The Tappan Riots break out in New York City. The riots were led by pro-slavery, Nativist New Yorkers who attacked the houses of abolitionists, as well as churches and homes of African-Americans. The riots lasted four days until they were quelled by the military.

1846: The United States annexes California. Commander J.D. Sloat of the U.S. Navy hoisted the American flag in Monterrey, California, after a Mexican garrison stationed there surrendered to his troops.

1876: Jesse James robs the Missouri-Pacific train. James, one of the most notorious criminals of the American West, made a fortune robbing trains. The Missouri-Pacific robbery was one of James' biggest scores, netting his gang $15,000.

1928: The best thing since... Sliced bread goes on sale for the first time in Chillicothe, Missouri. The Chillicothe Baking Company had acquired the first automatic bread-slicing machine, invented by Otto Frederick Rogwedder, earlier that year. His machine not only sliced the loaf of bread but also wrapped it. His friend, Frank Bench, was a baker at the Chillicothe Baking Company, and bought the first machine Rogwedder ever sold. Bench installed the machine at the company and began marketing sliced bread in the local community. It was a hit: sliced bread was widely popular, and Rogwedder sold his machine to other bakeries around the country. In 1930, the Continental Baking Company rolled out Wonder Bread, the first sliced bread to be sold nationwide.

1937: The Peel Report is published. The document was published by a commission in the United Kingdom tasked with investigating the situation in the British Mandate of Palestine. The report was the first to recommend a two-state solution for Palestine, one Arab and one Jewish.

1944: The largest banzai charge in history takes place. After more than three weeks of intense fighting, American forces had pushed the Japanese back until they had nowhere left to retreat. Their commander ordered a suicide attack, and nearly 3,000 men charged the American lines, followed by their wounded. More than 4.000 Japanese soldiers were killed in the battle, which lasted fifteen hours.

1946: Mother Frances Xavier Cabrini is canonized. Cabrini was the first American citizen to be made a Catholic saint.

1981: Sandra Day O'Connor is appointed to the Supreme Court. O'Connor was the first woman to be appointed to the Court, and she served until 2006. She usually sided with the Court's conservative side initially, but later in her career was a swing vote.

2005: London train bombing kills 56. Four suicide bombers detonated bombs on separate train lines around London. More than 700 commuters were wounded in the attacks.

July 8

52 BCE: Julius Caesar captures a village named Lutetia Parisiorium, in Gaul. The city's inhabitants were called the Parisii, and the city would go on to become one of the Roman Empire's most important cities in the region. It was eventually renamed Paris.

1099: The Siege of Jerusalem begins. Fifteen thousand Christian Crusaders marched around the walled city in a religious procession to begin the siege, while Muslim defenders watched. The siege lasted nearly a month, and then the Crusaders assaulted the walls and captured the city, massacring its inhabitants.

1497: Vasco da Gama departs from Lisbon. The Portuguese navigator and explorer was searching for a new sea route to India, and he discovered it by sailing around the southern tip of Africa. The King of Portugal, Emmanuel I, conferred the title of Admiral of the Indian Ocean for the accomplishment.

1608: Samuel de Champlain arrives in Canada. He establishes the first French settlement there, naming it Quebec.

1776: The Liberty Bell rings. The Bell was rung to call citizens of Philadelphia to the State House to hear a reading of the Declaration of Independence by Colonel John Nixon.

1800: The first cowpox vaccination is administered. Dr. Benjamin Waterhouse administered the vaccine to his son Daniel to prevent him from becoming infected by smallpox. The vaccine worked, and ushered in the beginning of a new era of medicine.

1905: The crew of the *Potemkin* surrender. The battleship was part of the Imperial Russian Navy's Black Sea Fleet, and the crew mutinied after having been served rotten meat by their officers. The mutiny was initially intended to protest the terrible conditions those serving in the Navy were under, but quickly grew to something bigger. After having their lives threatened by the ship's first mate, the crew killed him and several other officers. They established a revolutionary committee and sailed the ship to Odessa, flying a red flag. A general strike was occurring in the city, and the mutineers refused to put the ship's soldiers ashore to quell the strike. They then sailed to Romania, where they sought asylum from a friendly foreign government, but were unable to secure any. The ship was captured and the mutiny's ringleaders were sentenced to death, but the firing squad refused to shoot them. The mutiny is generally regarded as the first action in the Russian Revolution that eventually overthrew the Tsar and led to the October Revolution of 1917.

1936: The Spanish Civil War erupts. The fascist general Francisco Franco led an uprising against the government after the Popular Front, led by a coalition of socialists and communists, won popular elections. Although volunteers came from around the world to defend the democratically elected government of the Spanish Republic, no foreign governments other than the Soviet Union provided aid to the government. Franco was supported by fascists in Germany and Italy, and his forces won the Civil War after three years of bloody fighting.

1994: Kim Jong-Il takes power. Kim assumed supreme leadership of the country, which he would rule for nearly two decades, after his father Kim Il-sung died.

July 9

118: Hadrian enters Rome. As Emperor, Hadrian oversaw the expansion of Rome's holdings to its farthest Northern reaches in Great Britain, where a wall named for him was built. Portions of Hadrian's Wall are still there to this day. Hadrian was also ruthless over the rule of Palestine, where he issued a decree forbidding Jews from entering the city of Jerusalem.

1595: Kepler describes the universe in the *Cosmic Mystery*. In this book, German astronomer Johannes Kepler laid out a geometrical description of the heavenly bodies. He proposed a cosmological theory for the motion of the six planets known at the time. His theory applied Pythagorean shapes, called polyhedral, to the Copernican model. Kepler proposed that the geometric elegance of the system proved that it had a Divine plan. His model was also the first scientific attempt to prove that the Sun was at the center of the cosmos since Copernicus had proposed the idea.

1846: Captain John Montgomery lands in California. Montgomery sailed his ship into a large bay near the tiny Mexican village of Yerba Buena. He encountered no Mexican troops there, and ordered the American flag to be hoisted over the village, claiming it for the United States. The Americans later changed the name of the town and the Bay to San Francisco, and it grew to be one of the biggest metropolitan areas in Northern California.

1877: Wimbledon is inaugurated. The All England Croquet and Lawn Tennis Club organized its first tournament at Wimbledon, England, outside London. The competition included 21 amateur tennis players. It would go on to be one of the world's premiere tennis events, and is part of the suite of tournaments that make up competitive tennis's Grand Slam of trophies.

1893: The first open-heart surgery. Dr. Daniel Hale Williams performed the operation successfully, and without using anesthesia, at the Provident Hospital in Chicago, Illinois.

1900: The Commonwealth of Australia is founded. The British Parliament formally established the Commonwealth, uniting what had been several autonomous colonies under a single federal government. The Commonwealth took effect the following New Year's Day, in 1901.

1922: Johnny Weissmuller breaks the minute. Weissmuller, an American swimmer, was the first person to swim the 100-yard freestyle in less than a minute. He went on to set more than fifty world records and win five Olympic gold medals for swimming. He later played Tarzan in motion pictures.

1941: Enigma is broken. British cryptologists had labored for months to break the German military code. The Germans used a sophisticated computing machine to produce the code. The code-breakers, working in secrecy at Bletchley Park, were led by Alan Turing, who invented a machine of his own to break the code. That machine was the forerunner of modern computers.

2011: South Sudan gains independence. The country, bordered by Sudan, Ethiopia, Kenya, Uganda, the Congo, and the Central African Republic, was established following a referendum in which nearly 99 percent of the population voted for independence. In 2013 a civil war broke out after an attempted coup.

July 10

988: Dublin is founded. The Irish city became the seat of High King Mael Sechnaill mac Domnaill after the Norse King Gluniarn pledged loyalty to him and promised to accept Brehon Law, the system of statues that governed Ireland well into the 17th century.

1553: Jane Grey is proclaimed Queen of England. Grey, who was fifteen years old, was granted the throne by Edward VI following his death. Edward, a Protestant, passed it to her because she was an educated and devout Protestant. He passed over his Catholic sister Mary to do so. After just nine days, however, Mary claimed the throne, with wide popular support. Jane Grey was deposed, imprisoned, and ultimately executed at Mary's command.

1778: France declares war on England. The French King, Louis XVI opened hostilities against England in support of the American Revolution following negotiations with American leadership, represented by Benjamin Franklin. The French were primarily interested in weakening England's control of the Americas, but in joining the revolutionaries, they prevented their efforts from being crushed early on in the War. France is still considered the United States of America's oldest ally for the assistance they provided during the War.

1900: "On His Master's Voice": The Victor Recording Company, which went on to become RCA Victor, registered a trademark with the U.S. Patent Office that depicted a small dog staring into the amplifier of a gramophone record player. The trademark became one of the most famous and recognizable trademarks in the world.

1962: Telstar 1 is launched. The world's first communications satellite was part of an international agreement by telephone companies in the United States, United Kingdom, and France. It was launched from Cape Canaveral, Florida, at 4:30 in the morning, and by 7:30 A.M., it had relayed the first telephone call to be made by satellite, from the chairman of AT&T in Maine to U.S. Vice President Lyndon B. Johnson. Later that day, the satellite relayed the first transoceanic telecast, from Andover, Maine, to receivers in the United States and Europe. The satellite was able to relay messages for 20 minutes as it passed over the Atlantic ocean every 2.5 hours, due to its non-synchronous orbit. The experimental satellite remained in service until February 21, 1963. Today, there are more than 1,000 active satellites in orbit.

1987: The *Rainbow Warrior* is sunk. The ship, owned by the environmental activist group Greenpeace, was docked in the harbor at Auckland, New Zealand. The ship had been preparing to sail to a nuclear test site in the South Pacific where it intended to protest and interfere with the French Navy's testing of nuclear weapons. French intelligence agents planted underwater bombs that sank the ship, killing a photographer who was aboard. The agents were captured by New Zealand police and spent two years in prison.

1991: Boris Yeltsin is inaugurated. Yelstin was the first democratically elected president in Russian history. He served until 1999 and oversaw the later stages of the country's transition from a Communist government to a democratic one. He was succeeded by Vladimir Putin.

July 11

1533: Henry VIII is excommunicated. Pope Clement VII excommunicated the English King after he decided to divorce Queen Catherine of Aragon in order to wed Anne Boleyn. Unperturbed by the excommunication, Henry simply founded the Church of England and made himself its leader.

1807: Aaron Burr shoots Alexander Hamilton. Burr, the Vice President of the United States, had challenged Hamilton, the former Secretary of the Treasury, to a duel in New Jersey after a long and bitter rivalry between the two men. Burr had grown to hate Hamilton, who he considered a treacherous opportunist, after Hamilton had launched a series of personal attacks against him during the election of 1800. Although Burr had many critics, he felt that Hamilton was the only "gentleman" among them, and therefore the only one who would accept a challenge to a duel. Hamilton felt that he could not turn down the challenge, because he was unwilling to recant his repeated attacks on Burr, and because he did ascribe to the gentleman's code of honor. He intended to deliberately miss his shot, which could have ended the duel without bloodshed. But Burr did not miss, and mortally wounded Hamilton, who died the next day. Burr was indicted for murder, but never went to trial, and remained Vice President until 1805.

1914: Babe Ruth debuts in the Major Leagues. Ruth began his legendary career as a pitcher for the Boston Red Sox, earning $2,900 in his rookie season. Boston would later trade "the Bambino" to the New York Yankees, where he would go on to be one of the greatest hitters in the Modern Baseball era.

1938: *Mercury Theater on the Air* debuts. The radio program featured famed broadcaster and actor Orson Welles. It is best remembered for its broadcast later that year of a live reading of a radio play adaptation of H.G. Wells' novel *The War of the Worlds*. The broadcast, which purported to be of an actual alien invasion of a town in New Jersey, sparked widespread panic among listeners.

1974: The Terra Cotta Army is discovered. Consisting of more than 6,000 life-size figures of warriors, the Army was unearthed by Chinese archeologists near the site of the ancient capital, Xian. The warriors were created to guard the tomb of the first Ch'in Dynasty emperor in 206 BCE. The army and burial site covered more than three acres.

1979: Goodbye, Skylab: the first American space station came crashing down in a spectacular return as it showered the Indian Ocean and Australia in burning debris. Its final manned mission had been completed more than five years earlier.

1981: Neva Rockefeller is ordered to pay her husband alimony. It was the first case in United States history in which a woman was ordered to pay alimony to her former huband.

1995: The Srebrenica Massacre begins. Units of the Bosnian Serb Army, under the command of Ratko Mladic, carried out the massacre over the next eleven days, killing more than 8,000 Muslim Bosnians in the town of Srebrenica. The town had been declared a "safe area" under the protection of the United Nations, but the 370 U.N. soldiers in the town could not prevent Mladic's forces from capturing the town or carrying out the slaughter. Mladic was later found guilty of crimes against humanity by the International Criminal Tribunal on the former Yugoslavia and sentenced to life in prison.

July 12

1389: Geoffrey Chaucer is appointed Chief Clerk at Westminster by King Richard II of England. Chaucer was a middle-class son of a merchant, and he worked for the aristocracy in various positions throughout his adulthood. He would later become famous for writing *The Canterbury Tales*.

1630: The Governor of New Amsterdam buys Gull Island. The Governor purchased the island from Native Americans and renamed it Oyster Island. Years later, it would be called Ellis Island, and would be the site of entry to the United States by thousands of immigrants.

1679: Habeas Corpus is ratified. The Act was ratified by the British King Charles II, and was the first legal document to consider not just the guilt or innocence of the accused, but whether the act of imprisoning them was legal. It would become the basis for much of European and American law during the 18th and 19th centuries.

1817: The first recorded flower show is held. The show was held at Dannybrook, in County Cork, Ireland.

1861: Wild Bill Hickock draws down. The gunfighter coolly shot down three attackers in a shootout in Nebraska, establishing his reputation as a Wild West legend. One of the men Hickock killed, David McCanles, was married; Hickock visited his widow after he was tried and found not guilty of murder, apologized for the killing, and gave her all the money he had on him at the time, 35 dollars.

1948: David Ben-Gurion orders the expulsion of Palestinians. The event, which became known as the Palestinian Exodus from Lydda and Ramle, saw between 50,000 and 70,000 Palestinian Arabs forcibly expelled from those towns after they were captured by Israeli troops. The towns were renamed Lod and Ramla.

1960: The first Etch-A-Sketch is sold. The mechanical drawing toy went on sale for $2.99, and eventually sold more than half a million units that year alone. It became one of the 20th century's most iconic toys, and was inducted into the National Toy Hall of Fame in 1998.

1962: The Rolling Stones perform for the first time. The band played at the Marquee Club in London, England, and were billed as "The Rollin' Stones." After the performance, they went on tour around the United Kingdom, playing covers of Chicago Blues music and Chuck Berry songs. The Stones went on to be one of the greatest British rock 'n' roll bands in history, building a massive worldwide following and continuing to tour into the 21st century.

1967: The Newark Riots begin. The riots began after Newark police officers severely beat a Black motorist. Residents attacked police in response, and rioting lasted for five days. Twenty-six people were killed, hundreds were wounded, and businesses were burned.

1977: The *Enterprise* completes its first free-flight test successfully. The space shuttle would usher in a new era of low-earth-orbit flights. Named for the ship in *Star Trek*, the *Enterprise* remained in service until it was decommissioned in 2012. It was put on display at the Intrepid Museum in New York City following its retirement.

July 13

1568: Bottled beer is invented. The Dean of Saint Paul's Cathedral in London, Alexander Nowell, discovered the concept of conditioning beer in the bottle when he left his beer at his favorite fishing spot, and returned later to find that the beer became carbonated and fizzy after a few days.

1787: Congress passes the Northwest Ordinance. The Act permitted the Northwest Territory to be broken up into "three to five" states with a population of 60,000 settlers. The region later became the states of Ohio, Wisconsin, Illinois, Indiana, and Michigan.

1836: Patent No. 1 is issued. John Ruggles, an inventor from Thomastown, Maine, invented a traction wheel for use in locomotive steam engines. He was issued the first patent from the U.S. Patent Office under their then-new numbering system. Nearly ten thousand patents had been issued without numbers before Ruggles received his.

1837: Queen Victoria moves into Buckingham Palace. She was the first British monarch to live at the 600-room Palace, which was originally built to be the residence of the Duke of Buckingham. The estate was renovated in 1836–37, and became the home of the British ruler for the next two centuries.

1863: The Draft Riots break out. After the U.S. government announced it was instituting conscription to fill the ranks of the Union Army during the Civil War, riots broke out in New York City. The rioting continued for three days and is regarded as the worst such incident in U.S. history.

1923: HOLLYWOOD. The iconic sign in the Hollywood Hills area of Los Angeles was erected on Mount Lee. The sign, created as an advertisement for a local real estate development, features 44-foot-tall letters and originally read "Hollywoodland."

1925: Will Rogers debuts. The Ziegfeld Follies Burlesque Musical Debut was looking for a replacement for W.C. Fields, who was away attending his mother's funeral. This gave Rogers his break, and he would become one of the most famous comedians and showmen of the era.

1977: Blackout! Almost the entire city of New York experienced an electricity blackout that lasted for 25 hours, the longest such event in a major city anywhere in the world. The blackout came amidst a heat wave and while residents were already on edge because of a serial murderer, dubbed the Son of Sam, who was terrorizing the city. The blackout began in the evening, and under cover of darkness, looting of stores and acts of arson soon became widespread. More than seventy stores were looted in one five-block stretch of Crown Heights, and 25 fires were burning out of control in the Bushwick area of Brooklyn the morning after. Airports were closed, local television stations went off the air, highway tunnels connecting Manhattan to the mainland were shut down, and thousands of people were evacuated from the subway system, which ground to a halt. One person died as a result of violence during the blackout.

1985: The Live Aid concert is held. The concert took place in venues around the world, including Moscow, Sydney, and Philadelphia, and at Wembley Stadium in London. The concert was held to raise money to help famine-stricken areas in Africa. It was very successful, raising more than $125 million, the largest sum ever garnered in a benefit concert at the time.

July 14

1798: The Alien & Sedition Acts are passed. The Acts made it a crime to publish false, malicious, or scandalous material about the United States government. It also made it harder for new immigrants to vote in elections and allowed for deportation of foreigners.

1811: Lord Byron returns to England. The poet had been touring Europe and the Middle East for two years. Upon his return, Byron wrote *Childe Harold's Pilgrimage* based on his travels. The poem was highly successful and earned him his first acclaim.

1853: The first U.S. World's Fair opens. The Fair was held in Crystal Palace, New York, and featured the "Industry of All Nations." Among other exhibits, the Fair featured the demonstration of the first elevator equipped with a safety brake, a method to manufacture bromine, and the first pedal quadracycle, a rider-powered, four-wheeled contraption.

1865: Edward Whymper climbs the Matterhorn. The English mountaineer led the first successful expedition up the peak in the Swiss Alps. Four members of his party died on the descent.

The Matterhorn

1881: Billy the Kid is killed. The gunfighter William H. Bonney—also known as Henry McCarty—had been an outlaw in the American West since at least 1875. He became a wanted man after escaping jail in the New Mexico Territory and fleeing to the Arizona Territory. During the Lincoln County War, rival factions fought over the control of cattle ranching land there. Bonney joined the so-called Regulators faction of the war, and the group killed three men in 1878, including the Lincoln County Sherriff William J. Brady. Bonney was charged with his murder and became a national fugitive. He was captured by Sherriff Pat Garrett in the winter of 1880–81, tried for Brady's murder, and sentenced to hang. He escaped from jail in April of that year and killed two sheriff's deputies in the attempt. Garrett hunted Bonney down, and found him in Fort Sumner, New Mexico, where he shot the 21-year-old gunman on sight. Over the next twenty years, legends that Bonney survived the encounter spread, and several men claimed to be the famous outlaw.

1914: Robert H. Goddard patents the first liquid-fueled rocket. Until then, rockets operated using solid fuels. Goddard's invention revolutionized the science of rocketry and was the forerunner of all rocket technology that followed.

1972: Jean Westwood becomes head of the Democratic Party's National Committee. Westwood was the first woman to be appointed to such a high position in either of the two major parties in the United States.

2015: New Horizons reaches Pluto. The NASA probe performed a flyby of the dwarf planet and sent images back to Earth. In doing so, it completed the first full survey of the Solar System.

July 15

971: Bishop Swithun is exhumed. The Bishop originally had asked to be buried outside the west door of the ministry in Winchester, England, saying he wanted "the sweet rain of heaven to fall upon my grave." When the building was renovated, the grave was exhumed and the Bishop's remains were moved. It began raining and continued to do so for the next 40 days. This gave rise to the local legend that if it rains on Saint Swithun's Day, it will continue raining for weeks.

1149: The Church of the Holy Sepulchre is consecrated. The Church supposedly sits on the site where Jesus of Nazareth was crucified and is the location of his tomb. Several small chapels that had stood on the site for centuries were consecrated as a single holy site. It was destroyed in 1187 when Saladin captured the city, but later rebuilt, and remains on the site.

1799: The Rosetta Stone is discovered. French Captain Pierre-Francois Bouchard discovered the tablet while he was in Egypt with Napoleon's forces. The stone includes a decree issued in Egypt by King Ptolemy V, and is written in Ancient Egyptian and Ancient Greek. It was the first ancient text to be discovered that was written in both languages. This feature allowed archaeologists who were already familiar with Greek to begin deciphering Egyptian hieroglyphs for the first time. Bouchard took the stone to Alexandria, and when British troops captured the city in 1801, they took it with them to London. Several other decrees from the Ptolemaic era have been discovered since the Rosetta stone was found, but it remains the most significant key to translating Ancient Egyptian texts.

1815: Napoleon is exiled from France. He was brought from his home in France to the British ship HMS *Bellerophon*, where he surrendered to its Captain, Frederick Maitland. Napoleon believed the British would allow him to live in exile in Great Britain, but they took him instead to St. Helena, a tropical island in the South Atlantic Ocean, 2,500 miles east of Brazil. Napoleon died on the island on May 5, 1821.

1922: Platypus at the Zoo! The first duck-billed platypus, native to Australia, arrived in the United States, where it was put on display at the Bronx Zoo in New York City.

1933: Wiley Post takes to the sky. The American aviator piloted a Lockheed Vega airplane named the *Winnie May* in the first solo flight around the globe. Post landed in New York seven days, 18 hours and 49 minutes after he took off.

1965: *Mariner 4* flies by Mars. The NASA spacecraft was the first to send the first close-up images of the Red Planet—taken from an altitude of just 6,000 feet above the Martian surface—back to Earth.

1997: Gianni Versace is murdered. The former male escort and serial killer Andrew Cunanan murdered the fashion icon outside of his Casa Casuarina mansion. It led to a nationwide manhunt, with Cunanan placed on the FBI's Ten Most Wanted Fugitives list. Cunanan committed suicide before he was captured.

July 16

622: Muhammad flees Mecca. This date marks the traditional beginning of the Islamic Era. Muhammad, the religion's Prophet, fled Mecca with his followers, traveling to Yathrib, which he later renamed Medina ("The City"). The pilgrimage is called the *Hijrah*.

1439: England outlaws kissing. It was in an attempt to stop the spread of the Plague.

1769: San Diego is founded. Father Junipero Serra, a Spanish missionary, established a Catholic mission on the site, which he named San Diego de Alacala. He presided over a Catholic mass and unfurled the Spanish Royal Standard, declaring it a territory of Spain.

1790: The District of Columbia is established. The Residence Act, passed by the second session of Congress and signed by President George Washington, officially made the District the permanent seat of the United States government.

1861: Union troops invade Virginia. President Abraham Lincoln, frustrated by the cautious general George B. McClellan's unwillingness to take decisive action, personally ordered the invasion. The Army began a 25-mile march into the Confederate state that would later culminate in the First Battle of Bull Run.

1941: Joe DiMaggio hits for 56. DiMaggio, playing for the New York Yankees, successfully reached base in 56 consecutive baseball games. The record still stands as the longest consecutive hitting streak in Major League Baseball history.

1942: The Roundup at Vel' d'Hiv. The Vichy government, which was collaborating with Nazi occupiers, rounded up 13,152 Jews in France. The captives were held at the Winter Velodrome in Paris before being sent to the infamous concentration camp Auschwitz.

1945: The Manhattan Project succeeds. The atomic age is ushered in when the United States succeeds in a test detonation of the first plutonium-based nuclear weapon near Alamogordo, New Mexico. The U.S. would drop atomic bombs on Hiroshima and Nagasaki, Japan, later that summer, winning World War II and beginning the Cold War.

1969: Apollo 11 is launched. The mission took off from the Kennedy Space Center at Cape Canaveral, Florida, with mission commander Neil Armstrong and pilots Buzz Aldrin and Michael Collins. Three days later, the spacecraft entered lunar orbit, and Armstrong and Aldrin navigated the mission's lunar lander to the moon's surface. The landing was broadcast on live television to an audience around the globe. Upon alighting on the surface of the moon, Armstrong famously said, "That's one small step for [a] man, and one giant leap for mankind." Armstrong and Aldrin stayed on the lunar surface, in the Sea of Tranquility, for about 21 hours before rejoining Collins in the command module. They returned to Earth, landing in the Pacific Ocean, on July 24.

1979: Saddam Hussein takes power. Hussein replaced President Ahmed Hassan al-Bakr following the latter's resignation, and purged the Iraqi Ba'ath party of rivals in a bloody consolidation of power. He ruled Iraq in a ruthless dictatorship for more than two decades, until he was deposed by the United States invasion in 2003. Hussein was captured, tried, and executed in 2006.

1999: John F. Kennedy Jr. is killed. Kennedy was piloting a Piper Saratoga aircraft near Martha's Vineyard when it crashed, killing him, his wife and sister-in-law.

July 17

1453: The Battle of Castillon. The battle, between French and English forces, was the last one in the Hundred Years' War, which had been fought since 1337 between the two nations. The French won a decisive victory in the battle, ending English land claims in most of France.

1717: *Water Music* premieres. George Frideric Handel wrote the collection of orchestral pieces at the request of King George I. The King had Handel unveil the composition with a barge carrying 50 musicians who serenaded George I on a voyage down the River Thames.

1790: The first sewing machine is patented. English inventor Thomas Saint received a patent for his design of a mechanical sewing machine, although he apparently never had one manufactured. The design was a prototype for 19th-century sewing machines, which became widely popular.

1861: Congress authorizes printing of paper money. For the first time, the United States Congress approved the printing of paper banknotes by the Treasury. These "Demand Notes," so named because the bearer could remit them for silver or gold coins "on demand," were issued primarily to help finance the cost of fighting the Civil War. A year later, the Demand Notes were replaced by so-called "Legal Tender Notes," which would later become the Federal Reserve Notes still issued today.

1902: AC? OK! Willis Haviland Carrier, an American engineer, invents modern air conditioning. A publishing company in Brooklyn, New York, was experiencing an air quality problem in its warehouse, where humidity was causing paper and other materials to warp. Carrier was granted U.S. Patent 808,897 for his Apparatus for Treating Air. His design, tailored to fix this problem, was able to control temperature, humidity, ventilation, and filter the air. These factors were later adopted as the standard four requirements for an air conditioning unit.

1913: Pie to the face. In the silent film *A Noise from the Deep*, Mabel Normand hit Fatty Arbuckle in the face with a pie. It was the first time film audiences were treated to the gag, which became a staple in slap-stick comedy routines.

1938: Doug Corrigan goes the wrong way. Corrigan had intended to fly his 1929 Curtiss Robin airplane from Bennett Field in New York to Los Angeles, California. About 28 hours later, Corrigan landed at Baldonnel Field—in Dublin, Ireland.

1958: Discovery at the Olduvai Gorge. The husband-and-wife team of paleontologists George and Mary Leakey discovered one of the oldest known fossils of a hominid—one of the ancestors of the human species—in the dig site in northern Tanzania. The skull, estimated to be 1.75 million years old, is from an early hominid named *Australopithecus boisei*.

1968: *Yellow Submarine* premieres. The psychedelic animated film featured The Beatles in a fantasy landscape in which they must save Pepperland from the evil Blue Meanies.

1998: The International Criminal Court is established in The Hague. The Rome Statute was adopted by multiple nations to establish the Court, which prosecutes individuals for war crimes, genocide, and crimes against humanity.

July 18

64: Rome burns. A fire that broke out in the city quickly spread, and when the flames were finally put out more than a week later, over half of the city lay in ruins. Emperor Nero quickly accused Christians of setting the fire, and launched the first persecution of them by the Roman Empire.

1290: King Edward I issues his Edict of Expulsion, which banished all Jews from England. The Act was issued on Tisha B'Av, a holy day on the Hebrew calendar that commemorates many tragedies in Jewish history. At least 2,000 Jewish people were deported from England as a result of the Edict.

1863: The Battle of Fort Wagner. The 54th Massachusetts Regiment, an African-American detachment, had been recruited in February of that year in Boston to fight for the Union. Governor John Albion Andrew had advocated for the formation of an African-American regiment, and Frederick Douglass assisted in recruitment. Douglass's two sons were among the first soldiers to join the Regiment. At the Second Battle of Fort Wagner, the regiment, led by Colonel Robert Gould Shaw, attacked entrenched Confederate positions at the Fort, which protected the Charleston Harbor. The Regiment had to cross a long expanse of open beach in the attack, and were beset with artillery fire. They reached the Fort, but suffered heavy casualties: half of the 600 men who charged the Fort were killed, including Colonel Shaw. William Carney, a soldier with the 54th, single-handedly recovered the Regiment's flag under heavy fire, and was later awarded the Medal of Honor for his actions. The Confederates buried Colonel Shaw in a mass grave with his men, believing that doing so was an insult to him. His father later wrote to the Regiment's surgeon, saying that he could "imagine no holier place" for Shaw to be buried than with his men.

1921: The first tuberculosis vaccine is administered. Albert Calmette and Camille Guerin administered the vaccine in France.

1925: Hitler publishes *Mein Kampf*. The book, which Hitler wrote while he was imprisoned for his earlier attempts to seize control of Germany, became the bible of the Nazi Party. It detailed Hitler's ideas regarding German history and Aryan racial superiority, and Hitler used it as an organizing tool as he rose to power in the 1930s.

1969: Senator Ted Kennedy crashes at Chappaquiddick. After leaving a cocktail party on the Island, Kennedy accidentally drove his car off of a bridge and into a pond. Although he escaped the wreck, his passenger, Mary Jo Kopechne, drowned. Notably, Kennedy did not report the accident to police until the following morning; the Senator's first telephone call was allegedly to his lawyer. The resulting scandal influenced Kennedy's decision not to run for President in either the 1972 or 1976 elections. He later pled guilty to leaving the scene of an accident.

1976: A perfect 10. Fourteen-year-old Romanian gymnast Nadia Comaneci stunned the world when she executed a perfect score on the uneven parallel bars at the 1976 Olympic Games.

1994: The Rwandan genocide ends. Troops with the Rwandan Patriotic Front took control of the city of Gisenyi. The interim government was forced to flee to Zaire by the RPF troops, who were led by Paul Kagame. Between 500,000 and a million Rwandans were murdered over the three months that the genocide lasted.

July 19

1545: The *Mary Rose* sinks. The massive warship was part of King Henry VIII's Navy. The *Mary Rose* was leading an attack on the French fleet near the Isle of Wright when she sank with all hands aboard; nearly 500 sailors and soldiers were lost. The ship was raised from the depths in 1982 in one of the most complicated and expensive recoveries in the history of archaeology.

1848: The Seneca Falls Convention opens. The convention, held in Seneca Falls, New York, was the first to address women's rights in the United States. It was convened by Elizabeth Cady Stanton and Lucretia Mott, who had been at the World Anti-Slavery Convention in London, England, in 1840. At that event, the two had been barred from entering the convention floor, and this sparked their interest in organizing the Seneca Falls Convention. The convention lasted two days and featured six sessions that discussed the law and the role of women in society. It culminated in a preparation of a Declaration of Sentiments, which combined the social, civil, political, and religious rights of women. The Seneca Falls Convention is widely viewed as the beginning of the first wave of the movement for women's rights.

1900: The Paris Metro opens. The first subway line was opened during the Paris World's Fair. It was expanded over the next two decades and eventually spread to serve the entire Paris urban area. Its iconic Art Nouveau signs are still in use today.

1903: The first Tour de France is completed.
Maurice Garin won the first iteration of the now world-famous bicycle race. He completed the race nearly three hours before the second-place finisher.

1935: The Park-O-Meter is unveiled. The first parking meter was invented by Carlton Magee and installed in Oklahoma City by the Dual Parking Meter company. Parking spaces were laid out with paint every twenty feet on the pavement, and the Park-O-Meters were installed beside each one. The Park-O-Meters accepted nickels.

1943: The Allies bomb Rome. The bombing raid included more than 500 Allied aircraft and inflicted thousands of casualties on the civilian population.

1946: Marilyn Monroe takes her first screen test. The actress was signed to her first contract with Twentieth Century Fox Studios. Her first motion picture, titled *Scudda-Hoo! Scudda-Hey!*, debuted later that year.

1989: United Airlines flight 232 crashes. The plane crashed in Sioux City, Iowa, after a catastrophic failure of its rear engine. Of the nearly 300 passengers and crew on board the flight, 111 were killed in the accident and 185 survived—largely thanks to the quick actions of the crew.

1997: Ceasefire in Northern Ireland. The Provisional Irish Republican Army ends its 25-year guerilla campaign to end British rule in Northern Ireland. The PIRAs agreement to a ceasefire marked the end of the period of violence and unrest in Northern Ireland called the Troubles.

July 20

1773: The Jesuits are dissolved. Pope Clement XIV wrote the papal decree *Dominus ac redemptory noster*, in which he officially disbanded the Society of Jesuits. The suppression of the Jesuits was carried out across Europe, with the exception of Prussia and Russia, where Catherine the Great had overruled it. As the Society was largely responsible for foreign missions and Catholic education, the decree had a hugely negative impact on the Church. The Society was reinstated in 1804, and continues its work to this day.

1807: The Pyréolophore is patented. The device was the first internal combustion engine, and its inventor, Nicéphore Niépce Niepce, was granted the patent by Napoleon. The engine was first used to power a boat traveling upstream in the river Saone in France.

1903: The Ford Motor company ships its first automobile. The company would go on to manufacture and ship several hundred units of various Models—A, B, C, F, K, N, R, and S—before releasing the Model T, which sold millions between 1908–1930.

1904: The Trans-Siberian Railway is completed. The Railway, which spanned more than 4,600 miles, took 13 years to complete. It was later extended to connect Moscow with Vladivostok in 1916, and is still being expanded to this day.

1938: The U.S. Government sues Hollywood. The U.S. Department of Justice filed a lawsuit in New York City against members of the motion picture industry, charging them with violating antitrust laws. In 1948, the lawsuit resulted in the breakup of major production houses.

1968: The first Special Olympics are held. The Inaugural Games were held at Soldier Field in Chicago, and included more than 1,000 athletes from the U.S. and Canada with intellectual disabilities. At the one-day event, plans to hold the Games every two years were announced, and by 1988 the Games were officially recognized by the International Olympic Committee. The Special Olympics organization also provides athletic training to more than five million intellectually disabled athletes around the world and holds national and local competitions.

1973: Bruce Lee dies. The famous Chinese actor and martial-arts master died from a brain edema that was likely caused by a reaction to a prescription painkiller. He was only 32 years old.

1976: *Viking 1* lands on Mars. It was the first spacecraft to successfully land on the Red Planet, and held the record for the longest surface mission on Mars until the Mars rover *Opportunity* surpassed it in 2010.

1985: The Spanish galleon *Nuestra Senora de Atocha* is found. The galleon, which sank off the coast of Key West, Florida, in 1622, contained $400 million in coins and silver ingots. At the time, it was the most valuable underwater find in history.

1997: The USS *Constitution* weighs anchor. The 200-year-old ship, also called "Old Ironsides," had not taken to sea in over a century. Following a painstaking restoration, it celebrated its 200th birthday by setting sail from Marblehead, Massachusetts. The ship was later put on display in Boston's Charlestown Navy Yard, where it is open to the public year-round.

July 21

1733: John Winthrop is granted a Doctor of Law degree. Winthrop was granted the degree by Harvard College. It was the first issued in the United States by any institution.

1861: The First Battle of Bull Run. Many in the Union believed that the insurrection by slave-holding states would quickly be put down, and that the Confederate Army would be easily beaten in its first encounter with Union troops. The public in Northern states clamored for a quick march on Richmond, Virginia, the Confederate capital; they thought that capturing Richmond would bring an end to the uprising. To their surprise, the Union Army was routed at the First Battle of Bull Run (called the Battle of Manassas by Confederates), losing more than 3,000 men in a decisive defeat. The scale of the Confederate victory, and the disorganized retreat by Union troops back to Washington, D.C., demonstrated that the Civil War was going to last much longer and be far more costly than either side had anticipated.

1865: See you at high noon? In what may have been the first true gunslinger showdown in the American West, Wild Bill Hickock faced Dave Tutt in a duel in Springfield, Mississippi. They met in the town square. Tutt drew his revolver and fired the first shot, which went wide. Hickock then drew his six-shooter and shot Tutt square in the chest, killing him.

1873: Stick 'em up! The James-Younger gang, led by outlaw Jesse James, pulled off the first successful train robbery in the American West when they robbed the Rock Island Railroad outside of Adair, Ohio. The gang took $2,337 from the train after having derailed it. The train's engineer died in the crash.

1925: John T. Scopes is found guilty of teaching evolution. The high school teacher was fined $100 for doing so, but the verdict was later overturned on a technicality.

1959: *Lady Chatterley's Lover* has its day in court. A U.S. District Court Judge in New York City ruled that the steamy novel published by D.H. Lawrence in 1928 was not pornographic, in spite of its explicit descriptions and use of then-banned words. The book caused a sensation when it was released in the United States.

1960: Sir Francis Chichester crosses the Atlantic. The English adventurer and sailor crossed the Atlantic Ocean alone in just 40 days. He landed the *Gypsy Moth III* in New York, and set a new world record for a solo crossing of "the pond."

1983: Ice cold! The lowest temperature ever recorded in an inhabited location was documented at Vostok Station, Antarctica. The temperature was a bone-chilling -128.6 degrees Fahrenheit.

2011: The Space Shuttle Program ends. The Space Shuttle *Atlantis* landed at NASA's Kennedy Space Center following its last mission. The Shuttle Program was the first to use a launch vehicle that could be landed on Earth and re-used in additional missions. It had been the primary transportation method for bringing astronauts and cargo to low-Earth orbit beginning in 1981. During its existence, the Shuttles flew 133 successful missions.

July 22

1298: The English win the Battle of Falkirk. Under the command of King Edward I, the English army with its longbowmen and heavy cavalry defeat the forces fighting for Scottish independence with William Wallace. After the defeat, Wallace resigned as Guardian of Scotland in favor of Robert the Bruce, who would go on to become King of an independent Scotland.

1604: The King James Bible project is begun. King James I of England had convened a Conference at Hampton Court, where attendees proposed a new translation of the Bible be compiled. James sent a letter to Archbishop Bancroft asking him to request donations from all pious Englishmen to finance the translation of the Bible. James and Bancroft developed a set of fourteen instructions to the Bible's translators. The instructions included directives to keep biblical names as close as possible to the original ones and to ensure the Bible supported the theology of the Church of England. The new translation, called the King James Version of the Bible, was printed by the King's Printer Robert Barker. It was the third Bible ever to be translated into the English language. It quickly became widespread and is still one of the major works in Christianity.

1793: Mackenzie reaches the Pacific. Alexander Mackenzie, a young Scottish fur trader, set out from Montreal to cross the Canadian Rocky Mountains and reach the West Coast of Canada. The Mackenzie River in Canada is named for him.

1893: Katharine Lee Bates writes "America the Beautiful." The poet, a professor at Wellesley College, wrote the poem in Colorado Springs, Colorado, after being amazed by the beauty of Pike's Peak. The poem would later be put to music and become one of the United States' iconic patriotic songs.

1894: Gentlemen, start your engines. The first automobile road race in history was held between the French cities of Paris and Rouen. The fastest racer was Comte Jules-Albert de Dion, who broke the record for average speed at 13.6 miles per hour. The official victory was awarded to Albert Lemaitre for completing the 78-mile journey in a gas-powered automobile: Dion's motorcar was steam-powered, which disqualified him from winning the race.

1944: The Bretton Woods Conference is held. At the conference, held in New Hampshire, the International Monetary Fund was created. The IMF was designed to maintain a secure system for currency exchange so that payments made in foreign countries could be streamlined.

1995: The Hale-Bopp comet is discovered. The comet was named for Alan Hale and Thomas Bopp, its discoverers. It was visible to the naked eye for 18 months. It will not return to the inner Solar System for several thousand years.

July 23

1829: William A. Burt patents the typographer. The original patent for the machine, which was a precursor of the modern typewriter, and the working model Burt provided in order to secure the patent were destroyed in the 1836 Patent Office Fire. Burt's patent granted him exclusive rights over the typographer for 14 years.

1848: Henry David Thoreau refuses to pay a poll tax. Thoreau, a philosopher, pacifist, and opponent of slavery, refused to pay the poll tax as a protest of the Mexican-American War and slavery. He was arrested and jailed, and when he was told he was free to go the next day, he angrily demanded to remain in prison as a form of civil disobedience. Mohandas Gandhi later cited his account of the experience as an inspiration for his adoption of nonviolent tactics.

1914: Austria-Hungary issues an ultimatum to Serbia. Following the assassination of Austrian Archduke Franz Ferdinand by Serbian nationalists in Sarajevo, Austria demanded that its own representatives be allowed to conduct the investigation into the murder. When Serbia accepted all but one of the demands, Austria declared war, setting off World War I.

1961: The Sandinistas are founded in Nicaragua. The organization organized a long-term underground resistance movement to the Nicaraguan dictatorship that would eventually have international repercussions. Eighteen years after its founding, the paramilitary guerilla army overthrew Anastasio Somoza DeBayle, ending a dynasty that had ruled Nicaragua for decades.

1984: Miss America resigns. 21-year-old Vanessa Williams, the first African-American woman to be crowned the winner of the Miss America pageant, posed for nude photos in *Penthouse* magazine. The pageant's committee ruled that this was in violation of its rules and forced her to give up her crown.

1986: Andrew weds Fergie. Prince Andrew of England, fourth in line to the British throne and the Duke of York, married Sarah Ferguson, a commoner, at Westminster Abbey. The wedding was televised and viewed by about 300 million people around the world.

1996: Kerri Strug sticks the landing. The 18-year old American gymnast performed a dramatic vault with a badly sprained ankle, clinching the first Olympic gold medal for gymnastics ever won by an American.

2015: NASA announces the discovery of Kepler-452b. The planet, also known as Coruscant after the fictional planet in the Star Wars series, orbits a Sun-like star (Kepler-452) about 1,400 light-years from Earth. It was the first rocky planet to be found orbiting a star in the habitable zone—the range of orbits near a star in which a planet could have liquid water, one of the requirements for life to evolve.

July 24

1567: Mary Queen of Scots is forced to abdicate. While she was imprisoned at Lochleven Castle in Scotland, Mary was forced to give up her claim to the throne of England in favor of her son James VI, who was only one year old at the time. Twenty years later, Mary was beheaded for treason.

1701: Antoine de la Mothe Cadillac establishes a trading post. Cadillac was the administrator of French North America at the time. He established the trading post at Fort Pontchartrain du Detroit. The name was later shortened to Detroit, where centuries later Cadillac's name became synonymous with the luxury automobiles manufactured there.

1847: Brigham Young reaches the Valley of the Great Salt Lake. The leader of the Church of Latter-Day Saints, known as the Mormons, had led his followers to Utah to establish a new city. Upon gazing over the expanse of parched earth, Young simply declared, "this is the place." The settlement at the foot of the Wasatch Mountains would become Salt Lake City.

1866: Tennessee is readmitted to the Union. It was the first state to be readmitted following the Civil War. Andrew Johnson, who was Lincoln's Vice President and became President following Lincoln's assassination, was from Tennessee. Because Tennessee ratified the Fourteenth Amendment, it was the only Confederate state to be readmitted without a military governor during Reconstruction.

1911: Machu Picchu is rediscovered. The ancient "Lost City of the Incas" was rediscovered by Hiram Bingham III, an American professor and explorer. A guide named Melchor Arteaga, who lived in the nearby valley, led Bingham's expedition to the city, which is situated on a mountain ridge nearly 8,000 feet above sea level. Bingham incorrectly identified Machu Picchu as the Incan empire's capital, which was actually another city, named Vilcabamba. He returned three times, supported by the National Geographic Society. He came to believe that the city was a major religious site, but contemporary archaeologists have determined that it was most likely a summer estate for Incan royalty.

1915: The *Eastland* capsizes. The passenger ship was tied to a dock in the Chicago River when it sank suddenly. A total of 844 passengers and crew were killed in the tragedy, which remains the single largest maritime loss of life on the Great Lakes.

1943: Firestorm. The Allies firebombed Hamburg in what was the largest bombing raid conducted to that point during World War II. Some 42,000 German civilians were killed in the bombing and the subsequent fires. The Allies launched two bombing raids on the same day, dropping more than 2,000 tons of bombs on the city. The conflagration was described as a "fire storm," marking the first time that phrase was used. The following week, the Allies launched more bombing raids on the city.

1982: No more whaling. The International Whaling Commission voted to totally ban commercial whaling worldwide. Whaling still continues, largely under the guise of "research" by several nations.

July 25

1261: Alexios Strategopoulos recaptures Constantinople. Strategopoulos had a relatively small force, but was able to lead a detachment into the city at night by way of a secret passageway. Once inside, they overpowered the guards and opened the gates, allowing the rest of the army to enter. The capture of the city, which had been held by Latin Crusaders since 1204, marked the reestablishment of the Byzantine Empire.

1593: Henry VI of France converts to Catholicism. Henry had been a Protestant, and the powerful Catholic League had opposed his becoming King. His public conversion to the Roman Catholic faith allowed him to ascend the throne. He later declared the Edict of Nantes, which guaranteed religious liberties to Protestants in France. In 1610, a fanatical Catholic named Francois Ravaillac assassinated Henry.

1722: Dummer's War breaks out. The War was fought between Native American tribes allied with the French, known as the Wabakani Confederacy, and British settlers. The war was fought along the border between Acadia and New England and in Nova Scotia. It ended in 1725 with the signing of a peace treaty that gave control of western Maine to the British.

1788: Mozart completes his Great G Minor Symphony. It is one of only two symphonies written in a minor key by Mozart. Some historians believe Mozart never heard the symphony performed, although a letter discovered in 1802 asserted the composer was at a performance in Leipzig. The performance was so poor that Mozart walked out in frustration. Centuries later, physicist Albert Einstein called the symphony Mozart's "appeal to eternity."

1866: Ulysses S. Grant is promoted to four-star general. The U.S. Congress passed special legislation to establish the rank of General of the Army in order to promote Grant to the position. Grant would lead the Union to victory in the Civil War.

1943: Mussolini falls from power. The Italian Fascist dictator had ruled Italy with an iron grip since October 1922, when he marched on Rome with his shock troops—called the Blackshirts—ready to attack the government. The Italian King, wishing to avoid bloodshed, appointed Mussolini Prime Minister. In 1925, Mussolini dropped the pretense of democratic rule and declared himself Il Duce ("The Leader"). While in power, Mussolini signed an alliance with Nazi Germany, removed political opponents with the help of his secret police, outlawed labor strikes and openly aspired to establish a totalitarian state. Italy entered World War II on Germany's side in 1940, but its forces suffered multiple major setbacks during the war. On July 9 1943, American troops invaded Sicily and began preparing to assault the Italian mainland. The Fascist council voted on July 25 to remove Mussolini from power and imprison him. German commandos rescued him, and Hitler placed him at the head of a puppet government in the north of Italy. Less than two years later, Italian communist partisans captured the fascist dictator and summarily executed him.

1994: Peace in the Middle East. Yitzhak Rabin, the Prime Minister of Israel, and King Hussein of Jordan signed a peace agreement that ended a 46-year-long state of war between the two countries. As a result, Rabin was assassinated the following year by a nationalist Israeli named Yigal Amir who was vehemently opposed to the peace initiative.

1999: Lance Armstrong wins the Tour de France. In doing so, the cancer survivor was the first American to win the storied bicycle race. Years later, he had his multiple titles stripped when it was discovered that he had used performance-enhancing drugs.

July 26

657: The Battle of Siffin. The battle, fought in Mesopotamia, was between the forces of Mu'awiyah, an Islamic leader who later founded the Umayyad dynasty of caliphs, and the army of Caliph Ali. Mu'awiyah's forces won the battle, and he unified the Muslim empire under his rule, establishing its capital in Damascus.

1775: Neither rain nor sleet…. The United States Post Office was established in legislation passed by the Second Continental Congress. Benjamin Franklin was made the first Postmaster General.

1779: De Fleury is awarded a silver medallion. The medallion was granted to Lieutenant Colonel Francois Louis Treisseidre de Fleury by the U.S. Congress for his actions during the Revolutionary War. He led an attack at Stony Point, New York, that captured the British flag. De Fleury was the first foreign national ever decorated by the United States.

1847: Liberia declares independence. The nation on Africa's west coast was originally established as a colony by the American Colonization Society, which sought to repatriate African-American slaves there. The Republic of Liberia was recognized by the United States after England applied diplomatic pressure. It was the first democratic republic in African history.

1847: All aboard! Engineer Moses Garrish, of Dover, New Hampshire, built the first miniature train for children to ride.

1869: The Disestablishment Bill is passed. The Bill officially dissolved the Church of Ireland, and its opponents coined the longest word in the English language to describe their position: antidisestablishmentarianism.

1878: Black Bart robs his last stagecoach. An outlaw in the Old American West, Black Bart regularly left taunting poems behind for police to discover after he'd committed a robbery. In this robbery, he took a safe with $400 dollars and a diamond ring and watch from the stagecoach's passengers. He left a poem, signed "Black Bart," in the safe. He was captured soon after.

1887: Saluton! The Polish doctor L.K. Zamehof published the first book written in Esperanto. The language was artificially created in the 19th century in the hope that it would become a universal language that people from around the world could communicate in.

1941: FDR seizes all Japanese assets in the United States. The seizure was in retaliation for the Japanese invasion and occupation of French Indochina—modern-day Vietnam—during World War II. The Japanese later attacked Pearl Harbor in retribution, drawing the U.S. into the war.

1953: Fidel Castro attacks the Moncada Barracks. Castro was attempting to overthrow the government of dictator Fulgencio Batista. The attack was unsuccessful, and Castro was captured and imprisoned along with several of his comrades. He was later released by the Batista government, and traveled to Mexico, where he met Che Guevara and began putting together a guerilla force to invade Cuba.

July 27

1540: Thomas Cromwell is executed. Cromwell had been the chief adviser to King Henry VIII of England. He arranged a marriage between Henry and Anne of Cleves in order to secure an alliance with German nobles. But Henry found Anne unattractive and the alliance fell apart. The King blamed Cromwell for both, and had him executed for treason.

1586: Sir Walter Raleigh brings tobacco to England. The explorer was presented with tobacco by Native Americans in Virginia, and he brought it to England, where it was widely popular. He was later granted a charter to establish tobacco plantations in the New World.

1794: Maximillian Robespierre is arrested. Robespierre was the leader of the Reign of Terror during the French Revolution, and under his watch thousands of nobles and members of the French middle class—as well as French Revolutionaries who were denounced—were put to death by the guillotine. Robespierre suffered a fate of his own design: the day after he was arrested, he was guillotined before a cheering crowd.

1816: The deadliest cannon shot in history is fired. During the Battle of Negro Fort, fought between escaped slaves and their Choctaw and Seminole allies against the United States Army, a U.S. gunboat fired a shell at the Fort. It exploded in the Fort's powder magazine, causing a massive explosion that killed more than 250 defenders and ending the battle.

1866: The transatlantic cable is completed. The undersea cable was laid between Ireland and Newfoundland, and reduced the time it took to get a message across the Atlantic from ten days to seventeen hours.

1919: Race riots break out in Chicago. In the racially segregated city, a white man began throwing rocks at African-American boys who were swimming in Lake Michigan. One of the rocks struck Eugene Williams, who drowned. When a white police officer not only refused to arrest the rock-thrower, but also arrested an African-American, several African-American observers objected. A white mob attacked them, and the riot soon spread throughout Chicago's South Side. The riot lasted nearly a week, and only ceased when the National Guard arrived and were stationed around African-American neighborhoods to protect them from white mobs. During the violence, more than 30 fires were set by white mobs in a single day. Thirty-eight people, including an African-American policeman, were killed as a result of the rioting and more than a thousand African Americans were left homeless because of the fires.

1974: Baryshnikov premieres in the United States. The Russian ballet dancer had defected from the Soviet Union to

Canada that June. In his debut, he danced with the American Ballet Theatre in a production of *Giselle* in New York City.

1996: The Olympics are bombed. In Centennial Olympic Park in Atlanta, a pipe bomb exploded in a public park during the Summer Olympics. The bombing was committed by the American Christian terrorist Eric Robert Rudolph to protest abortion. A security guard, Richard Jewell, discovered the bomb and cleared most of the crowd out of the park. Jewell was initially hailed as a hero, but later became a suspect himself. In 2006 he was publicly thanked by the Governor of Georgia for his actions.

July 28

1586: Sir Thomas Harriot brings potatoes to England. The English scientist had traveled to the New World on an expedition financed by Sir Walter Raleigh. He was introduced to potatoes by indigenous people in Colombia, and brought some back with him. The tubers were quite popular, and spread around the globe, where they were introduced into local cuisines everywhere.

1858: The first fingerprints are made. William Herschel, an employee of the Indian Civil Service of Jungipur, took the fingerprints of Rajyadhar Konai to sign a contract. It was the first time in history that fingerprints were used as a means of identification. Herschel later established the first fingerprint register.

1868: The Fourteenth Amendment is certified. The Amendment guaranteed due process of law to all citizens and established citizenship for African Americans. Secretary of State William H. Seward certified the Amendment after it was ratified by Georgia.

1917: The Silent Parade marches through New York City. The parade was organized by the National Association for the Advancement of Colored People (NAACP) to protest the ongoing mob violence, riots, and lynchings perpetrated against African Americans by whites. It was precipitated by race riots in East St.

Louis, during which white mobs killed at least 40 Black people. The NAACP hoped the parade would pressure President Woodrow Wilson to sign an anti-lynching law, but he refused.

1932: Herbert Hoover orders the eviction of the Bonus Army. World War I veterans had gathered in Washington, D.C. to demand the promised cash payments for their service. They were so named because they had been given "Bonus" certificates guaranteeing the payments. Out of work since the beginning of the Great Depression and in dire financial straits, 20,000 veterans marched on the United States Capitol. When a solution was not reached, they camped on the National Mall with their wives and children in protest. Hoover ordered General Douglas MacArthur to forcibly remove the marchers. MacArthur, who declared the protest an attempt to overthrow the government, attacked the Bonus Army campsite with infantry and cavalry. The infantry used fixed bayonets and fired tear gas into the campsite, and the cavalry repeatedly charged the marchers. These troops were additionally supported by six tanks. Two Bonus marchers were killed in the ensuing melee, and more than a thousand were injured. Their shelters and belongings were burned by the soldiers. The incident is considered a deciding factor in Hoover's loss by a landslide to Franklin Delano Roosevelt later that year.

1933: The singing telegram is introduced. The service, in which a telegram is delivered by a singing courier, was first sent to the American singer Rudy Vallee for his 32nd birthday.

1945: The U.S. Senate approves the United Nations charter. The establishment of the United Nations, and the Senate's approval of the treaty, signaled the end of the longstanding U.S. policy of isolationism that had been established by the Monroe Doctrine.

1996: Kennewick Man is discovered. The remains of the prehistoric man were discovered near Kennewick, Washington. At the time it was one of the most complete skeletons ever found.

July 29

1565: Mary, Queen of Scots, weds Henry Stuart, Duke of Albany. Stuart was later murdered, and Mary left Edinburgh with the Earl of Bothwell, who was suspected of involvement in the killing. The widespread suspicion that Mary had colluded in the death of her husband eventually led to her downfall, loss of the crown, and execution.

1751: Touch gloves! The first International World Title Prize Fight was held in Norfolk, England. The competitors, Jack Slack of England and Jean Petit of France, did not wear boxing gloves, as they hadn't been invented yet. In the "bare-knuckle" era of boxing, competitors stood with their feet touching a chalk line—the origin of the term to "toe the line"—and struck each other with bare fists. Slack defeated Petit after 25 minutes.

1776: Silvestre de Escalante and Francisco Dominguez set out from Santa Fe, New Mexico. The Spanish Franciscan friars traveled throughout the American Southwest in a journey during which they attempted to spread the Gospel and convert as many Native people as possible to Christianity. They were the first Europeans to explore the Southwest so extensively, and Escalante's journal of the expedition later became a guide for other explorers of the region.

1845: The New York Yacht Club is founded. It is the oldest yacht club in the United States that is still in operation.

1935: *Seven Pillars of Wisdom* is published. The book was T.E. Lawrence's account of his exploits during the Arab Revolt. Lawrence advised Arab revolutionaries in their war for independence from the Ottoman Empire, and fought with them in multiple battles. The book solidified his reputation as a daring soldier and has since become a classic of war literature.

1976: David Berkowitz commits his first murder. The serial killer became known as the "Son of Sam" because Berkowitz signed letters taunting the police with the moniker. He terrorized New York City for nearly a year, murdering six people and wounding eight more, before being captured in July 1977.

1981: Lady Diana weds Prince Charles. In what was perhaps the most anticipated wedding of the 20th century, Prince Charles, the heir to the British throne, and Lady Diana Spencer, an English schoolteacher with royal blood, were married at St. Paul's Cathedral. More than 2,500 guests attended the ceremony, and some half a million spectators lined the route to the Cathedral to see the couple. Nearly one billion people tuned in to watch the event from 74 countries around the world. In the wedding, Diana became the first Englishwoman to marry an heir apparent to the English throne in more than three centuries.

2015: Windows 10 is released. The operating system was the first version of Windows to receive ongoing updates, including security patches and bug fixes.

July 30

657: Saint Vitalian begins his rule as Pope. At the time, a centuries-old schism in the Christian Church had left the rulers of Rome at odds with those of Constantinople. Vitalian was successful in reestablishing good relations with Constantine VI, leading to a long peace between the rivals.

1619: The House of Burgesses convenes. The body was the first representative assembly established by European settlers in North America. It was convened at Jamestown, Virginia, where it remained the representative branch of the Virginia legislature until 1776. When Virginia joined the other colonies in declaring its independence, it became the House of Delegates.

1811: Father Miguel Hidalgo y Costilla is executed. Less than a year earlier, the Catholic priest had delivered a speech called "The Cry of Dolores" after the town where he gave it. The speech triggered an uprising against Spanish rule; Hidalgo then marched across Mexico, rallying others to the cause of Independence. Although the rebels won a few victories through sheer numerical superiority, an army loyal to the Spanish Crown soon regained the territory Hidalgo's troops had captured. The royalist army attacked Hidalgo's rebels at the Battle of Calderon Bridge, in Mexico City. Although they were outnumbered more than ten to one, the royalists were better trained and defeated the rebels in a decisive victory. Hidalgo fled, but was captured by royalists. He was formally expelled from the priesthood and then excommunicated from the Catholic Church before a military court found him guilty of treason. The rebellion continued, however, and in 1821 Spanish rule of Mexico was overthrown.

1894: The Kellogg brothers invent Corn Flakes. John Harvey and Will Keith Kellogg, who jointly ran a health spa and hospital in Battle Creek, Michigan, came up with the plain, no-frills cereal as part of their program for "clean living." They stressed a diet that eschewed meat, alcohol, tobacco, and caffeine. The cereal was a hit, and eventually became one of the best-recognized breakfast cereals in America.

1930: The first FIFA World Cup is held. Uruguay won the title match over Argentina, 4–2, in front of a crowd of more than 68,000 spectators. The World Cup of football, or soccer, depending on where you are, grew to be the world's premiere international sporting event. It was held every four years since, with the exceptions of 1942 and 1946, due to World War II. Over 200 nations are eligible to compete, with 32 entering the final tournament.

1956: "In God We Trust" is adopted as the U.S. national motto. The phrase was eventually added to U.S. currency. It was popularized as part of the Cold War, and meant to distinguish pious Americans from Soviet communists, who were officially atheist.

1965: LBJ establishes The Great Society. As part of his War on Poverty, President Lyndon B. Johnson signed the Social Security Act of 1965 into law. The Act established Medicare and Medicaid, which provided federal health insurance for the elderly and poor families.

1975: Jimmy Hoffa disappears. Hoffa, the president of the International Brotherhood of Teamsters, had been on his way to meet with Anthony Giacalone, a member of the mafia, at a restaurant in a Detroit suburb. Hoffa was never seen again.

July 31

1703: Daniel Defoe is pilloried—literally. Defoe, an English author, had published a pamphlet titled "The Shortest Way with Dissenters." It was an attack on officials of the High Church. Defoe was found guilty of seditious libel and sentenced to be locked in a pillory. Years later, he published his now-famous book *The Life and Strange Adventures of Robinson Crusoe.*

1777: Marquis de Lafayette is made a General in the Continental Army. The 19-year-old French nobleman and military officer traveled to the United States to fight for the colonists in their Revolutionary War. He was instrumental in securing French support for the American cause. His troops blockaded British troops at the Siege of Yorktown, an action that proved instrumental in securing a victory in that battle, which was the last major engagement of the War. Upon his return to France, Lafayette helped write the Declaration of the Rights of Man and of the Citizen during the French Revolution. That document was largely inspired by the Declaration of Independence drafted by the American patriots.

1790: The first U.S. patent is issued. Although it was not officially Patent Number One, because the Patent Office did not begin numbering the patents it issued until years later, the first patent was issued to Samuel Hopkins. It was for a means of manufacturing potash.

1792: The cornerstone of the U.S. Mint is laid. The Mint, located in Philadelphia, was the first official building of the United States Federal Government to be erected.

1812: Spain reconquers Venezuela. A year earlier, Francisco de Miranda had led an insurrection and Venezuela declared its independence from Spain. Nine years later, Simon Bolivar led Venezuela to permanently attain independence.

1874: Dr. Patrick Francis Healy is made president of Georgetown University. Upon his inauguration, Healy became the first African-American to hold the position of president of a mostly white university.

1928: The MGM lion debuts. The lion appeared on the silver screen for the first time to introduce the first talking motion picture made by the Metro-Goldwyn-Mayer film studio. The picture, *White Shadows in the South Seas*, featured a doctor who is marooned on an island in the South Pacific.

1941: Herman Goering orders a plan for the Holocaust. Acting under orders from Adolf Hitler, the high-ranking Nazi wrote an order to Reinhard Heydrich, a general in the SS and his second-in-command, to draft a general plan for the systematic extermination of all Jewish people living in "all the territories of Europe under German occupation." Before it was over, more than 6 million Jews would be murdered in the Holocaust.

1964: Ranger 7 sends back the first close-up images of the Moon. The space probe was the first successful mission during the Ranger program. It carried six television cameras and transmitted more than 4,300 photographs of the lunar surface during its final 17 minutes of flight, before crash-landing into the Moon. Its landing site was named *Mare Cognitum* ("The Sea that has Become Known").

1991: George Bush and Mikhail Gorbachev sign the long-range Strategic Arms Reduction (START) treaty. It mandated reductions of nuclear arsenals by the two nations.

Chapter 8

August

August 1

1498: Christopher Columbus lands at the Paria Peninsula in Venezuela, setting foot on the continent of South America for the first time. Columbus thought he had landed on an island, and named the peninsula Isla Santa. He claimed the "island" for Spain.

1715: Thomas Doggett, an Irish theater producer, stages the first rowing race to commemorate the ascension of King George I to the British throne. The race has been held every year since, making it possibly the oldest sporting event in the world. The winner's coat and trophies are funded by a bequest Doggett left in his will.

1774: Joseph Priestley discovers oxygen. The English scientist called the gas "dephlogisticated air" because he adhered to the popular theory of phlogiston, an imaginary flammable element. The theory was discredited during the chemical revolution, but Priestley refused to abandon it, and was ultimately shunned by much of the scientific community as a result.

1831: London Bridge opens. King William VI and Queen Adelaide formally opened the Bridge, which replaced the one that had stood since 1209. After nearly 140 years, the "new" Bridge began to sink into the Thames because the weight of modern cars and trucks was more than it had been built to withstand. Rather than demolish the historic span, the city of London put it up for auction. Robert McCulloch, an oil baron and the founder of Lake Havasu City, a retirement community in Arizona, purchased the Bridge. He moved it to Lake Havasu City piece by piece and had it rebuilt there, where it opened in October 1971.

1834: The British Parliament passes the Abolition Act. The Act outlawed slavery throughout the British Empire, freeing three quarters of a million slaves. The Industrial Revolution, growing humanitarian protests, and slave revolts in the British colonies all contributed to the passage of the Act.

1941: The Jeep is born. The U.S. Army had issued requests for proposals for a fast, all-terrain vehicle that had to be finished in under 50 days. The Willys Truck Company successfully answered the call, and over the course of World  War II produced about 640,000 jeeps for the war effort. The lightweight truck would become popular after the war, eventually becoming a luxury automobile.

1956: The polio vaccine is made publicly available. Dr. Jonas Salk and a research team at the University of Pittsburg developed the vaccine. Salk refused to profit from the vaccine and his efforts ensured that it was widely distributed. As a result, polio was virtually eradicated worldwide.

1966: Charles Whitman opens fire from the clock tower at the University of Texas. Whitman killed 22 people and wounded 31 more before he was shot by Austin police officers.

1981: MTV goes on the air. The Music Television station's first music video was "Video Killed the Radio Star," by the Buggles.

August 2

1769: Gaspar de Portola and Juan Crespi explore "Alta California." Portola, a Spanish explorer, and Crespi, an Franciscan priest, described the area as a suitable place to establish a settlement. That settlement would later become Los Angeles, California.

1790: The first U.S. census is completed. The census was mandated by Article I, Section 2 of the U.S. Constitution, and counted just under 4 million Americans in the thirteen states and Southwest Territories. The census included the name of the male head of each household, and categorized Americans at the time as free white males, free white females, all other free persons, and slaves. Secretary of State Thomas Jefferson and President George Washington both thought the resulting figure was too low, and believed that the actual population of the United States had been undercounted. The Congress passed legislation based on the first census to increase the number of seats in the U.S. House of Representatives from 69 to 105. The census has been performed every ten years since.

1830: Charles X abdicates. Charles was the last king of France from the House of Bourbon. A revolution broke out during his reign because he dissolved an elected representative congress and severely restricted the right to vote.

1861: Congress passes the first income tax law. The law was passed to raise money for the Union Army during the Civil War. While it was not enacted, the law established a precedent for income taxation and contributed to the expansion of the federal government.

1865: Lewis Carroll publishes *Alice's Adventures in Wonderland*. The first printing had problems with the illustrations, so Carroll had the entire print run withdrawn. Only 22 copies of the first edition remain in existence.

1939: Albert Einstein writes to FDR. The physicist and discoverer of the Theory of Relativity urged the President to take "quick action" to develop an atomic weapon. He feared Nazi Germany had already begun working on an atomic bomb, and that the U.S. had to catch up in the nascent nuclear arms race.

1945: The Potsdam Conference concludes. The Conference, between the leaders of the Soviet Union, Great Britain, and the United States, was held in part to determine the fate of Germany following World War II. Churchill and Truman could not reach an agreement with Stalin, which set the stage for the later Cold War.

1974: John Dean, White House Counsel to Richard Nixon, is sentenced to four years in prison. Dean was ultimately placed in a safe house and spent the next several months testifying against other Watergate conspirators. He served only four months of the sentence.

1990: Iraqi forces invade Kuwait. Within two days, the Iraqi Republican Guard had taken control of the small, oil-rich kingdom, which it would occupy for the next seven months. After negotiations reached an impasse, an international coalition of troops drove the Iraqi Army out. As they retreated, the Iraqi Army set hundreds of Kuwaiti oil wells on fire.

August 3

1460: King James II of Scotland is killed. The king was leading an invasion of England, and his army had besieged Roxburgh Castle. During the siege a Scottish cannon exploded, killing him outright.

1769: Father Juan Crespi sees the La Brea Tar Pits while exploring an area near modern-day Los Angeles, and notes them in his journal. The Tar Pits would later become a valuable archaeological site that preserved the remains of dozens of prehistoric animals.

1900: Harvey Firestone founds the Firestone Tire & Rubber Company. Based in Akron, Ohio, Firestone's company supplied his friend Henry Ford's factories with equipment for his automobiles. In 1911, the winner of the Indianapolis 500 used Firestone tires, and so did all of the winners from 1950–1974. Firestone eventually grew to be one of the giants of tire manufacturing in the 20th century along with Goodyear.

1914: Germany declares war on France and invades neutral Belgium. These actions drew the French into World War I. Great Britain sent an ultimatum to Germany to withdraw, which was ignored.

1933: The first Mickey Mouse watch is sold for $2.75 in Waterbury, Connecticut. The struggling Waterbury Clock Company patented a Mickey Mouse watch after Herman Kamen, an ad man from Kansas City, sold the idea to Walt Disney. Disney thought the watch would never sell. By 1957, the company had sold 25 million.

1948: Cold War charges: Whittaker Chambers, an admitted former member of the Communist Party, accuses Alger Hiss, a former official at the U.S. State Department, of being a communist and a spy for the USSR in testimony to the House Un-American Activities Committee. Hiss denied the charges and sued Chambers for defamation. During the trial Chambers produced more evidence that Hiss was a spy, and while he was not convicted of espionage, a jury later found Hiss guilty of perjury. He served 3 years in prison, and maintained his innocence for the rest of his life. Whether Hiss was in fact a Soviet agent has never been definitively established.

1958: The USS _Nautilus_, the world's first nuclear submarine, completes the first undersea voyage to the North Pole. It had submerged at Point barrow, Alaska, and travelled under the polar ice cap for almost 1,000 miles to reach the Pole.

1977: Radio Shack releases the TRS-80 desktop computer.
The machine included a full keyboard, 1.77-megahertz processor, and 64-character per line black & white video monitor. It is considered the first completely assembled computer to be sold at retail. It had a price tag of $599.95 and four kilobytes of random-access memory (RAM).

1981: Thirteen thousand air traffic controllers go on strike to protest obsolete equipment and overly long working shifts. The strike, organized by the U.S. Professional Air Traffic Controllers Organization, grounded flights nationwide. In response, President Ronald Regan demanded the controllers return to their jobs. Only about 1,300 did, and Regan fired 11,345 who refused to do so. The union was decertified later that year.

August 4

1693: The Benedictine monk Dom Perignon invents champagne. Perignon was the cellar master of the Benedictine Abbey at Hautvillers, France. He invented many new winemaking techniques beginning in the 1670s, such as using corks to seal the bottles, strengthening bottles, blending different grapes to create new flavors, and developing sparkling wine. The practice continued to be perfected for a century after his death in 1715. The monk and cellar master is buried at the Abbey.

1735: John Peter Zenger, the publisher of the *New York Weekly Journal,* is acquitted of libel by a colonial jury. The British royal governor, William Crosby, had had Zenger arrested after he refused to follow the governor's orders to stop publishing articles that were critical of the Crown. The case was the first court decision in favor of a free press in the British colonies in America.

1821: The first *Saturday Evening Post* is published. The weekly newspaper became the most widely circulated magazine in the United States under the guidance of George Horace

Lorimer. In the 1920s, the *Post* began a relationship with illustrator Norman Rockwell after Lorimer discovered him in New York. Rockwell's iconic cover illustrations defined the look of the magazine through much of the 20th century.

1899: The Great Fire of Spokane, Washington, burns 32 city blocks, mainly in the city's commercial district. When the fire began, there was no water pressure in the city because a pumping station had failed. By the time it was finished the fire had destroyed the downtown area and with it Spokane's hopes of becoming the state capital.

1914: Britain enters World War I by declaring war on Germany. President Woodrow Wilson officially proclaimed America's neutrality in the conflict. Woodrow's position would become increasingly difficult to maintain as U.S. ships en route to England were damaged and sunk by German mines.

1922: All the telephones in North America are silent for one minute. Switchboards and switching stations across the United States and Canada shut down to commemorate Alexander Graham Bell, who had died two days earlier. Bell patented the first telephone and is widely credited as being its inventor.

1944: Anne Frank and her family are arrested in Amsterdam. The family had been in hiding for nearly two years from the Nazis. They were taken to concentration camps, where all but Anne's father died.

1962: Marilyn Monroe dies. The actress, model, and singer was one of the most recognizable celebrities of the 1950s. As a "blonde bombshell," she embodied the decade's attitudes about sexuality and was one of the 20th centuries iconic figures in popular culture. Monroe—whose original name was Norma Jeane Mortenson—began her career as a pin-up model and began acting in motion pictures with 20th Century Fox in 1951. At age 36 Monroe overdosed on barbiturates at her Los Angeles home.

1964: Three bodies are discovered in an earthen dam in Mississippi. The FBI later identified the bodies as Michael Schwerner, James Chaney, and Andrew Goodman, three civil rights workers who had gone missing two months earlier while they were registering African American voters for the Congress of Racial Equality. Eighteen men were eventually indicted but only seven were convicted and they received light sentences. The murder of the activists helped contribute to the passage of the Voting Rights Act of 1965.

August 5

1305: William Wallace is captured by English troops near Glasgow. Wallace had led the Scottish resistance against English rule, winning several major battles and invading England before being defeated at the Battle of Falkirk in 1298. He was taken to London, tried for treason, and executed.

1620: The *Mayflower* departs from Southampton, England, heading for America for the first time. Its sister-ship the *Speedwell* soon sprang a leak, forcing both ships to return to England. They would attempt the voyage again in September.

1864: Damn the torpedoes, full speed ahead! At the Battle of Mobile Bay in Alabama during the Civil War, Union Admiral David Farragut ordered his own ship to lead a naval assault through a Confederate minefield and past entrenched cannons on shore. The daring move succeeded, and led to the capture of crucial forts that defended the Bay and completed the Union naval blockade of the Confederacy east of the Mississippi River.

1884: Workers lay the cornerstone of the foundation for the Statue of Liberty on Bedloe's Island (now called Liberty Island), New York. The statue itself would arrive two years later, when President Grover Cleveland would accept it as a gift from the nation of France.

1888: Bertha Benz makes the first long-distance automobile trip. Benz, the business partner and wife of Karl Benz, drove from Mannheim to Pforzheim and back, a distance of about 130 miles (105 km), with just her young sons to accompany her. Before the trip automobiles had only made very short journeys. During the trip, she stopped at a cobbler's to install leather pads on her failing wooden brakes, thus inventing the world's first brake pads.

1921: "On the Road to Moscow," a political cartoon drawn by Rollin Kirby, is published in the *New York World*. It depicted the Grim Reaper leading a line of starving people to their deaths. It would be the first cartoon to be awarded a Pulitzer Prize.

1957: *American Bandstand* debuts on the ABC television network. The program played popular music by rock 'n' roll acts and featured dancing baby boomers.

1962: Nelson Mandela is imprisoned in South Africa. He would be found guilty of sabotage and conspiracy to overthrow the government for his activities with Mkonto We Sizwe, the armed wing of the African National Congress. Mandela was not released until 1990. He later became the first Black president of South Africa.

1989: The Sandinista National Liberation Front wins sweeping victories in elections in Nicaragua. This brought about an end to a decade-long civil war with American-backed Contras.

August 6

1181: Chinese and Japanese astronomers observe a supernova lighting up the sky in the constellation Cassiopeia. In the 20th century, astronomers observing that area of the sky discovered that it was an exploding neutron star.

1762: The sandwich is invented. The story goes that John Montagu, the Fourth Earl of Sandwich, England, was playing cards when he ordered a valet to bring him meat assembled between two pieces of bread. Sandwich hit upon the design so he could continue to play while eating without using a fork or getting the cards dirty. The other card players asked the valet to bring them "the same as Sandwich," and the name stuck. The story was first reported in *A Tour to London*, published by Pierre-Jean Grosley in 1772. Sandwich's official biographer claimed that the Earl had actually invented the finger food while he was working at his desk.

1787: At the Constitutional Convention in Philadelphia, Pennsylvania, delegates begin to debate the first draft of the Constitution of the United States. Debate continued until September 10.

1824: Simon Bolivar defeats a Spanish army at the Battle of Junin during the Peruvian War of Independence. Bolivar's victory cut off the Spanish retreat towards Cuzco. The Spanish never recovered, and four months later Bolivar's army would effectively win the war at Ayacucho.

1889: The Savoy Hotel opens in London. It was the first luxury hotel opened in Great Britain. Richard D'Oyly Carte opened the hotel with profits he'd amassed from producing operas by Gilbert and Sullivan.

1890: William Kemmler is executed by the electric chair at Auburn Prison in New York. Kemmler had been convicted of murdering his lover with an axe, and he was the first convicted criminal to be put to death using electrocution. The method was initially proposed by Dr. Albert Southwick in 1881 as a humane alternative to hanging. Southwick developed the chair over the ensuing decade.

1912: The Bull Moose Party meets at the Chicago Coliseum. Former President Teddy Roosevelt formed the progressive third party after he lost the presidential nomination to the Republican Party to William Howard Taft. Roosevelt was defeated in the 1912 presidential election, and the party dissolved soon after.

1932: Richard Hollingshead Jr. registers a patent for a drive-in movie theater. He never developed the concept, and in 1950 the patent was invalidated, leading to a boom in drive-in theater construction in the United States.

1945: The United States Air Force drops an atomic bomb on Hiroshima. It is the first use of nuclear weapons in human history. The bomb killed about 75,000 civilians in the explosion and ensuing firestorm. US President Truman called for Japan to surrender later that day.

1965: President Lyndon B. Johnson signs the 1965 Voting Rights Act into law. The legislation, passed during the height of the Civil Rights Movement, prohibited racial discrimination during voting.

1991: Tim Berners-Lee first makes the World Wide Web publicly available. Berners-Lee released a set of files describing the Web on an online newsgroup called alt.hypertext.

2012: NASA's *Curiosity* rover lands on Mars. Originally designed to operate for two years on the Red Planet, the rover continued operating for years, logging thousands of Martian sols and investigating the Martian atmosphere, mineralogy, and chemistry.

August 7

1782: George Washington orders the creation of a Badge of Military Merit for American soldiers who have been wounded in battle. It was awarded to only three soldiers who fought in the Revolutionary War. The Badge was later renamed the Purple Heart after Elizabeth Will was chosen to redesign the medal in 1931 by Douglas MacArthur. Close to 2 million have been awarded since then.

1789: The U.S. Congress establishes the War Department. It was later renamed the Department of Defense.

1794: During the Whiskey Rebellion in western Pennsylvania, President George Washington invokes the Militia Acts of 1792. The Acts allowed the President to take command of state militias to put down the anti-tax insurrection.

1944: The Harvard Mark I is dedicated at Harvard University. Weighing five tons, it was the first calculator that was controlled by a computer program. It could store up to 72 numbers, do three additions or subtractions per second, and perform trigonometry and logarithmic operations.

1947: *Kon Tiki* reaches the Tuamotu Archipelago. The large balsa-wood raft was designed by Thor Heyerdahl who led a six-man expedition that had left Peru 101 days earlier. Heyerdahl was attempting to prove his theory that ancient inhabitants of Peru could have colonized the Polynesian islands by drifting on ocean currents.

1955: The first transistor radios are sold. Texas Instruments and Raytheon each developed transistor radios as research and development projects following the invention of transistors—semiconductors that amplify electronic signals—in 1947. It took nearly a decade before Tokyo Telecommunications Engineering, which later became the multinational corporation Sony, designed one for sale to the general public, however. In doing so they became the first company to manufacture transistors and other components for the radio, which they dubbed the TR-55. The compact, portable radios eventually became the most popular electronic device ever sold, with billions manufactured over the next two decades. The popularity of transistor radios declined in the 1980s with the invention of boomboxes and later, portable CD players, mp3 players, and smart phones.

1959: The Lincoln Memorial penny design goes into circulation. The penny, which featured an image of Abraham Lincoln on the front side and one of the Memorial on the back, replaced the Wheat Penny. The last one was minted in 2008.

1964: Congress passes the Gulf of Tonkin Resolution. The legislation was passed in response to a reported attack on a U.S. destroyer by North Vietnamese torpedo boats, although the truth of those reports was later disputed and remains unclear. The Resolution gave President Lyndon Johnson broad powers to send military forces into Vietnam without a declaration of war, and led directly to a major escalation of American involvement in the Vietnam War.

1987: Lynne Cox swims across the Bering Strait. In so doing, she became the first person to swim from the United States to the Soviet Union, crossing from Alaska to Big Diomede Island.

August 8

117: Hadrian becomes emperor of Rome following the death of his father Trajan. Hadrian ruled until his death at age 62 in the year 138. He oversaw the completion of the Pantheon and the construction of the Temple of Venus and Roma in Rome, but is perhaps best known for ordering the construction of Hadrian's Wall along the northern border of the Roman province of Britannia. Portions of the wall are still standing.

1709: Father Bartolomeu de Gusmao, a Brazilian priest and naturalist, makes the first ascent in a hot-air balloon in Lisbon, Portugal. For his invention, he was awarded a professorship at a university in Coimbra, Portugal, by the king.

1860: Queen Emma of Hawaii arrives in New York City, becoming the first queen to visit the United States.

1876: Thomas Edison patents the mimeograph. The device automatically copied documents, and was licensed by Albert Blake Dick, who founded a copier manufacturing and office supply company that dominated the industry for much of the 20th century. The mimeograph was widespread in American offices until the 1980s, when electronic copy machines took over.

1907: The Rolls-Royce Silver Ghost is named the "Best Car in the World." The British-made automobile completed a 15,000-mile trial easily, allowing it to be called the Best. Only 6,173 Silver Ghosts were ever manufactured by the Rolls Royce company.

1963: Bandits make off with more than 2.5 million pounds after robbing a mail train in Buckinghamshire, England. The heist would become  known as "The Great Train Robbery," and became an international sensation. An initial press conference announced the robbers had stolen only about a quarter of a million dollars, but the real sum was more than ten times that. Most of the money was never recovered by authorities, although ten men were imprisoned in 1964. One of them, Ronnie Biggs, led an escape and remained in hiding for nearly forty years before being captured. The train conductor, Jack Mills, died in 1970 from injuries sustained during the robbery.

1974: Richard Nixon resigns. The President was facing impeachment by the US Congress for his involvement in covering up the Watergate burglary of the DNC by White House operatives during Nixon's 1972 reelection campaign. An FBI investigation of the break-in, combined with reporting by Bob Woodward and Carl Bernstein of *The Washington Post*, had uncovered a conspiracy to cover up the White House connection to the burglars. Mark Felt, the associate director of the FBI, famously provided Woodward and Bernstein with leads as the anonymous source they dubbed "Deep Throat." Nixon fought to remain in office, but after it was revealed that Oval Office conversations between him and his deputies had been recorded, he was forced to release the tapes. One of them clearly implicated him in the cover-up. Congressional Republicans informed him he would almost certainly be impeached and removed from office, and he resigned instead, the first and only US President to do so.

August 9

378: The Battle of Adrianople. A Roman Army consisting
of more than 20,000 soldiers attacked a force of Gothic rebels
about 15,000 strong during the Gothic War. The Goths defeated
the larger Roman force during the battle, and the Roman
Emperor Valens was killed. The defeat is considered the beginning of the fall of the Roman Empire during the fifth century. It
also established the supremacy of cavalry over infantry.

1173: Construction of the Leaning Tower of Pisa begins.
Construction of this campanile, or bell tower, took nearly two
centuries to complete. The tower's foundation was too soft to
support its weight, and it began to tilt while it was still being
built. Due to its shift, the construction was put on hold when
it was only two stories tall, and it was not resumed for 99 years
to allow the soil underneath it to settle. In 1272, Giovanni di
Simone took over construction, and ordered that subsequent
floors be built with one side taller than the other to correct
for the angle. This resulted in the tower becoming curved.
Construction was halted again in 1284 when Genoa defeated
Pisa in the Battle of Meloria. The penultimate seventh floor was
finished in 1319, but the bell-tower was not completed until
1372. There have been many efforts to fix the tower's tilt, but
none have been successful. In 2008, engineers were successful in
stopping the tower from leaning further.

**1842: The border between the United States and Canada
is formalized.** The Webster-Ashburton Treaty, signed by the
US and Great Britain, established the boundary from the Great
Lakes to the East Coast.

1854: Henry David Thoreau publishes _Walden_. The book
described Thoreau's time spent living in a small cabin on the edge
of Walden Pond for two years. A transcendentalist, Thoreau was
exploring self-reliance and spiritual discovery as he lived in near-
solitude in the woods.

1892: Thomas Edison patents the two-way telegraph. The device allowed operators to send messages over a single wire simultaneously, greatly cutting down on the time required to do so. Telegraphs remained the primary means of long-distance communication until the telephone became widespread in the early 20th century.

1902: King Edward VII is crowned King of Great Britain and Ireland at Westminster Abbey in London, England. Edward was originally supposed to be coronated on June 26, but his appendix had burst, requiring the ceremony to be postponed.

1910: A. J. Fisher, an inventor from Chicago, patents the first electric clothes washing machine.

1930: Betty Boop appears for the first time. The iconic cartoon "star" debuted in Max Fleischer's animated film *Dizzy Dishes*. A caricature of a Jazz Age "flapper," she became one of the most well-known cartoon characters of the 20th century.

1969: Members of the Manson Family murder actress Sharon Tate, who was pregnant at the time, and four others in a house in Los Angeles. The gruesome killings shocked the nation and resulted in a murder conviction for Charles Manson, who while not at the scene had ordered the murders in an attempt to start an apocalyptic race war.

August 10

1776: News of the American Declaration of Independence reaches London more than a month after it is signed by revolutionaries in Philadelphia. The Declaration was published in British newspapers starting in mid-August, and copies made it to Switzerland and Warsaw by September. The British government did not officially acknowledge the Declaration, but began mobilizing troops.

1793: The Louvre Museum opens in Paris. The museum is housed in the Louvre Palace, which had been the site of the

King's royal collection of art and sculpture. During the French Revolution it was repurposed to display the masterpieces of the French Republic. Napoleon later had the museum renamed in his own honor and expanded its collection. In the latter half of the 19th century, Louis XVIII and Charles X both further increased its holdings, and the museum continued to expand its collection during the twentieth century. It eventually became one of the largest art museums in the world.

1846: The Smithsonian Institution is granted a charter.
James Smithson, an English chemist, left half a million dollars in his will to fund the establishment of the Smithsonian, and the US Congress granted it a charter. Smithson's scientific notes, mineral collection, and library were sent to the United States to serve as the core collection of the museum. The Smithsonian grew to be a group of museums and research institutions with a collection of more than 150 million items.

1907: An eight-thousand-mile automobile race begins in Beijing, China. Racers drove across Asia and Europe to finish the race 62 days later in Paris, France. Prince Borghese of Italy won the race after driving through a swamp, narrowly avoiding a brush fire, and being pulled over by a Belgian policeman who didn't believe he was involved in a race, rather than simply speeding.

1920: The Treaty of Sevres divides the Ottoman Empire between the Allies following World War I. The partitioning of the Empire also saw the Allied takeover of much of modern-day Turkey as well as the creation of Mandatory Palestine and the French Mandate for Syria and Lebanon.

1948: *Candid Camera* debuts. The program originally began as a radio series called *Candid Microphone*, and expanded to include theatrical short films. The show, which featured hidden cameras recording practical jokes, continued to air until the 1970s. It was revived on CBS from 1996–2004 and again for a single season on TV Land in 2014.

1961: US forces use Agent Orange in the Vietnam War for the first time. The chemical defoliant was used to kill thick vegetation that the United States said guerillas were using for cover. Its use in Vietnam resulted in millions of civilian health issues and extreme environmental damage.

1985: *Like A Virgin,* by the pop singer Madonna, reaches sales of five million copies. It was the first solo album by a female artist to reach that number of sales.

2003: A temperature of 101.3 degrees Fahrenheit is recorded in Kent, England. It was the first time a temperature above 100 degrees Fahrenheit was recorded in the United Kingdom.

August 11

3114 BC: The Mesoamerican Long Count Calendar begins. The calendar is a base-20 counting system used by multiple pre-Columbian cultures in Central America, including the Maya. It counts from the mythical creation date, and the "long count" ended on December 21, 2012.

1711: The first Royal Ascot race is held in England. Queen Anne had bought land near the village of East Cote for 558 pounds to build a racetrack. During the race, horses competed for the prize Her Majesty's Plate. The Royal Ascot race is now held in June, and is famous for its high fashion.

1929: Babe Ruth hits his 500th home run. Ruth had begun his career in Major League Baseball with the Boston Red Sox in 1914. They infamously traded him to the New York Yankees after Ruth broke the single-season home run record in 1919. Ruth spent the rest of his career in New York, solidifying his reputation as "The Sultan of Swat" who remains one of the greatest baseball hitters of all time. During his 15 years there, Ruth led the Yankees to win seven American League pennants and four World Series championships.

1934: The first prisoners arrive at Alcatraz Island. Classified as "most dangerous," the federal prisoners were the first civilians to be housed on the rocky island that sits a mile and a half from San Francisco Bay. The prison, called "the Rock," saw about three dozen escape attempts, all unsuccessful. It closed in 1963. From 1969–1971 activists with the American Indian Movement occupied the island to protest government treatment of their people.

1942: Dimitri Shostakovich's Seventh Symphony is performed in Leningrad. Later known simply as "The Leningrad Symphony," the piece was originally dedicated to Vladimir Lenin but later given to the city of Leningrad upon its completion. The symphony was performed while the city was besieged by Nazi forces and became a popular symbol of resistance to fascism.

1952: Prince Hussein becomes King of Jordan. His father, King Talal, had been declared unfit to rule due to mental illness. Hussein became the third constitutional monarch of Jordan and on his death in 1999 was the longest-serving head of state in the 20th century.

1968: The Beatles launch Apple Records. The label produced much of the band's music as well as former members' solo projects after they broke up. Other artists such as James Taylor, Mary Hopkin, and Jackie Lomax were also produced by the label.

1972: The last US ground combat troops leave Vietnam. The Third Battalion, 21st Infantry, which guarded the Da Nang airbase, left for the United States. American involvement in the war had begun more than 19 years earlier, and at the war's height more than half a million American servicemen were deployed to Vietnam. Of them, 58,318 did not return alive, and another 300,000 were wounded.

1984: "We begin bombing in five minutes." While running for re-election, President Ronald Reagan jokingly said these words while preparing to give a weekly address on National Public Radio.

August 12

1099: During the First Crusade, crusaders defeat an Egyptian army at Ascalon. Godfrey of Bouillon launched a surprise attack and defeated an army nearly twice the size of his own in the battle, which lasted less than an hour.

1492: Christopher Columbus lands in the Canary Islands on his journey west. Columbus thought that he could reach Asia by sailing across the Atlantic, but was stopped short when he found the Americas in his way.

1508: Juan Ponce de Leon lands in Puerto Rico for the first time. The conquistador had recently crushed an uprising by native Taino people on the nearby island of Hispaniola, and he was dispatched to explore Puerto Rico for the Spanish crown. He was appointed the first Governor of Puerto Rico in 1509 and extracted a great deal of wealth from mines and plantations.

1831: "The Glorious Twelfth" is officially enshrined in English law for the first time. The annual event opens the grouse-hunting season in Great Britain, and the date is a traditional one whose origins are unknown. The Twelfth starts a race to shoot the birds during the hunt on the moors of Scotland and Northern England and bring them to dinner tables in restaurants of London.

1865: Joseph Lister first uses disinfectant during an operation. The English surgeon was the first to use antiseptic methods during surgery, sterilizing his instruments with phenol and using the chemical to clean his patients' wounds. Lister had observed workers using phenol to reduce the stench coming from fields irrigated with sewage water, and his further realization that the livestock grazing on the fields did not get sick. His methods soon became widely used as other surgeons saw they led to fewer infections arising from operations. Lister is widely considered the "father of modern surgery" for his efforts. The mouthwash Listerine is named after him.

1898: An annexation ceremony takes place in Hawaii. The flag of the Republic of Hawaii was lowered from ʻIolani Palace in downtown Honolulu and replaced with the United States flag, signaling the American takeover of the island nation.

1941: FDR meets Winston Churchill in Newfoundland. The pair met to discuss issues ranging from how to best support the Soviet war effort to how to deal with the impending Japanese threat. They also discussed how to structure the world following the war, jointly declaring a promise to free the world of "aggrandizement, territorial or other." The agreement, which became known as the Atlantic Charter, comprised the foundation of the United Nations mission.

1944: At Sant'Anna di Stazzema, German Waffen-SS troops massacre 560 civilians. The massacre was carried out as a collective reprisal against Italians in response to actions by partisans fighting for the Italian resistance to Nazi occupation.

1994: The North American Free Trade Agreement is completed. NAFTA reduced barriers to trade between Mexico, Canada, and the United States, but generated a great deal of controversy almost as soon as it was proposed.

August 13

1521: Conquistadors led by Hernan Cortes capture Tenochtitlan. The Spanish invaders had besieged the Aztec capital city for three months before defeating the Aztec army. The Aztecs had been weakened by a smallpox epidemic that the Spanish were immune to. The Spanish also prevented any food or water from reaching the inhabitants of the city. Upon being resupplied, the conquistadors attacked the city directly, and after several days of fighting they captured Tlatoani Cuauhtemoc, the Aztec emperor. He was tortured and kept prisoner until Cortes had him executed in 1525. The fall of Tenochtitlan ultimately caused the destruction of the Aztec civilization and brought the initial phase of the Spanish conquest of Mexico to a close.

1779: The British Royal Navy defeats the Penobscot Expedition during the American Revolutionary War. The British attacked a flotilla of nineteen American warships that were accompanying 25 other vessels carrying troops and artillery. All the American ships were destroyed, making it the greatest naval loss by the American military until the Attack on Pearl Harbor 162 years later.

1784: The English Parliament passes the India Act. The Act provided for the British East India Tea Company and the British Crown to jointly control the Indian subcontinent. It remained in effect for the next 75 years.

1889: William Gray patents the coin-operated telephone.

1906: Black infantrymen with the US Army's 25th Regiment are accused of murdering a white bartender and wounding a white police officer in Brownsville, Texas, in spite of the fact that there is evidence that clears them. They were all later dishonorably discharged in relation to the incident.

1907: Taxicabs first appear in New York City. The automobiles had already debuted on European streets in the 1890s. They were named for the devices that recorded the distance travelled in order to calculate the fare, called taximeters. The word "cab" came from the cabriolet, a horse-and-buggy that could be hired during the 19th century.

1918: Opha May Johnson enlists in the US Marine Corps, making her the first woman to do so. Johnson served in the Marine Corps Reserve during World War I along with 300 other women. She worked as a clerk in the War Department and later chartered the first women's post of the American Legion.

1942: Eugene Reybold authorizes the construction of facilities that would house the Manhattan Project. The code name was initially the Development of Substitute Materials. It grew to employ 130,000 people and led directly to the invention of the first nuclear weapons.

1961: East Germany closes the border between East and West Berlin. The border closing, meant to prevent East Germans from escaping to the West, eventually led to the construction of the Berlin Wall. The infamous symbol of repression and militarism stood for nearly three decades until it was torn down in 1989.

1996: The Galileo space probe sends back data that indicates there may be water on Europa, a moon of Jupiter. It was the first water found elsewhere in the solar system by a space probe from Earth.

August 14

1040: Soldiers loyal to Macbeth kill King Duncan of Scotland in battle. Macbeth became king of Scotland upon Duncan's death, and ruled until 1057, when he was killed in the Battle of Lumphanan by forces loyal to Malcolm III, the son of Duncan. Macbeth's life was immortalized in 1606 by William Shakespeare.

1678: William of Orange surprises the French army near Mons. The French were able to quickly regroup and counterattack, however, and William's forces lost several thousand soldiers and were forced to retreat.

1784: Grigory Shelikhov founds Three Saints Bay on Kodiak Island. Initially a small encampment, it was the first permanent Russian settlement established in Alaska. Russian adventurers led by the Danish explorer Vitus Bering, for whom the waterway between Alaska and Russia is named, had originally mapped much of Alaska in 1741.

1834: Richard Henry Dana sets sail with the Royal Navy. He would later recount his experiences at sea in the autobiographical account *Two Years Before the Mast*. The book was hugely popular and influenced public perception of the treatment of sailors in the Navy.

1932: Guglielmo Marconi, an Italian inventor and engineer, completes his work on the short-wave radio.  The invention would radically change communications worldwide in the coming years, eventually allowing messages to be transmitted around the globe faster than ever before.

1936: The United States defeats Canada in the first basketball competition to take place at the Olympic Games. The US won the game 19–8.

1945: News of Japan's unconditional surrender in World War II is broadcast to the Japanese public for the first time. In response to the news, more than 1,000 Japanese soldiers who refused to give up the cause stormed the Imperial Palace to try to prevent the message from being carried abroad.

1975: *The Rocky Horror Picture Show* opens in London. The campy musical horror-comedy went on to become the longest-running release in the history of film.

1980: Workers in Poland take over the shipyard in Gdansk. The workers were demanding pay raises and the right to organize a labor union that would be autonomous from Communist party control. The strike, dubbed "Solidarity," gained international attention and brought attention to labor leader Lech Walesa.

2013: The Egyptian government declares a state of emergency. Security forces commanded by General Abdel Fattah el-Sisi raided protest camps of the Muslim Brotherhood, killing nearly 1,000 protestors.

August 15

1805: The expedition led by Lewis and Clark crosses the Continental Divide in Montana. They followed the Missouri river to its headwaters at Lemhi Pass, where they crossed over. The Continental Divide is the line that divides the East and West watersheds in North America.

1848: Milton Waldo Hanchett patents the dentist's chair. Hanchett, a manufacturer and businessman from Connecticut who was also the choirmaster and organist at his church, added a pedal taken from a grand piano to the dental chair. This allowed the dentist to manipulate the chair's height and orientation, and has been a crucial feature in all dental chairs since.

1852: The first game of croquet is played in the United States. Originally an English lawn game, croquet became widely popular as a social game rather than an athletic one. It was one of the first games that American men and women played together.

1914: The *Ancon,* an ocean liner, passes through the Panama Canal. It was the first ship to do so and inaugurated the Canal with its passage.

1945: Victory in Japan is declared by the Allied forces in WWII. The Allies had decimated the Japanese Navy, American planes had dropped two atomic bombs on major Japanese cities, and an Allied ground invasion of the Japanese home island was imminent. Soviet forces had recently entered the Eastern theater of the war, attacking Japanese troops in Manchuria. Faced with these factors, Emperor Hirohito overruled his generals and ordered his Supreme War Council to accept the Allies' demand of unconditional surrender. On the same day, Korea gained its independence from the Empire of Japan.

1969: The Woodstock Music and Art Festival opens. Nearly half a million people converged on Max Yasgur's 600-acre dairy farm in the Catskill Mountains in upstate New York for the festival. It was advertised as "An Aquarian Exposition of Three Days of Peace and Music." Thirty-two performers whose music defined the generation, including Janis Joplin, Jimi Hendrix, Carlos Santana, Joe Cocker, the Grateful Dead and Jefferson Airplane, performed, rain or shine, in the outdoor venue. The iconic event is considered a focal moment in the counterculture of the 1960s and the overall history of American pop music.

1970: Patricia Palinkas is the first woman to play in a professional American football game. Palinkas, who attended Northern Illinois University, went on to play as a holder for the Orlando Panthers in the Atlantic Coast Football League. In her first game against the New York Jets, she was attacked by a Jets lineman who was trying to injure her. Nevertheless, she persisted in playing, holding for three successful extra point conversions and one field goal that was blocked.

1977: Astronomers using the Big Ear radio telescope hear a radio signal from deep space. The telescope was part of the Search for Extraterrestrial Intelligence (SETI) project. The radio signal was named the "Wow!" signal because of a written note on the report. The signal came from the constellation Sagittarius and had the appearance of being extraterrestrial in origin.

2013: The Smithsonian announces the discovery of the olinguito. A member of the raccoon family that lives in the Andes mountains, it was the first carnivorous species found in the Americas in 35 years.

August 16

1792: Robespierre presents demands from the Paris Commune to the Legislative Assembly. The demands included the establishment of a revolutionary tribunal, extending the right to vote to everyone, and sending delegates to a national convention. Robespierre was elected first deputy to the convention, which was held in September and declared a French Republic.

1858: The Transatlantic Telegraph Cable is inaugurated. US President James Buchanan sends greetings to Queen Victoria of the United Kingdom, who replies in kind. The line had to be shut down after only a few weeks because the signal was too weak, however.

1896: The Klondike Gold Rush begins. George and Kate Carmack discovered gold in a tributary of the Klondike river in Alaska. When the news of the discovery spread to San Francisco and Seattle, it sparked a migration of about 100,000 gold prospectors to the harsh climate.

1916: Canada and the United States sign the Migratory Bird Treaty. The treaty provided for both countries to protect migratory birds from hunting and trapping as they pass from their nesting grounds in the Arctic to their winter grounds in Central and South America. It was the first purely environmental treaty in history.

1929: Rioting breaks out in Mandatory Palestine between Jewish settlers and Arabs. The rioting continued for two weeks and left 133 Jews and 116 Arabs dead. Seventeen Jewish communities were evacuated during the violence. The incident resulted in an official inquiry by the British government.

1954: The first issue of *Sports Illustrated* is published. The cover of the inaugural issue showed Eddie Mathews of the Milwaukee Braves at bat in Milwaukee County Stadium, with We Westrum catching for the New York Giants behind the plate. The magazine was the first to use color photos widely, and was the first to include in-depth sports journalism and scouting reports. *SI* eventually grew to more than 3 million subscribers and is read by more than 20 million people, including 18 million men, each week. It annually awards the title of "Sportsperson of the Year" to an athlete. The award was first given to Roger Bannister for breaking the four-minute mile. In 1999 the magazine named boxer and human rights advocate Muhammad Ali as "Sportsman of the Century."

1960: Joseph Kittinger parachutes 102,800 feet over New Mexico. Kittinger set three world records for his stunt: the highest-altitude jump, longest free fall, and fastest speed by a human without an aircraft.

1966: The House Un-American Activities Committee (HUAC) begins its investigations. The Committee was convened to investigate charges of Americans aiding the Viet Cong during the Vietnam War, but soon included antiwar activists and movement leaders.

1977: Elvis Presley dies in his home in Memphis, Tennessee. The "King of Rock 'n' Roll" was 42 years old. Presley, one of the iconic pop culture figures of the 1950s and 60s, was in poor health due to prescription drug abuse. Elvis was buried at Forest Hill Cemetery two days later, but his body was reinterred at Graceland after grave robbers attempted to steal the body.

1999: Vladimir Putin is confirmed as Prime Minister of Russia by the Parliament. He was the nation's fifth prime minister since the early months of 1998, and would hold onto power for decades.

August 17

1585: Colonists arrive on Roanoke Island, North Carolina. The colonists were sent by Sir Walter Raleigh and under the command of Ralph Lane. The colony was intended to be the first permanent settlement in the part of North America claimed by Great Britain. They initially built a small fort on the island and waited for supplies to arrive. The colonists had destroyed a native village down the coast, and this resulted in an attack on the fort that was repulsed. In 1587, when a contingent dispatched to relieve them arrived, they found no survivors at the fort, and only a human skeleton. They reestablished the colony, but in 1590 when Walter Raleigh returned, he found it was deserted, with the only clue of the more than 100 inhabitants' disappearance being the word "Croatoan" carved into a tree.

1807: Robert Fulton's steamboat leaves New York City. The steamboat was initially ridiculed by laughing crowds as they watched its maiden voyage, and newspapers called the adventure "Fulton's Folly," believing it wouldn't catch on. But the North River Steamboat successfully completed its first trip to Albany, and Fulton was soon running passengers and cargo up and down the river on a regular schedule.

1835: Solyman Merrick of Springfield, Massachusetts, patents the first wrench.

1896: Bridget Driscoll becomes the first pedestrian killed by an automobile. Driscoll was crossing the street near the Crystal Palace in London when a motorcar struck her, crushing her skull and killing her instantly.

1907: Pike Place Market opens in Seattle. It is one of the oldest continuously operating farmers' markets in the United States and one of the most popular tourist destinations in Seattle.

1943: The US 7th Army captures Messina, Italy. Led by General George S. Patton, the Seventh Army was soon joined by the British 8th Army under Bernard Montgomery. The capture of Messina was the final piece in the conquest of Sicily by the Allies.

1945: Indonesia declares its independence from the Dutch Empire. The struggle for independence was waged for four years, with Dutch forces holding urban centers and revolutionary guerillas controlling the countryside. The revolutionaries were successful, and the Netherlands recognized Indonesia's independence in 1949.

1962: East German soldiers kill Peter Fechter. Fechter, an 18-year-old bricklayer, was shot while trying to flee to West Berlin across the Berlin Wall. The guards left him bleeding to death in the 'no-mans-land' between East and West Berlin in full view of hundreds of Western bystanders and journalists.

1998: US President Bill Clinton admits under oath that he had an "improper physical relationship" with White House intern Monica Lewinsky. As a result, he was impeached by the House of Representatives later that year.

August 18

1686: Giovanni Domenico Cassini first reports observing a satellite orbiting Venus. Cassini, an Italian astronomer and engineer, had originally observed the satellite in 1672 but did not report it until he saw it again fourteen years later. Although numerous other astronomers also thought they observed the satellite in the 18th century, which they dubbed "Neith," it was later determined to have been a case of mistaken identity. The satellite was most likely explained by stars that were nearly in alignment with Venus at the time.

1825: Alexander Gordon Laing reaches Timbuktu. He was the first European to reach the fabled city by crossing overland from Tripoli, traveling more than 2,200 miles to do so. Laing was killed on the return journey by Tuareg raiders that ambushed his expedition.

1900: The word "television" is coined. During a presentation at the Paris International Electricity Conference, Constantin Perskyi, a Russian scientist, first used the word in a paper presented on the use of photoelectric selenium to transmit images.

1914: President Woodrow Wilson issues the Proclamation of Neutrality. The Proclamation was meant to keep the United States out of World War I, but after months of German attacks on American shipping vessels that were resupplying England, the US decided to enter the war.

1920: The 19th Amendment to the US Constitution is ratified. When Tennessee ratified the Amendment, which extended the right to vote to women, it gave it the two-thirds majority required to add it to the Constitution.

1932: Jim Mollison makes the first east-to-west solo flight across the Atlantic Ocean. Mollison, a Scottish aviator, flew from Portmarnock, Ireland, to New Brunswick, New Jersey.

1943: Betty Smith publishes *A Tree Grows In Brooklyn*. The book, about life in the tenements of Brooklyn, New York, went on to become an American classic.

1963: James Meredith graduates from the University of Mississippi. Meredith, who earned a degree in political science, was the first African-American graduate from the university. When he enrolled, white segregationists rioted for days, killing two men. Meredith endured harassment and ostracism from other students, and had to be protected by federal troops during his time at Ole Miss.

1991: Communist hard-liners attempt to overthrow Mikhail Gorbachev. The leaders of the coup were opposed to the reforms Gorbachev had implemented in the Soviet Union, as well as his policy of decentralizing government power. Russian president Boris Yeltsin, an ally of Gorbachev, led a campaign of civil resistance over the following days, condemning the coup and encouraging a general strike. Tanks arrived in the Red Square in Moscow but were held back from attacking the Parliament by crowds of civilians. After a few days, the coup was over and its leaders placed under arrest by soldiers loyal to Yeltsin. The coup left Gorbachev politically ineffective, and he resigned in December of 1991. The following day, the Soviet Union ceased to exist. Yeltsin retained his position as President of Russia until 1999, when he was succeeded by Vladimir Putin.

August 19

1692: Several "witches" are executed in Salem, Massachusetts. After Elizabeth Hubbard and other teenaged girls in Salem accused a number of people of practicing witch-craft, a mass hysteria soon gripped the town and many more were also accused. Local clergy and magistrates arrested and tried more than thirty townspeople. George Burroughs, Mary Eastey, Martha Corey, and George Jacobs were convicted and hanged in August, and eight more were executed in September. The following year a superior court found several more accused witches not guilty at trial, which ended the rash of accusations and executions. The Salem witch trials ultimately resulted in twenty people being executed and is widely considered to have influenced the separation of Church and State by American patriots following the Revolution.

1812: The USS *Constitution* defeats the British frigate *Guerriere* in a fierce battle off the coast of Nova Scotia during the War of 1812. Observers later claimed that cannonballs fired by the British ship bounced off the hull of the *Constitution* as if the ship were built of iron instead of wood. The rumor led to the ship being nicknamed "Old Ironsides." By the war's end it had destroyed or captured seven more British ships. The *Constitution* remains the oldest commissioned naval vessel still at sea. Its homeport is in Boston.

1839: Louis Daguerre announces his photographic process. The method allowed for an image to be chemically fixed as a permanent picture, and was named the "daguerreotype" after its inventor. The French government called it a "free gift to the world," and the announcement marked the beginning of photography.

1848: The *New York Herald* breaks the news of a gold rush in California. The California Gold Rush had begun the previous January, but the *Herald*'s announcement was the first to spread the news to the East Coast of the United States. It spurred countless hopefuls to "Go West" in search of fortune.

1909: The first automobile race is held at a dirt racetrack in Indianapolis, Indiana. The racetrack would become the Indianapolis Motor Speedway, home to the Indy 500 and Brickyard 400, two of the premiere events in American auto racing.

1919: Afghanistan declares independence from the United Kingdom. Amanullah Khan, the first emir of Afghanistan, won independence when he quickly fought the British to a stalemate during the third Anglo-Afghan war. He introduced social reforms during his rule including the establishment of co-educational schools for both men and women.

1934: The first All-American Soap Box Derby is held in Dayton, Ohio. The race was later moved to Akron, Ohio, because it had hillier terrain that was better suited to the race. It has been held every year since.

1953: A CIA-backed coup overthrows the President of Iran and installs the Shah. Mohammad Reza Shah ruled with the help of a ruthless state security apparatus that imprisoned and tortured dissidents until 1979, when he was overthrown in the Iranian Revolution.

1991: The Soviet Union is dissolved after almost 70 years. With the failure of the August coup, maintaining a Communist government became untenable, and the Union of Soviet Socialist Republics, which had dominated world politics for nearly a century, ceased to exist.

August 20

1191: The Massacre ay Ayyadieh. King Richard I of England ordered his troops to slaughter some 3,000 Muslim soldiers and civilians in full view of the armies of Saracen leader Saladin. Although the Saracens attacked the Crusaders upon seeing the massacre, Richard's army was able to hold them off.

1667: John Milton publishes *Paradise Lost*. The epic poem about the fall of Adam and Eve filled ten volumes and was more than ten thousand lines long. Its publication cemented Milton's reputation as one of the great English poets of all time, and it remains a classic.

1858: Charles Darwin first publishes the Theory of Evolution. Darwin had initially conceived his theory—that species arise from differences in survival and reproduction based on variations in their physical traits—while voyaging to South America in 1838 on the HMS *Beagle*. He was reluctant to publish his theory without thoroughly explaining it and spent the next two decades quietly revising his ideas. But when Alfred Russell Wallace, a Welsh explorer and naturalist, wrote to Darwin to describe a very similar theory, Darwin realized he had to publish. The two jointly published their findings in *The Journal of the Proceedings of the Linnean Society of London*. Darwin later published a comprehensive argument for evolution by natural selection in *On the Origin of Species* the following year. The theory underlies the work in the life sciences that has come since and is one of the most important discoveries in human history.

1882: Peter Tchaikovsky's *1812 Overture* is first performed in Moscow. The work commemorated the Russian defense against Napoleon's invasion of 1812. It would go on to become one of the composer's best-known works, along with the ballets *The Nutcracker* and *Swan Lake*.

1910: Jacob Fickel fires a rifle from an airplane at a target on the ground in Brooklyn, New York. Fickel, a lieutenant in the US Army, was the first person to fire a gun from an aircraft.

1922: The first Women's Olympic Games open in Paris, France. After it was discovered that women would be barred from competing in track and field events at the 1924 Olympic Games, the Women's Olympics were organized in response. The International Olympic Committee protested the use of the term "Olympic," and the name was changed to the World Games.

1940: Winston Churchill addresses the nation during the Battle of Britain. In reference to the Royal Air Force pilots who were defending the island nation from attacks by the German Luftwaffe, Churchill uttered the famous line "never in the field of human conflict was so much owed by so many to so few."

1968: Soviet forces invade Czechoslovakia. More than 200,000 troops from Warsaw Pact nations allied with the USSR accompanied by 5,000 Soviet tanks invaded the nation following a short period of liberal reforms known as the "Prague Spring." Czech citizens protested the invasion with public demonstrations and civil disobedience, and 137 were killed. The invasion put an end to the reform efforts.

1977: Voyager 2 is launched. The space probe spent the next three decades on a "Grand Tour" of the Solar System before travelling beyond it in 2007. It was the first spacecraft to pass by both Saturn and Jupiter.

August 21

1770: James Cook claims eastern Australia for Great Britain. Cook named the territory New South Wales. In 1788 a penal colony was founded in the territory, and in 1901 it joined the new federation of Australian colonies as the state of New South Wales.

1831: Nat Turner launches a revolt in Virginia. Turner was born into slavery in 1800. By the time he was in his twenties, he was experiencing frequent visions that he interpreted as messages from God to lead his people to freedom. After a solar eclipse on August 7 that he took as a sign from the Almighty to begin his rebellion, Turner organized an uprising. Free and enslaved African-Americans attacked plantations in the vicinity, killing sixty men, women, and children, but sparing the homes of poor whites. Turner was captured in October and executed on November 11.

1858: Abraham Lincoln and Stephen Douglas debate one another. The debate was the first of seven the two held over the next two months across the state of Illinois during the 1858 race for US Senate. The encounters became known as the "Lincoln-Douglas debates" and were a precursor to the issues that would define Lincoln's later his time in office. All of the debates mainly centered on a discussion of slavery in the United States. Lincoln took the position that the federal government should limit the spread of slavery to new territories, while Douglas argued that the decision should be left to individual states. The debates raised Lincoln's national political profile and helped propel his candidacy for President in 1860.

1863: Quantrill's Raiders attack Lawrence, Kansas. The pro-Confederate guerillas, who were organized by William Quantrill in neighboring Missouri, attacked and burned the town because it was known to be abolitionist. They killed 150 men.

1911: Vincenzo Peruggia steals the *Mona Lisa*. Peruggia, an employee at the Louvre Museum in Paris, walked out of the museum with the painting concealed under a smock. He kept it for two years, avoiding suspicion until he contacted an Italian art dealer who alerted the police. He served a short jail sentence for the theft.

1912: Arthur R. Eldred becomes the first Eagle Scout. The highest rank in the Boy Scouts of America has been attained by more than 55,000 scouts since then.

1935: The Swing Era is born. Benny Goodman, a jazz clarinet player, performed at the Palomar Theater in Los Angeles, California. The show was broadcast nationwide and drew a large audience, popularizing the Swing musical genre.

1957: The ICBM is born. The Soviet Union successfully tested the long-range R-7 Semyorka, the first Intercontinental Ballistic Missile (ICBM). The invention of this system, which could deliver nuclear payloads virtually anywhere in the world, ushered in a new era of the Cold War.

1961: British authorities release Jomo Kenyatta from prison.
Kenyatta had served almost nine years in prison for his efforts
to secure independence for his native Kenya. Two years later, he
would become the first prime minister of Kenya when the nation
won its freedom.

August 22

**1485: Henry Tudor defeats Richard III at the Battle of
Bosworth Field.** The battle signified the end of the Wars of
the Roses, a civil war between rival English factions that had
lasted thirty years. Henry's victory put an end to the Plantagenet
dynasty that had ruled England for over three centuries. During
the battle, Richard divided his army, which was larger than
Henry's, allowing Henry's men to command the battlefield.
When Richard led his knights in a charge towards Henry's position in an attempt to kill Henry, they were trapped and killed by
Henry's men instead. The battle was later immortalized in the
final act of William Shakespeare's play *Richard III*, in which the
titular king delivers the famous line "A horse, a horse, my kingdom for a horse!" Henry ruled England until his death in 1509.
The House of Tudor existed until 1603, when the House of
Stuart succeeded it.

1642: The English Civil War breaks out. Charles I raised
the royal standard at Nottingham in defiance of the English
Parliament. Supporters of Charles I, called "Royalists" or
"Cavaliers," and supporters of the English Parliament, called
"Roundheads," met in several skirmishes over the next few
months. By 1645 Charles' forces were defeated and he was captured. He was executed in 1649 for treason.

1791: The Haitian Revolution begins. Hundreds of enslaved
persons rose up and killed their masters in Saint Domingue,
Haiti, and by the end of the month they had taken control of
the island's entire Northern province. Toussaint Louverture, a
free Black man, took command of the rebellion and led until his

death in French custody in 1803. Haitian rebels won their independence the following year.

1851: The US schooner *America* wins the first America's Cup race. The schooner beat out fifteen other ships in a race around the Isle of Wight, finishing eight minutes ahead of the next ship. When Queen Victoria, who was watching the race, inquired who was in second place, the famous reply "Ah, Your Majesty, there is no second," was uttered.

1864: The first Geneva Convention is signed by 12 nations. The Convention was proposed after Henry Dunant visited wounded solders after the Battle of Solferino in 1859. It governed the treatment of sick and wounded soldiers on the battlefield, and was the basis for all international laws overseeing armed conflict that have come since.

1906: The first Victrola record player is manufactured. The Victor Talking Machine Company of Camden, New Jersey, developed the Victrola, which retailed for $200 and was hand cranked.

1922: Anti-Free State nationalists assassinate Michael Collins. Collins, who had led the Irish Republican Army to fight the British to a stalemate, supported a treaty that allowed the British to keep six counties in Northern Ireland and designated the rest of the Island a Free State. The treaty sparked a civil war. Collins, leader of the pro-treaty faction, was gunned down by nationalists as he traveled through County Cork.

1989: Nolan Ryan reaches 5,000 strikeouts. Ryan, a pitcher for the Texas Rangers at the time, struck out Rickey Henderson of the Oakland Athletics. After the game, Henderson was quoted as saying that if Ryan "ain't struck you out, then you ain't nobody."

2004: *The Scream* and *Madonna*, paintings by Edvard Munch, are stolen at gunpoint. Four men were later convicted for the robbery, and the paintings were recovered in 2006.

August 23

1305: William Wallace goes on trial for treason. Wallace argued in his defense that he could not have been a traitor because he had never sworn allegiance to King Edward I, called the "Hammer of the Scots." He was summarily found guilty, hanged, drawn, and quartered, as was the prescribed punishment for traitors in England at the time.

1617: The first official one-way streets are created. In London, England, seventeen narrow and congested streets were specified as allowing traffic in only one direction to reduce the "disorder and rude behavior of Carmen, Draymen, and others using Carts."

1775: King George III acknowledges the American Revolution. Following the British defeat at the Battle of Bunker Hill, King George delivered a Proclamation of Rebellion to the Court of St. James, stating that the American colonies were engaged in "open and avowed rebellion" against the British Crown.

1898: The Southern Cross Expedition departs from England. It 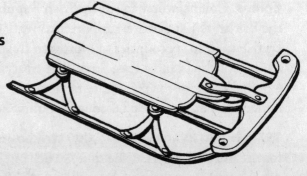 was the first mission in the Heroic Age of Antarctic Exploration, the first to use sled dogs to travel, and the first to spend the winter on the Antarctic mainland. Carsten Borchgrevink led the voyage, which traveled farther south than anyone had done before and opened the way for later expeditions to the South Pole.

1922: Chitty Bang Bang wins the Southsea Speed Carnival in England. The car, which featured a Mercedes chassis and a 6-cylinder Maybach engine, was driven by Count Louis Zborowski, an English automobile engineer and amateur racer. The car inspired Ian Fleming, author of the James Bond novels, to write a children's book titled *Chitty Chitty Bang Bang: The Magical Car*. In the novel, an inventor named Commander Caractacus Pott renovates the racecar so that it can fly. The novel was later adapted to a musical film starring Dick Van Dyke.

1926: Rudolph Valentino dies. The film star and heartthrob's wake drew throngs of more than 100,000 fans in a crowd lined up for over a mile in Manhattan to pay their respects. Valentino's remains were transported by train to California, where they were interred at the Hollywood Forever Cemetery.

1942: The Luftwaffe begins bombing Stalingrad. Six hundred bombers attacked the Soviet city to begin a siege that lasted until February 1943. Although the city was reduced to rubble, Soviet defenders were able to stop the German offensive at the Volga River by November, when they counterattacked and trapped the German army in the city. After three more months of heavy fighting, the remaining Germans surrendered.

1970: The Salad Bowl strike begins. Organized by Cesar Chavez and the United Farm Workers, it was the largest strike by farmhands in the history of the United States, and led directly to the California Agricultural Labor Relations Act of 1975.

2011: Muammar Qaddafi is overthrown. Qaddafi had ruled Libya since he seized power in 1969. Assisted by NATO airstrikes, a coalition of rebels overthrew the government. Qaddafi was captured and executed by rebels in October.

August 24

455: The Vandals sack Rome. Pope Leo I asked Genseric, King of the Vandals, that they spare the ancient city and the lives of Roman citizens. Genseric agreed, and the gates of the city were opened for the Vandals, who looted a great deal of treasure.

1215: Pope Innocent III declares the Magna Carta invalid. King John of England was forced to sign the Magna Carta by rebellious barons. It limited the power of the Crown and established a system of justice that still remains in British and American legal systems. Pope Innocent relented in his opposition when John reached a compromise with him.

1456: The Gutenberg Bible is completed. The printing took two years to complete, and the Gutenberg Bible was the first major book produced using movable metal type in Europe. It ushered in the "Gutenberg Revolution" and the era of printed books in the West. About forty copies of the Bible still survive.

1780: King Louis XVI of France outlaws the use of torture to obtain confessions from suspects accused of a crime.

1814: The British burn Washington, D.C. Following a victory over American troops at the Battle of Bladensburg during the War of 1812, General Robert Ross led British troops into the American capital. They attacked the city in retaliation for the American destruction of Port Dover, Canada, earlier that spring. The British burned the Capitol and the White House during the 26 hours they occupied the city.

1857: The Panic of 1857 begins. A worldwide economic crisis led to a decline in the stock market, and several banks and railroad businesses failed. On the morning of August 24, the president of the Ohio Life Insurance and Trust Company announced it was suspending payments. Bank runs ensued in major American cities, and the American economy did not recover until after the Civil War.

**1938: Clark Gable reluctantly agrees to play Rhett Butler in
Gone with the Wind.** Gable was concerned that the high-profile
production would set unreasonable expectations on leading
actors going forward. Metro-Goldwin Mayer, the production
company, offered him a large bonus, and he agreed. The film was
the most successful ever made up to that point.

**1950: Edith Sampson becomes the first African-American
US delegate to the United Nations.** Sampson, a prominent
attorney, served in the post until 1953. In 1961 she became the
first African-American representative to NATO.

1995: Microsoft's Windows 95 is released. The operating
system merged two products, MS-DOS and Windows, and
included a graphical user interface for the first time. It was succeeded three years later by Windows 98.

August 25

1609: Galileo Galilei demonstrates his first telescope.
Galileo, an Italian astronomer, had invented and improved
existing telescopes based on descriptions by Hans Lippershey.
Galileo's designs eventually improved the telescopes' magnification by as much as thirty times, fundamentally changing their use
from terrestrial observation to astronomical observations. Using
his telescopes, Galileo would go on to discover the moons of
Jupiter in 1610 and the phases of Venus in 1611. These discoveries revolutionized astronomy in that they directly challenged
Aristotle's theory of cosmology, which dictated that all heavenly
bodies orbited the Earth. In 1632, convinced Aristotle's theory
was incorrect, Galileo published *Dialogue Concerning the Two
Chief World Systems*. In it he presented the argument that the
Sun was the center of the universe. He was tried and convicted of
heresy in 1633, but he was ultimately proven to be correct.

**1823: Hugh Glass is mauled by a grizzly bear in South
Dakota.** Glass, a fur trapper and mountain man, was wounded
so badly by the attack that his companions left him for dead. He

survived, set his own broken leg, and crawled overland more than 200 miles to the nearest European settlement, Fort Kiowa. His journey was later immortalized in the film *The Revenant*.

1894: Kitasato Shibasaburo discovers the infectious agent of bubonic plague. The Japanese bacteriologist published his findings in *The Lancet*.

1910: Walden Shaw and John Herts found a livery company in Chicago. Originally called the Walden Shaw Livery Company, it would go on to become the Yellow Cab Company when its first brightly-colored taxicab debuted in 1915. It remained one of the iconic giants of the cab industry for the 20th century.

1940: The Bombing of Berlin begins. The British Royal Air Force dispatched 95 aircraft to attack the airport in the center of Berlin. As a result, Hitler ordered the Luftwaffe to shift from targeting air defenses and airfields to attacking British cities during the Battle of Britain.

1991: Linus Torvalds announces the creation of Linux. He developed the operating system kernel while he was a student at the University of Helsinki. Linux is the first and most widely used open-source operating system. Torvalds did not conceive of the name, and originally called it "Fre-ax," but it became known as "Linux" when it was made widely available.

2001: R&B singer Aaliyah is killed in a plane crash. The 22-year-old singer and actress sold more than fourteen million albums during her short career. She was filming a music video for her hit single "Rock the Boat" when her plane crashed on takeoff in the Bahamas, killing everyone aboard.

2012: Voyager 1 leaves the Solar System. As it entered interstellar space, it became the first human-made object to do so.

August 26

1346: English longbowmen help win the Battle of Crécy.
The battle, fought during the Hundred Years' War, was the first
of three famous English victories. It signaled the rise of the
longbow as the chief weapon on the European battlefield; more
than 4,000 French soldiers were killed in the fighting, while only
about 300 English died. The victory prevented the French from
breaking the siege at Calais, which remained under English rule
for the next two centuries.

1498: Michelangelo is commissioned to sculpt the Pietá.
The statue, of the Virgin Mary cradling the dead body of Jesus,
was carved from a single clock of marble over a period of ten
years. It is housed in St. Peter's Basilica in the Vatican, and is the
only work of art Michelangelo ever signed.

1768: James Cook sets sail from England. Captain Cook
sailed the HMS *Endeavor* on a voyage to the south Pacific Ocean
that lasted three years. The aim of the voyage was to observe the
transit of Venus across the face of the Sun and to find evidence of
Terra Australis, a hypothetical continent in the Pacific. Cook and
his crew also mapped New Zealand, were the first Europeans
to land on the east coast of Australia, and nearly crashed on the
Great Barrier Reef.

**1873: The first public kindergarten in the United States is
authorized to operate in St. Louis, Missouri.**

1883: The final eruption of Krakatoa begins. The Indonesian
island and its surrounding archipelago were completely
destroyed, and at least 36,000 people died in the eruption and
ensuing tsunamis. Eruptions of steam and ash began in May and
continued sporadically for the next three months before dramati-
cally intensifying. On the 26th a black cloud of ash seventeen
miles high could be seen over the island, with hot pumice landing
on ships up to 12 miles away. The volcano would erupt over the
next 24 hours, with a colossal, final explosion on the 27th. That

explosion caused a pressure wave that was recorded by barometers around the world and ruptured the eardrums of sailors some 40 miles away. Tsunamis rocked ships as far as South Africa and destroyed coastal towns along the coasts of Indonesia and Asia. The eruption left a veil of ash around the world that dimmed the sun for years afterwards. The eruption was one of the deadliest and most powerful in recorded history.

1944: Charles de Gaulle enters Paris. As he proceeded along the Place de la Concorde, his motorcade was attacked by Vichy militia with machine guns. The militia were overwhelmed by French troops, and de Gaulle addressed a crowd after the battle, proclaiming the Third Republic was resurgent. The German Wehrmacht, furious and humiliated at losing Paris, bombed the city overnight, killing thousands. De Gaulle would go on to serve as Prime Minister and later President of France.

1978: Albino Luciani is elected Pope John Paul I. He served until he died suddenly 33 days later, making his reign one of the shortest in the history of the Vatican and resulting in The Year of Three Popes.

1997: F.W. de Klerk resigns as leader of the National Party. De Klerk was the last president of South Africa elected during the Apartheid era. During his presidency he negotiated the dismantling of the Apartheid regime, freed freedom fighter Nelson Mandela, and helped bring about the first elections that allowed everyone, white and black, to vote.

August 27

1776: British forces win the Battle of Brooklyn Heights during the American Revolution. It was the largest battle of the war. As a result, George Washington was forced to evacuate the Continental Army from New York City and retreat to Pennsylvania. Washington's nighttime retreat surprised the British and is considered one of his greatest military actions, as it saved the Continental Army from destruction or surrender.

1832: The Black Hawk War ends. Chief Black Hawk, leader of the Sauk tribe of Native Americans, led a group of allied tribes into Illinois to settle near the Mississippi in April. US forces attacked them, sparking the conflict. Black Hawk eventually surrendered to American authorities and was imprisoned for a year.

1893: The Sea Islands Hurricane makes landfall in Georgia. The hurricane killed between one and two thousand people in the storm surge and was one of the deadliest hurricanes in American history.

1918: Mexican militia and American soldiers clash in the Battle of Ambos Nogales. Fought in Nogales, a city that straddled the Mexican-American border, it was the only fighting that occurred in the United States during the Mexican Revolution.

1928: The Kellogg-Briand Pact is signed. The Pact, signed by representatives from more than 30 countries, including France, Germany, the United Kingdom, Japan, and the United States, renounced the use of war to settle disputes. Just a decade later, Japan invaded China and Germany invaded Poland, starting World War II.

1968: Order breaks down at the Democratic National Convention. Following days of escalating demonstrations by anti-war activists in nearby Grant Park and Lincoln Park, tension between protesters and police had reached a breaking point. Ordered to clear Clark street near the convention, police charged demonstrators, beating them mercilessly without distinguishing between protesters and bystanders. TV footage of the police riot was shown around the country, with protesters famously chanting "the whole world is watching" as police attacked. Democratic Senator Abraham Ribicoff denounced the attack during a speech to Convention, accusing the police of using "Gestapo tactics." Chicago Mayor Richard J. Daley defended the tactics, famously misspeaking when he stated, "the policeman isn't there to create disorder; the policeman is there to preserve disorder."

1979: The IRA assassinates Lord Mountbatten. The Irish Republican Army, a nationalist paramilitary organization, bombed Mountbatten's boat while he was vacationing in Sligo, Ireland. Later that day, IRA guerillas attacked a British Army convoy with bombs and gunfire. It was the deadliest attack on British forces during the four-decade war known as "The Troubles." Mountbatten was given a state funeral at Westminster Abbey.

1984: President Ronald Reagan announces a schoolteacher will be the first "citizen astronaut" in NASA's history. Christa McAuliffe, who was chosen for the mission, died when the *Challenger* space shuttle exploded on liftoff in January 1986.

August 28

1565: Pedro Menéndez de Avilés sights land in what is now Florida. Upon landing, he founded St. Augustine, which is the oldest continuously occupied European settlement in the continental United States. De Avilés, a Spanish conquistador, was later appointed governor of Cuba. He died in Spain in 1574.

1609: Henry Hudson sails into Delaware Bay. Hudson was the first Englishman to sight the Bay, which he noted in his journal but did not explore, choosing to continue exploring the coast north of it.

1789: William Herschel discovers Enceladus, a moon of Saturn. Herschel was a German-born British astronomer who later discovered infrared light, determined the rotation period of Mars, discovered Martian polar ice caps, and discovered two moons of Uranus and another of Saturn.

1833: Britain adopts the Slavery Abolition Act of 1833. The Act of Parliament, upon being granted royal assent, abolished slavery throughout much of the British Empire. An exception was made for "Territories in the Possession of the East India Company," as well as present-day Sri Lanka and Saint Helena.

1862: The Second Battle of Bull Run begins. Fought in the same location as the First Battle, the Second involved far more troops and had many more casualties. General Robert E. Lee's Army of Northern Virginia won the battle after three days of fighting, defeating a Union Army commanded by General John Pope. The victory was secured by a joint attack led by "Stonewall" Jackson and James Longstreet that crushed the Union forces between them.

1898: Caleb Bradham renames "Brad's Drink" as "Pepsi-Cola." Bradham originally developed the soft drink in 1893. The new name referenced dyspepsia, or indigestion, which it claimed to relieve, and the kola nuts used in the recipe. In 1961 "Cola" was dropped, and it has been known as Pepsi ever since.

1955: Emmett Till is brutally murdered in Mississippi. Till, an African-American teenager visiting from Chicago, was kidnapped from his bed by white racists after he was accused of whistling at a white woman. He was beaten to death and left in a swamp. After rail porters smuggled his body back to Chicago, his mother insisted on an open-casket funeral to show what had been done to her son. The funeral and media coverage helped galvanize the fledgling Civil Rights Movement.

1963: The March on Washington takes place. At the largest demonstration of the Civil Rights Movement, Dr. Martin Luther King, Jr. delivers the now-famous "I Have a Dream" Speech from the steps of the Lincoln Memorial to a crowd of more than 250,000.

1984: The Jackson Family's Victory Tour breaks the record for concert ticket sales. In just two months the tour sold 1.1 million tickets, and ultimately topped 2 million ticket sales.

1988: At an air show in Rammstein, Germany, three aircraft collide. The wreckage fell into the crowd, killing 75 and seriously injuring more than 300.

August 29

708: The first copper coins minted in Japan. The Wado-kaichin was the oldest official Japanese coinage, and was in circulation for over two centuries.

1756: Frederick the Great of Prussia attacks Saxony. The attack sparked the Seven Years' War, which involved every great European power at the time and was fought on five continents, making it the first global conflict in human history.

1831: Michael Faraday, an English scientist, demonstrates his electrical transformer to the Royal Institution in London. The transformer was the first to use the principle of electromagnetic induction, in which electricity is produced by moving an electrical conductor through a magnetic field.

1835: John Batman buys 600,000 acres of land from Australian Aborigines. In exchange for the land, Batman gave the elders blankets, tomahawks, knives, flour, and clothing. He initially named the new settlement "Batmania" but later changed the name to Melbourne after the British prime minister. It eventually grew to become Australia's second-largest city.

1885: Gottlieb Daimler patents an internal combustion motorcycle. Dubbed the *Reitwagen*, or "riding car," it is widely considered the first true motorcycle ever invented. Two-wheeled vehicles that used steam engines were invented in the 1860s, but the *Reitwagen* is the first to use a gasoline-powered internal combustion engine. The motorcycle had a single-cylinder four-stroke engine that had an output of 0.5 horsepower with 600 RPM. It had a top speed of about seven miles per hour. Daimler's son took the bike on its inaugural ride, traveling about five miles. During the trip, the motorcycle's leather seat caught fire from the heat generated by the engine. Daimler and his business partner Wilhelm Maybach continued to pioneer the development of early automobiles, developing a *Stahlradwagen* ("steel-wheeled car") in 1889.

1949: The Soviet Union tests its first atomic bomb in Kazakhstan. The explosion of the bomb, dubbed *First Lightning*, yielded 22 kilotons of TNT, similar to early American atomic weapons. The test surprised American intelligence, which did not expect the Soviets to enter the nuclear era for another four years. It accelerated the growing divide between the Soviets and the West during the Cold War.

1966: The Beatles perform at Candlestick Park in San Francisco. It was the band's last commercial concert and was the end of a four-year period of near constant touring that included more than 1,400 shows.

1991: The Communist Party's 75-year rule in the Soviet Union comes to an end. The Supreme Soviet, the parliament of the Soviet Union, officially suspended all activities of the Soviet Communist Party.

2005: Hurricane Katrina devastates southern Louisiana. The storm had been gathering strength over the Gulf of Mexico and made landfall as a Category 5 hurricane. It caused catastrophic damage to much of the Southern coast of the United States and overwhelmed New Orleans. Eventually 80% of the city was flooded. There were 1,833 fatalities as a result of the storm.

August 30

1146: European leaders agree to outlaw the use of the crossbow. The ban was agreed to on the pretense that it would prevent future wars, but the weapon continued to be used in battle until the longbow became the superior weapon.

1682: William Penn sails from England for America. Upon his arrival, Penn would found the colony of Pennsylvania, which later became one of the thirteen original States.

1909: Charles Doolittle Walcott discovers fossils in the Burgess Shale. The fossils dated to the Cambrian Period and included imprints of soft parts of animal remains, rather than only bones. Located in the Canadian Rockies in British Columbia, the Shale was the source of thousands of specimens; Walcott himself collected more than 65,000. The diversity of the fossils significantly impacted human understanding of ancient life on Earth.

1918: Fanni Kaplan shoots and wounds Vladimir Lenin. Kaplan was a member of the Socialist Revolutionary Party, which Lenin's Bolsheviks had banned following the October Revolution. The attempt, along with the successful assassination of Moisei Uritsky, a prominent Bolshevik, prompted the Red Terror, in which thousands of political enemies were arrested and executed.

1935: President Franklin D. Roosevelt signs the Revenue Act of 1935. Popularly called the Wealth Tax or the "Soak the Rich" tax, it raised income taxes on the ultra-rich to 75% on their highest earnings.

1967: Thurgood Marshall is confirmed to the Supreme Court of the United States. President Lyndon Johnson nominated Marshall, the first African-American to serve on the Court, saying it was "the right thing to do, the right time to do it, the right man and the right place." Marshall had been appointed to the Second Circuit Court of appeals by President John F. Kennedy in 1961. Prior to that he had been the chief counsel for the National Association for the Advancement of Colored People (NAACP) and had successfully argued cases before the Supreme Court. Marshall served on the Court for 24 years, and compiled a liberal voting record. He upheld the rights of the accused, supported a woman's right to choose, and opposed the death penalty.

1984: The Space Shuttle *Discovery* lifts off on its first voyage. *Discovery* went on to fly 39 times over 27 years of service, flying 149 million miles and completing 5,830 orbits of Earth. Among other notable achievements, it carried the Hubble Space Telescope into orbit in 1990.

1992: An 11-day standoff at Ruby Ridge, Idaho, ends.
Randy Weaver was arrested after surrendering to federal agents on firearms charges. When agents initially attempted to surround Weaver's remote cabin, a shootout left one agent and the Weaver's son dead. During the siege, an FBI sniper killed Weaver's wife, Vicki. The event, along with the 1993 siege at Waco, galvanized radical right-wing extremists, including Timothy McVeigh, who would later bomb the Federal Building in Oklahoma city in revenge.

August 31

1798: Irish rebels establish the Republic of Connacht with French assistance. The Republic was short-lived. Following a victory by the rebels at the Battle of Castlebar a few days prior, French general Jean Joseph Humbert declared John Moore as President of the new Republic. At the Battle of Ballinamuck, the British defeated the rebels and captured Moore, who was arrested and exiled. He died later that year.

1864: William Tecumseh Sherman's Union forces attack Atlanta. Sherman was able to draw Confederate forces under General John Hood out of their defensive fortifications and into a battle north of the city. After two days of fighting, the Confederates were defeated, and Hood ordered his troops to evacuate the city. Sherman's troops occupied Atlanta on September 2.

1895: Count Ferdinand von Zeppelin patents a "navigable balloon." The German count had originally hit upon the idea for a rigid airship in the 1870s and developed the idea in the ensuing years. Soon the word "zeppelin" became associated with all rigid airships, which were widely used until the passenger airship the *Hindenburg* exploded in 1937, killing 36 people.

1949: The Greek Civil War ends. The Democratic Army of Greece, the military wing of the Greek Communist Party, was defeated after three years by government forces that were supported by the United Kingdom and United States. The civil war is considered the first proxy war of the Cold War, in which the West and the Soviets supported antagonists in a local conflict.

1962: Trinidad and Tobago gains its independence. The Caribbean island nation had been a Spanish colony since Columbus arrived in the Americas, and was under British control beginning in 1802. As the British Empire's power waned in the 20th century, Trinidad and Tobago gained its independence, although it remained a member of the Commonwealth.

1985: Serial killer Richard Ramirez is captured. Ramirez, dubbed the "night stalker" by the new media, went on a crime spree that terrorized Los Angeles residents for over a year. Ramirez murdered 14 victims between June 1984 and August 1985. After police found a fingerprint belonging to him at a crime scene, they released a mug shot to the media. Ramirez was then identified and subdued by bystanders in East LA who saw his picture in the newspapers. He died in prison in 2013.

1997: Diana, Princess of Wales, is killed in a car crash in Paris. Her companion Dodi Fayed, and the car's driver, Henri Paul, also died in the collision. Diana's bodyguard survived the crash. The media initially blamed paparazzi who were following Diana's car for the high-speed collision, but a French investigation two years later determined that her driver was responsible. Paul had been intoxicated at the time of the crash. Diana's death sparked an outpouring of grief around the world. Thousands of flower bouquets were left outside of Kensington Palace in tribute to her, and her funeral was watched by about 2.5 billion people.

2016: Dilma Rousseff, the President of Brazil, is impeached and removed from office. Rousseff was found guilty of administrative misconduct in relation to a scandal at the national petroleum company, Petrobras, of which she was formerly on the board of directors.

Chapter 9

September

September 1

1715: King Louis XIV dies. Louis had reigned as King of France for 72 years, making him the longest-serving European monarch in history.

1773: Phillis Wheatley publishes *Poems on Various Subjects.* It was the first volume of poetry published by an African-American writer. Wheatley, who was enslaved at the time of the book's publication, was set free soon after.

1873: Cetshwayo kaMpande ascends the throne of the Zulu Kingdom. In 1879, Cetshwayo led his people in defending their territory against the British. Cetshwayo's Impi troops were initially victorious, defeating them in the Battle of Isandlwana. The British won the war after less than a year, though, forcing the annexation of much of the Zulu territory.

1878: Emma Nutt joins the Boston Telephone Dispatch Company. Recruited to the job by Alexander Graham Bell, Nutt was the first female telephone operator in history. Before then, the company had employed teenaged boys in the job. After a few hours on the job, the company hired Emma's sister Stella, the second female operator in history.

1897: The first subway in North America opens in Boston. The Tremont Street Subway initially served just five stations, but the line was later expanded. The original tunnel later became part of Boston's Green Line system and is still in use.

1914: The passenger pigeon, *Ectopistes migratorius*, goes extinct. When Europeans first arrived in North America, passenger pigeons were so numerous that their migrating flocks would darken the sky for days at a time. As demand for their meat grew in the 19th century, hunting caused their decline over several decades until the last known wild one was shot in 1901. The last surviving passenger pigeon, named Martha, died in captivity at the Cincinnati Zoo. After the pigeon died, its body was frozen in a 300-pound block of ice at the Cincinnati Ice Company and sent to the Smithsonian Institution's National Museum of Natural History to be photographed and mounted. Martha and her species have become a symbol of the dangers posed by human-caused extinction.

1923: The Great Kanto earthquake destroys Yokohama and much of Tokyo, Japan. The quake killed at least 105,000 people and destroying more than 2.5 million buildings.

1959: Elizabeth Taylor signs a contract to star in *Cleopatra*. The English-born actress was offered $1 million for the role by 20th Century Fox, and its total budget of $31 million made it the most expensive film to date.

1969: Muammar Gaddafi seizes power in Libya. The 27-year-old Army officer seized power in a bloodless coup against King Idris and abolished the monarchy. During his four decades in power, Gaddafi championed African socialism and supported anti-Western terrorist attacks. He was overthrown in 2011.

1983: Soviet fighters shoot down Korean Airlines Flight 007. The passenger jet had strayed into Russian airspace. All 269 passengers and crew, including US Congressman Larry McDonald, were killed.

1985: The wreck of the HMS *Titanic* is found. A joint British-French expedition located the wreckage on the floor of the North Atlantic 73 years after it struck and iceberg and sank off the coast of Newfoundland.

September 2

1666: The Great Fire of London begins. The fire started as a blaze in the house of a baker in Pudding Lane, and eventually spread across the city. It burned for four days and completely devastated London's medieval old city, destroying approximately 13,000 homes, 90 churches including St. Paul's Cathedral, and hundreds of other buildings along the north bank of the River Thames. The fire mostly was contained in the walls of the old city, although it spread outside of the wall along Fleet Street to the west. Despite the intensity of the conflagration and extent of its destruction, fewer than 16 people were confirmed to have died in the blaze. Sir Christopher Wren, an English architect and scientist, was commissioned by King Charles to rebuild more than 50 churches for London including St. Paul's. The cathedral was completed on Christmas Day 1711.

1789: Congress establishes the United States Department of the Treasury. George Washington appointed Alexander Hamilton the first Secretary of the Treasury, and he was sworn in on September 11.

1885: Rock Springs massacre occurs. White miners in Rock Springs, Wyoming, attacked the city's Chinatown district during a labor dispute. The rioters killed 28 Chinese miners and injured 15 more, burned dozens of homes, and drove hundreds of Chinese miners out of the city. Federal troops had to be deployed to Rock Springs to put down the riot, but copycat violence against Chinese workers soon spread as a result of it.

1901: Speak softly and carry a big stick. Then-Vice President Theodore Roosevelt uttered the famous phrase in a speech delivered at the Minnesota State Fair. Roosevelt was describing his policy of negotiating peacefully but maintaining a strong military. Roosevelt would go on to become President less than a week later after President McKinley was shot, and the foreign policy became a hallmark of his administration.

1939: Nazi Germany annexes the Free City of Danzig. Now Gdansk, Poland, Danzig was a semiautonomous state that was protected by the League of Nations. Local Nazis took over the city government in 1933, and Germany abolished the Free City upon invading Poland. Thousands of the city's defenders were executed upon surrendering to the Nazis.

1945: Japan formally surrenders to the Allies in World War II. Japanese Foreign Minister Mamoru Shigemitsu signed the Japanese Instrument of Surrender aboard the USS *Missouri* in Tokyo Bay. Admiral Chester Nimitz accepted the surrender on behalf of the United States.

1963: *CBS Evening News* switches to a half-hour format. The station's flagship news program was the first to use the half-hour broadcast structure. It was anchored by Walter Cronkite until 1981.

1969: Ho Chi Minh dies. The Vietnamese revolutionary led an independence movement against French colonizers beginning in 1941, and was Chairman of the Workers' Party of Vietnam from 1951 until his death. Following the Communist victory in the Vietnam War in 1975, Saigon was renamed Ho Chi Minh City in his honor.

1993: The United States and Russia announce plans for the International Space Station. The Station's first components were launched in 1998 and it was completed in 2011.

September 3

301: San Marino is founded. Located on the Italian Peninsula, the 24-square-mile nation is one of the world's smallest. It is also the world's oldest sovereign state still in existence and the oldest constitutional republic.

1189: Richard I is crowned King of England at Westminster. The "Lionheart" king ruled until his death in 1199. He spent much of his reign leading Crusaders in their conquest of Egypt and Jerusalem. Captured in 1192 by the Duke of Austria, he was imprisoned until 1194. He was shot by a crossbowman while besieging a rebel stronghold in France and died of gangrene. His heart was buried at Rouen Cathedral, and his body at Fontevraud Abbey in Anjou.

1658: Oliver Cromwell dies. Cromwell had risen in the ranks of the New Model Army during the English Civil War, and his defeats of Royalist Armies in several key battles gained him political influence following the war. He co-signed King Charles I's death warrant after his trial and became a member of the "Rump Parliament" that governed England after Charles' execution. In 1649 Cromwell led a brutal invasion of Ireland to put down an alliance between Irish Catholics and English royalists. Cromwell oversaw massacres of the civilian populations of the Irish cities of Drogheda and Wexford during the invasion, which consolidated English rule of Ireland. In 1653 Cromwell dissolved the Rump Parliament by force and had himself appointed Lord Protector of England, a post he held until his death. He was succeeded by his son Richard, but Richard was unable to hold power and Charles II restored the monarchy in 1660. Cromwell's body was exhumed from its tomb in Westminster Abbey, hanged, and beheaded.

1777: At the Battle of Cooch's Bridge, Maryland, Continental Army troops display the American flag for the first time. The flag featured 13 red and white stripes and 13 stars in a circle on a blue background.

1861: Confederate troops invade Kentucky. The state was neutral in the Civil War at the time, but by leading his soldiers into the state, General Leonidas Polk prompted the State Legislature to request federal troops. As a result, the Union Army gained vital strategic positions and Kentucky remained in the Union.

1895: The first professional American football game takes place. The Latrobe Athletic Association paid David Berry

$10 to play for them in a game against the Jeanette Athletic Association. Latrobe won the game, 12–0.

1935: Sir Malcolm Campbell breaks the 300-mph land speed barrier. Campbell, an English auto racer, reached a speed of 304.331 mph on the Bonneville Salt Flats in Utah in a Rolls-Royce Railton named the *Blue Bird*.

1943: The Allied invasion of Italy begins. The Kingdom of Italy signed an armistice with the Allies in Sicily, which the Allies had just finished occupying. Germany retaliated by attacking Italian forces in Italy and France and occupying much of the Italian Peninsula.

2017: North Korea conducts a nuclear test. It was the sixth test of a nuclear weapon by the nation, and set off a 6.3-magnitude earthquake in the process.

September 4

1781: Felipe de Neva and 44 Spanish colonizers establish Los Angeles. The city's original name was "El Pueblo de Nuestra Señora La Reina de los Ángeles de Porciúncula," or "The Village of Our Lady, the Queen of the Angels of Porziuncola." It would go on to become the most populous city in California and the second-most populous in the United States with a population of nearly 4 million.

1862: Robert E. Lee and the Army of Northern Virginia invade Maryland. The Union Army of the Potomac, under General George B. McClellan, intercepted Lee's Army at the Battle of Antietam and repulsed the invasion. Lee would try again in 1863, invading Pennsylvania before being turned back at the Battle of Gettysburg.

1888: George Eastman trademarks *Kodak*. On the same day, Eastman also patented the first camera to use roll film. The Kodak 1888 Black Camera and roll film helped popularize photography and make it a mainstream hobby that was accessible to amateur photographers. Kodak went on to become a giant of the photography business. By 1976, the company was responsible for 85% of camera sales and 90% of film sales in the United States.

1928: George Wood sets the first water speed record. Wood piloted the speedboat *Miss America VI* down the Detroit River, reaching a speed of 92.838 mph.

1941: The USS *Greer* is the first US ship to fire on a German ship during World War II. When the German U-boat *U-652* fired on the *Greer*, it returned fire and drove the submarine away. President Franklin D. Roosevelt condemned the attack as an "act of piracy" and ordered American ships to shoot Nazi subs on sight.

1957: Arkansas governor Orval Faubus orders the National Guard to prevent African-American students from entering Central High School in Little Rock. In response, President Dwight D. Eisenhower sent the 101st Airborne Division of the US Army to Little Rock and federalized the Arkansas National Guard. By the end of the month, the students were allowed to attend the school under heavy guard.

1970: Salvador Allende is elected President of Chile. As a Marxist, Allende nationalized industry and collectivized agriculture. The US Central Intelligence Agency helped General Augusto Pinochet overthrow Allende in 1973.

1972: US swimmer Mark Spitz wins a seventh gold medal at the 1972 Summer Olympics in Munich, Germany. Spitz's record for Olympic gold medals stood until Michael Phelps won eight in the 2008 Summer Olympics in Beijing.

1998: Larry Page and Sergey Brin found Google. At the time, the founders were PhD candidates at Stanford University. They built the search engine as part of a research project. It would go on to be one of the leading internet-technology companies in the world, with more than 80,000 employees and a market share of over 90% by 2018.

September 5

1774: The First Continental Congress convenes. Notable attendees included George Washington, John Adams, Patrick Henry, and John Jay. The Congress assembled in Philadelphia to draft a response to the British Parliament's passage of the Intolerable Acts. The Acts gave British military officers immunity from prosecution, forced American colonists to house British troops in their homes, and established martial law in Massachusetts. One faction, led by conservatives such as Jay, argued the Congress should put pressure on Parliament; the other, led by Patrick Henry, argued the colonies should assert their liberty. After debating for weeks, the Congress decided upon a boycott of British goods beginning on December 1 and a list of grievances that were sent to King George III. The Congress also decreed that a Second Continental Congress would meet in May of the following year. By April 1775, the colonies would be in open rebellion in the breakout of the American Revolution.

1781: The Battle of the Chesapeake. Also known as the Battle of Virginia Capes, it proved to be a decisive confrontation in the American Revolution. The French Navy defeated the British Navy, stranding British General Cornwallis and his troops on the Yorktown Peninsula in Virginia. Cornwallis would surrender in October following a month-long siege, ending the War.

1793: The Reign of Terror begins in France. Bertrand Berére declared at the French National Convention "Let us make terror the order of the day!" Maximilien Robespierre headed the infamous Committee of Public Safety, which oversaw the arrest of hundreds of thousands of political and class enemies, and the executions of more than 16,000 at the guillotine. Robespierre himself would fall victim to the guillotine after he was denounced as a tyrant and arrested by his comrades-in-arms in 1794.

1877: Crazy Horse is killed by US soldiers. The Oglala Sioux chief who helped defeat Custer's troops at the Battle of Little Bighorn had surrendered at Fort Robinson, Nebraska. When a guard attempted to force him into a prison cell, Crazy Horse resisted, and the solider bayoneted him to death.

1882: Fridtjof Nansen reaches the top of Greenland's ice cap. Nansen was leading an expedition to the Arctic, and after summiting, he and five other skiers began making their way to Greenland's west coast.

1914: The Battle of the Marne begins. French forces counterattacked German troops that were advancing on Paris during World War I. In a six-day battle, the French repulsed the German push to the capital, inflicting more than 100,000 casualties.

1945: Igor Gouzenko defects to Canada. Gouzenko was a clerk at the Soviet embassy, and he provided Canadian authorities with documents that revealed Soviet espionage in North America. The "Gouzenko Affair" is considered the start of the Cold War.

1960: Muhammad Ali, boxing as Cassius Clay, wins gold at the Rome Olympics in the light heavyweight boxing competition.

1972: Black September terrorists take 11 Israeli athletes hostage at the Munich Olympics. The terrorists killed two athletes during the initial attack. The other nine, as well as the terrorists, were later killed in a failed rescue attempt.

September 6

1522: Ferdinand Magellan's sailors complete the first round-the-world journey. Magellan had set out three years before with a fleet of five ships in order to circumnavigate the globe. Of the 270 men who set out with him, only 22 returned, with just one ship. Magellan was killed in the Philippines in April 1521.

1861: Ulysses S. Grant leads Union troops in the capture of Paducah, Kentucky. Grant's troops seized the strategically vital river city without firing a shot, ensuring Union control of the mouth of the Tennessee River.

1879: Great Britain's first telephone exchange opens in London, on Lombard Street.

1901: Leon Czolgosz shoots US President William McKinley. Czolgosz, an unemployed steelworker from Michigan, became interested in anarchism after seeing a speech by Emma Goldman earlier that year. In 1900, when the Italian anarchist Gaetano Bresci assassinated King Umberto I, Czolgosz heard news of the attack and decided to act. He attended the Pan-American Exposition in Buffalo, New York, armed with a .32 caliber revolver. He stood in line to meet the President, and when McKinley extended his arm to shake Czolgosz's hand, the anarchist fired twice at pointblank range. McKinley died of his wounds on September 13, and Vice President Theodore Roosevelt was sworn in as President the following day. Roosevelt would serve until 1909. Czolgosz was convicted of murder and sentenced to death in a speedy trial. He was then executed on October 29.

1943: At age 16, Carl Scheib pitches for the Philadelphia Athletics, becoming the youngest baseball player to appear in an American League game. Scheib came in as a reliever in the ninth inning of the game, which the Athletics lost to the New York Yankees, 11–4. He pitched five more games that year as a reliever.

1970: The Dawson's Field hijacking begins. Members of the Popular Front for the Liberation of Palestine seized five aircraft, forcing three to land at Dawson's Field in Jordan. The hijackers demanded the release of Palestinian prisoners. The incident ended with the death of one hijacker and the arrest of several others.

1975: Martina Navratilova, a Czech tennis star, requests political asylum in New York. She became a US citizen in 1981. During her tennis career she won 18 Grand Slam singles titles, including nine at Wimbledon, and 31 doubles titles.

1991: Leningrad's name is changed back to St. Petersburg. It had been renamed in 1924 after the Bolshevik Revolution swept communists into power.

1997: Diana, Princess of Wales, is laid to rest in London. More than a million people lined the funeral procession's route, and 2.5 billion more watched on television around the world.

2007: Israeli fighter jets bomb a suspected nuclear reactor in Syria. Israeli intelligence agencies identified the site as a military nuclear reactor, but did not acknowledge the attack had occurred until 2018.

September 7

1695: The English Parliament passes a series of draconian laws against Irish Catholics. The legislation prohibited them from teaching, sending their children overseas for education, owning firearms, and owning horses worth more than five pounds.

1776: The first use of a submarine in warfare. During the American Revolution, a submersible craft powered by its pilot crept along the side of a British warship in New York Harbor. The submariner, Ezra Lee, steered his craft, named the USS *Turtle*, alongside the ship, where he attempted to plant a time bomb, but was unsuccessful.

1812: Napoleon's army wins the Battle of Borodino. The battle was the bloodiest in the Napoleonic Wars, involving a quarter of a million troops and resulting in more than 70,000 casualties. The French victory over the Russian army opened the way to capturing Moscow, which was only 70 miles to the east. Napoleon's forces approached the city the following week, but Russian civilians looted and set fires rather than turn its food stores over to the French invaders. The fires soon spread, and more than 80 percent of the city was destroyed. The Russian army continued its retreat to the east. By November, the French Army was too deep in Russian territory to maintain supply lines, and not prepared for the harsh winter. Napoleon was forced to abandon his campaign and retreat from Russia.

1860: Giuseppe Garibaldi's forces capture Naples. Garibaldi, an Italian nationalist and general, led a military campaign to unite the several states on the Italian Peninsula into a single kingdom. With the capture of Naples, he declared himself the "Dictator of the Two Sicilies."

1864: Union General William T. Sherman orders the evacuation of Atlanta. Sherman's army had captured the city five days prior. In preparation for his march to the sea, Sherman ordered the burning of all government and military buildings in the city, although the fire spread to private businesses and houses as well.

1876: The James-Younger gang robs the First National Bank of Northfield, Minnesota. The robbers were unprepared for the wrath of the townspeople, who soon surrounded the Bank, intent on protecting their savings. Although Jesse and Frank James escaped, the rest of the gang was nearly wiped out in the ensuing gun battle.

1921: The First Miss America Pageant is held. The two-day competition took place in Atlantic City, New Jersey. Margaret Gorman, of Washington, D.C., was crowned the first Miss America.

1936: The last existing thylacine dies in a zoo in Tasmania. Commonly called the Tasmanian Tiger, the thylacine, a carnivorous marsupial, was driven to extinction by intensive hunting, the introduction of dogs to its environment, and disease.

1986: Bishop Desmond Tutu becomes the Archbishop of Cape Town, South Africa. Tutu, who had won the Nobel

Peace Prize in 1984 for his nonviolent opposition to apartheid, was the first Black clergyman to head the Anglican Church in South Africa.

2008: The US government takes control of Fannie Mae and Freddie Mac. The two mortgage financing companies lost billions in the subprime mortgage crisis.

September 8

1504: David is unveiled. The statue, by Italian sculptor Michelangelo, was put on display in the Piazza della Signoria in Florence. It took the master sculptor three years to carve the statue from marble.

1565: An Ottoman army withdraws from Malta. The army had besieged the city for four months, and withdrew after the Knights Hospitaliers defended it heroically in the face of overwhelming odds.

1760: Montreal surrenders to the British. With the surrender of the city during the French and Indian War, the French abandoned all claims to the Canadian territories.

1810: The *Tonquin* departs from New York. The merchant ship belonged to the Pacific Fur Company and was bound for the coast of Oregon. It was the first American trade vessel to sail to the Pacific Northwest, and the company hoped to establish a fur trading base of operations there.

1883: The Northern Pacific rail line is completed. It was the second transcontinental railroad to be completed in the United States. Former president Ulysses S. Grant drove the final golden spike home to mark the completion.

1892: The Pledge of Allegiance is recited for the first time.
The Pledge was composed by George Thatcher Balch, a school-teacher and former Captain in the US Army.

1944: Nazi Germany fires the first V-2 missile. Aimed at Paris, it was the world's first guided ballistic missile and damaged buildings but resulted in no casualties. Germany would launch more than 3,000 over the next year at targets in Europe and the United Kingdom, killing 2,754 civilians in London alone.

1988: For the first time, Yellowstone National Park is closed.
Multiple small wildfires in the Park had spread out of control and combined into a single huge conflagration, forcing the Park's closure. Ultimately the wildfires had a positive effect on much of the Park's plant life, as it had evolved over millennia to respond well to such disasters.

1998: Mark McGwire hits his 62nd home run of the season.
In doing so, the right-handed slugger broke Roger Marin's single-season record of 61 homers. McGwire had hit 52 home runs for the Oakland Athletics in 1997, and 58 in 1997. McGwire raced Chicago Cub Sammy Sosa to set the single-season record in 1998, and he tied the record in a game against the Cubs in St. Louis on September 7. The following night, facing the Cubs yet again, McGwire hit a 341-foot home run to break Maris' record. Sosa finished the season with 66 homers, and McGwire with 70. Both players later had their records tainted by allegations that they had used performance-enhancing drugs during the 1998 season. McGwire later testified before Congress on the subject, but refused to confirm or deny the charges. In 2010, however, he admitted in an interview with Bob Costas that he had used steroids during the home run race.

2004: Genesis, an autonomous spacecraft designed to collect samples from the solar wind and return to Earth, crash-lands when its parachute fails to deploy.

September 9

1000: The Battle of Svolder is fought in the Baltic Sea.
Norwegian King Olaf Tryggvason's fleet was sailing home after a raid on Northern Europe when they were attacked by an armada of Swedish and Danish ships. Olaf's fleet comprised just 11 warships, while the attackers had more than 70. The alliance ships sank or captured each of Olaf's until the King's was the only one remaining. Olaf threw himself into the sea rather than be captured. The Kingdom of Norway was divided among the Swedish and Danish victors.

1499: Vasco da Gama returns to Lisbon. The Portuguese navigator had just completed his first voyage to India, which opened up a sea route to the East by going around the Southern Cape of Africa and launched a new era of exploration.

1739: Cato's Rebellion begins. It was the largest slave uprising in North America prior to the American Revolution. An enslaved man named Jemmy, often called Cato after the family that held him, organized about 60 other enslaved people in South Carolina in the months leading up to the rebellion. They armed themselves and set out for Florida, which was under Spanish rule at the time. The Spanish had promised freedom to anyone who escaped from slavery in British territory. Cato's band killed several whites before the South Carolina militia intercepted them. Over the next week, the group was pursued by and fought pitched battles with the militia before they were killed or captured. As a result of the uprising, the South Carolina legislature passed the Negro Act of 1740, which prohibited enslaved Africans from meeting, learning to write, or earning money. It also allowed owners to kill rebellious slaves.

1791: The capital of the United States is named Washington after the nation's first President. The three commissioners who chose the name also decided to call the federal district where it is located Columbia, after the poetic name for the Americas.

1839: John Herschel produces the first glass plate photograph. Herschel was an English mathematician, astronomer, chemist, and inventor. His photograph, of a 40-foot telescope, used a new chemical process to take the image on a plate of glass. Hershel was the first to use the term "photography" to describe the process.

1901: The first long-distance car race begins. Over five days, drivers raced from New York City to Buffalo, New York, a distance of 464 miles. The winning car posted an average speed of 15 miles per hour.

1919: The Boston Police go on strike. The police were seeking recognition of their union and better wages and working conditions. During the strike, the city descended into lawlessness in the first few nights until then-governor Calvin Coolidge sent in the state militia to restore order.

1956: Elvis Presley makes his first appearance on *The Ed Sullivan Show*. The singer performed "Don't Be Cruel" and "Hound Dog," and became a household name overnight.

1971: A prison riot breaks out in Attica. Hundreds of inmates took guards hostage and sent demands for better conditions to Governor Rockefeller. The Governor responded by sending in State Police, who opened fire indiscriminately, killing prisoners and guards alike.

September 10

1509: An earthquake hits Constantinople. The 7.2 magnitude earthquake occurred along a fault off the coast of the city, devastating buildings and causing a tsunami in the Marmara Sea. More than 10,000 people died and over a thousand homes, as well as 109 mosques, were destroyed. Aftershocks continued for over a month after the quake, which became known as "The Lesser Judgement Day."

1608: John Smith is elected council president of Jamestown. Smith, an English explorer, had managed the town's efforts to survive the harsh first year.

1776: Nathan Hale volunteers to spy for the American Revolution. Hale, a 21-year-old schoolteacher, had joined a militia in Connecticut at the outbreak of the war, and later accepted a commission as a first lieutenant. In the Fall of 1776, the Continental Army was occupying Manhattan, and General George Washington needed to gather intelligence about where the British intended to attack the island. Hale was the only person to volunteer for the mission. He traveled behind enemy lines on September 12, and while he was undercover, the British captured New York City. Hale continued his mission, but Robert Rogers, a Loyalist double-agent, recognized Hale at a tavern and approached him, pretending to be a spy for the Continentals as well. Hale fell for the ruse and was captured. On September 22, Hale was executed by hanging. At the gallows, he famously said "I only regret that I have but one life to lose for my country."

1813: Oliver Hazard Perry's flotilla wins the Battle of Lake Erie. Perry's nine ships defeated a British naval group of six warships and forced the British to evacuate Detroit and give up control of the American Northwest. It was the first unqualified victory of an American force over a squadron of British naval vessels in history.

1846: Elias Howe patents the hand-cranked sewing machine. Howe's machine was the first to use a needle with an eye at its point, a shuttle, and an automatic feeder mechanism. He donated most of the money he earned from his invention to the 17th Connecticut Volunteer Infantry during the American Civil War.

1861: Union troops drive Confederates from the bluffs overlooking Carnifex Ferry, seizing control of the Kanawha River Valley in Virginia. Two future US presidents, Rutherford B. Hayes and William McKinley, were among the combatants.

1913: The Lincoln Highway opens. It was the first paved highway in the United States that went all the way from the East Coast to the West Coast.

1960: Abebe Bikila wins gold at the Rome Olympics. Bikila, an Ethiopian long-distance runner, placed first in the marathon event, running it in his bare feet because his new shoes gave him blisters. He was the first person from sub-Saharan Africa to win Olympic gold.

1990: Pope John Paul II consecrates the basilica of Our Lady of Peace. The basilica, in Yamoussoukro, Cote d'Ivoire, is the largest Christian church in the world.

2008: The Large Hadron Collider begins operations at CERN. The particle collider was built over a ten-year period and involved more than 10,000 scientists from around the world. It is used for particle physics experiments, and has performed such tasks as measuring the Higgs boson.

September 11

1609: Henry Hudson arrives at Manhattan Island and meets its inhabitants. The next day he began sailing up the river that now bears his name.

1777: The British are victorious in the Battle of Brandywine. General George Washington threw the Continental Army against the British near Wilmington, Delaware, during the American Revolution. A British army led by Generals Howe and Cornwallis repulsed the attack and nearly destroyed Washington's army.

1847: "Oh! Susanna" is performed for the first time in a tavern in Pittsburgh. The song, written by American composer Stephen Foster, became wildly popular and catapulted him to fame.

1857: Mormon settlers massacre 120 pioneers in Mountain Meadows, Utah. The pioneers' wagon train was headed to California when they camped at the site to rest. The attackers besieged the pioneers for five days. When they surrendered, the attackers killed them all.

1903: The Milwaukee Mile opens at the Wisconsin State Fair Park. It is the oldest motor speedway in the world.

1945: Former Japanese Prime Minister Hideki Tojo attempts suicide following the surrender to the Allies. Tojo survived the attempt, but was later tried for war crimes and hanged.

1973: Augusto Pinochet overthrows Salvador Allende in Chile. Pinochet, backed by the Central Intelligence Agency, seized power after the democratically elected Marxist instituted a series of reforms. He ruled as dictator, engaging in a ruthless "dirty" war against dissidents, until a referendum in 1988.

1985: Pete Rose breaks the record for most career hits in baseball. With his 4,192nd hit, Rose surpassed the record that Ty Cobb had set in 1928.

1997: The Mars Global Surveyor reaches the Red Planet. The NASA orbiter spent the next four years mapping the planet's atmosphere and surface. It continued operating until November 2006, when it failed to respond to commands.

2001: The 9/11 attacks. Terrorists affiliated with Al-Qaeda hijacked four planes in the United States, crashing two into the north and south towers of the World Trade Center, a third into the Pentagon, and a fourth in a field in Pennsylvania. Within two hours of the planes striking the towers, both collapsed due to the catastrophic damage. Part of the Pentagon was also destroyed. The attacks killed 2,996 people and injured more than 6,000 more. Al-Qaeda's leader, Osama bin Laden, took credit for the attacks and said they were in retaliation for the US stationing military troops in Saudi Arabia, where Mecca and Medina are located, as well as US support for Israel. As a result of the attacks, the United States declared a War on Terror, and invaded Afghanistan, where Al-Qaeda had been granted safe haven by the Taliban government. Bin Laden was killed by a US Special Forces raid in May of 2011.

September 12

490 BC: The Battle of Marathon (traditional). A Persian force of more than 25,000 attacked the city-state of Athens, landing near Marathon. The Athenians, numbering only about 10,000, routed the Persian army after the Athenian general Miltiades trapped them in a pincer move. More than 6,000 Persians were killed, while the Athenians lost less than 3,000 men. The defeat ended the first Persian invasion of Greece. According to legend, a runner named Pheidippides was sent to Athens to announce the victory to the king. Upon arriving, the exhausted runner could only say the words "Joy, we won," before collapsing and dying. The account later inspired the creation of the 26.2-mile race called the marathon, which was one of the events in the original modern Olympics, held in 1896.

1772: A French court in Aix finds the Marquis de Sade and his servant Latour guilty of sexual abuse. As the pair were tried in absentia, the court ordered effigies of them to be executed.

1846: English poet Elizabeth Barrett elopes with Robert Browning. Barrett's father disapproved of Browning during their courtship, so the couple absconded to Italy, where they lived until her death fifteen years later.

1890: Colonizers with the British South Africa Company arrive in Salisbury, Rhodesia. The new arrivals began staking prospecting claims almost as soon as they landed. They did not ask permission of the Ndebele residents who were already there.

1919: Adolf Hitler joins the German Workers' Party. The organization would go on to become the Nazi Party.

1935: Howard Hughes sets an airspeed record in a plane of his own design. Hughes reached a speed of 352.46 miles per hour during the flight.

1940: Prehistoric paintings are discovered in a cave in Lascaux, France. An 18-year-old found the cave entrance and returned with friends later that day. The paintings depict horses, deer, and aurochs, an extinct species of bulls. There are over 6,000 figures painted on the cave walls, and the paintings have been estimated to be about 17,000 years old.

1943: German forces rescue Benito Mussolini from prison during World War II. The Germans set Mussolini up as a puppet ruler of Italy.

1959: *Bonanza* airs for the first time. The series was the first regularly scheduled TV show in color. It ran on NBC until 1973.

1974: A military coup overthrows Emperor Haile Selassie of Ethiopia. He had ruled for 58 years. A group of officers and enlisted men in the army called The Derg seized power and had Selassie imprisoned. He died the following year.

1983: Puerto Rican nationalists rob a Wells Fargo bank in Connecticut. Called *Los Macheteros*, the group escaped with $7 million. The money was spirited out of the United States and used to fund the group's later operations.

September 13

1577: Matteo Ricci arrives on the west coast of India. The Italian Jesuit missionary was originally headed to China, but stayed in India for 30 years, where he introduced Christianity to locals.

1788: The Philadelphia Convention establishes New York City as the temporary capital of the United States. The Convention also set the date for the first presidential election, which took place from December 15, 1788 to January 10, 1789.

1848: Phineas Gage, a railroad worker, has an iron rod accidentally driven through his head. Gage survived the incident, but it destroyed most of his frontal lobe. As a result his personality and behavior were changed profoundly, indicating the brain's role in both.

1814: Francis Scott Key pens "The Star-Spangled Banner." Originally titled "Defense of Fort McHenry," the poem described the bombardment of the fort by the British fleet in Baltimore Harbor during the War of 1812. Key was a captive on the British Navy's HMS *Minden* during the battle. He watched while the British ships pounded the fortress with shells and rocket fire, and saw that the fort's small "storm flag" was flying during the rainy night, illuminated by the explosions. After the barrage had finished, Key would not know until the next morning what the outcome of the battle was. That morning, the defenders raised the fort's larger American flag, and Key was inspired by the sight. He wrote the poem on the back of a letter, and later it was published in a national magazine. In the 19th century it was set to music, and officially adopted by the US Navy. In 1931, the song was officially adopted as the national anthem.

1826: For the first time, a rhinoceros is exhibited in the United States, in New York City.

1847: Troops under General Winfield Scott storm Chapultepec Castle during the Battle of Chapultepec in the Mexican-American War. The Castle was defended by just six teenaged cadets, who died in the fighting. With the capture of the city, the last significant Mexican resistance was defeated.

1862: While resting in a meadow, Union soldiers find a copy of General Robert E. Lee's battle plans. The intelligence was passed up the chain of command to General George McClellan, who failed to act on it. The discovery foreshadowed the upcoming Battle of Antietam.

1899: Henry Bliss is killed in an automobile accident. He was the first passenger in a car to die in a motor vehicle accident in the Western Hemisphere.

1922: The highest ambient temperature ever measured in the shade, 136.4 degrees Fahrenheit, is recorded in a Libyan village south of Tripoli.

1953: Nikita Khrushchev assumes power in the USSR. Khrushchev, the first General Secretary of the Communist Party after Stalin's death, would go on to disavow the former leader's purges in a "secret speech" delivered to the Party Congress in 1956.

1985: Super Mario Bros. is released in Japan for the Nintendo Entertainment System. The game was a successor to the 1983 arcade game Mario Brothers. It sold more than 40 million copies.

1993: The Oslo Accords. Israeli prime minister Yitzhak Rabin shook hands with Palestinian Liberation Organization leader Yasir Arafat. The accord granted autonomy to part of Palestine, and the PLO acknowledged Israel's right to exist.

September 14

1224: Saint Francis of Assisi has an epiphany. The Italian Catholic friar had been a soldier as a young man but had left the military to pursue a religious calling. He went on a pilgrimage to Rome and would often spend time praying and fasting in solitude. Upon seeing a vision of Jesus, Francis renounced his family inheritance and took a vow of poverty, begging for alms in Assisi and working to rebuild churches in the local countryside. Later in his life, he traveled to Egypt during the Fifth Crusade in an attempt to convert the Sultan to Christianity. He was not successful. Back in Italy, Francis founded a religious order that still exists today. Its followers take a vow of poverty and serve the poor. On this day in 1224, Francis was praying when he reportedly had a vision of an angel and was struck by the stigmata, a condition of wounds on the hands and feet. He died in 1226 and was canonized as a saint two years later.

1752: Great Britain changes from the Julian calendar to the Gregorian one. September 3, 1752 becomes September 14, 1752 in an instant.

1781: The Continental Army, led by General George Washington, arrives in Williamsburg, Virginia. They began laying plans to besiege the British Army, which was trapped on the Yorktown Peninsula. The arrival in Williamsburg marked the beginning of the end of the American Revolution.

1854: A combined force of French, Ottoman, and British troops land near Sevastopol during the Crimean War. The 175,000-man army besieged the Russian city for nearly a year before finally capturing the strategically vital port city. The capture of Sevastopol ended the Crimean War with an Allied victory.

1886: George Anderson patents the typewriter ribbon. Before his invention, typewriters used carbon paper and embossing techniques. The ribbon design standardized the typewriter and remained in use until the advent of the word processor a century later.

1911: Dmitry Bogrov assassinates Russian Prime Minister Pyotr Stolypin. The Prime Minister was attending an opera with the Tsar when Bogrov, a left-wing revolutionary, shot him in the chest.

1960: The Organization of the Petroleum Exporting Companies (OPEC) is founded in Baghdad. The original five founding nations—Iran, Iraq, Kuwait, Saudi Arabia, and Venezuela—were later joined by ten more companies. OPEC was founded to wrest control of global oil companies from multinational companies and establish national control of natural resources.

1994: Baseball is cancelled. A strike that began on August 12 resulted in the cancellation of the remainder of Major League Baseball's regular season, as well as the postseason and World Series. The strike lasted until April 2, 1995, making it the longest work stoppage in MLB history.

2015: Gravitational waves are detected. The Laser Interferometer Gravitational-Wave Observatory (LIGO) discovered the phenomenon, which had been postulated by Albert Einstein.

September 15

1440: Gilles de Rais is arrested. The French knight and lord had fought with Joan of Arc in the war to force the English out of France. But after an accusation by the Bishop of Nantes, it was discovered that the war hero was a serial killer who preyed on children. He was later executed for his crimes.

1616: Joseph Calasanz opens the first free public school in Europe in Frascati, Italy. Calasanz, a Spanish priest, believed education should be extended to the poor. Before then it was exclusively the domain of the wealthy. A Catholic, Calasanz accepted Jewish and Protestant children in his schools without prejudice, which was also uncommon for the day. The school taught Latin, mathematics, and hygiene to its pupils.

1752: The first professional theater production in the American colonies is staged. Lewis Hallam, an English theater director, and his company produced Shakespeare's play *The Merchant of Venice* in Williamsburg, Virginia. The play included a musical accompaniment with a harpsichord.

1789: Congress creates the US Department of State, transferring powers previously held by the Department of Foreign Affairs to it.

1812: The Attack of the Narrows. During the War of 1812, an American garrison under the command of future president Zachary Taylor was besieged at Fort Harrison, Indiana, by a combined force of Native Americans. A supply wagon heading to the fort was ambushed by a group of Potawatomi warriors near modern-day Fairbanks. Only two soldiers survived the attack.

1862: Confederate forces capture Harpers Ferry, Virginia, during the American Civil War. General Stonewall Jackson led a three-day attack on the Union garrison stationed there. They took more than 12,000 Union prisoners. After capturing the garrison, Jackson's troops had to rush to Sharpsburg in advance of the Battle of Antietam.

1916: Tanks are used in war for the first time. At the Battle of the Somme in World War I, 32 British tanks attacked German lines. Nine made it across the no-man's-land between the German and Allied lines. Over the next half-century tank warfare developed into a mainstay of modern combat.

1963: Klansmen bomb the 16th Street Baptist Church. During the Civil Rights Movement, the church in Birmingham, Alabama was an organizing hub for activists. Martin Luther King, Jr. and other civil rights leaders used the church for meetings, and marches including the Children's Crusade of 1963 were staged there. In response, four members of the Ku Klux Klan, a racist terrorist organization, planted a dynamite time-bomb at the church. When it exploded, it blew a seven-foot-wide hole in the church's wall and killed four girls who were preparing to sing in the choir. They were all fourteen years old or younger. Twenty other people were injured in the blast. The bombing was a pivotal moment in a momentous year of the Civil Rights Movement, and helped catalyze support for the passage of the Civil Rights Act of 1964.

1981: The Senate approves Sandra Day O'Connor to become the first female justice of the Supreme Court of the United States.

September 16

1252: William of Rubrouck, a French Franciscan monk, leaves the camp of Batu Khan, leader of the Mongols. Rubrouck's account of his journey to Karakorum, the Mongol capital, would become a classic of travel writing and a major historical document.

1498: Tomás de Torquemada dies. The head of the Spanish Inquisition, Torquemada had been responsible for the torture and execution of thousands of supposed heretics, apostates, and nonconformists, as well as numerous Jewish persons.

1620: The Pilgrims set sail for America aboard the *Mayflower*. The Pilgrims were fundamentalist Puritan Christians who believed they should be separate from the Anglican church. A party of 102 sailed to America. One Pilgrim and one crewman died on the journey. A child, named Oceanus, was also born during the trip. They sighted land on November 9 near Cape Cod.

1776: The Battle of Harlem Heights is fought. While retreating from the British Army, George Washington's men fought a rearguard action to cover their escape. The battle took place in an area that is now between 103rd and 120th streets in Manhattan. The battle was the first success of Washington's command during the Revolutionary War.

1893: The largest land run in American history begins. More than 100,000 people poured into the Cherokee Outlet. The Outlet was a 60-mile by 225-mile strip of land in northwestern Oklahoma that had been reserved for the Cherokee Nation after they were pushed out of other parts of the state by the US Government. Boomers, as settlers who tried to enter Indian Territory in Oklahoma to seize land for homesteading were known, had begun pressing the government to officially open these lands for white settlers. In response, President Benjamin Harrison signed an executive order that prevented Cherokee settlers from leasing their land for grazing. As a result, the Cherokee sold the land to the government. One hundred thousand boomers gathered in Kansas in preparation for the land being opened to claims. At noon on September 16, a cannon boomed to signal the opening of the Outlet, and boomers raced in to stake claims. Many settlers had sneaked in ahead of the boomers to secure the best parcels; they were called "Sooners."

1959: The Xerox 914 is introduced. It was the first photocopier in history, and was shown in a demonstration on live television. The machine could make seven copies per minute, but it tended to catch fire when it overheated. As a result, Xerox provided a fire extinguisher with the copier.

1979: Eight people escape East Germany in a homemade hot air balloon. Peter Strelzyk and Gunter Wetzel, workers in a plastics factory, designed and built the balloon over a two-year period. After two unsuccessful attempts, they finally succeeded in floating over the border at 2 A.M., and turned themselves in to a Bavarian police officer.

1982: The Sabra and Shatila massacre is committed. During the Lebanese Civil War, the Israeli Defense Forces (IDF) ordered right-wing Phalangist militias they were allied with to clear out the Sabra neighborhood and Shatila refugee camp in Beirut. The militias murdered between 760 and 3,500 civilians during the operation.

September 17

1577: Henry III of France signs the Treaty of Bergerac. The treaty was a defining moment in the French Wars of Religion between Catholics and Huguenots, a French Protestant sect. The treaty permitted Huguenots to practice their faith on an extremely limited basis.

1787: Members of the Constitutional Convention sign the newly drafted document. Thirty-eight of the 41 delegates to the convention in Philadelphia signed the Constitution, which would later require ratification by nine of the 13 states.

1849: Harriet Tubman escapes from slavery. Tubman, who was born into slavery in 1822, was horrifically mistreated by slave masters. When she was a young girl, she was badly injured when a slave owner hit her in the head with a heavy metal weight. The injury left her with dizziness, spells of hypersomnia, and visions, which she interpreted as messages from God. After escaping to Philadelphia, Tubman immediately returned to the Maryland plantation where her family was held to free them as well. She ultimately made 13 missions back to the South to free a total of about 70 enslaved people. In 1858, she met abolitionist John Brown and assisted him in planning his raid on the federal garrison at Harpers Ferry. During the Civil War, she served as a scout and spy for the Union Army. She led an assault by Union forces on plantations along the Combahee River, becoming the first woman to lead a detachment of US soldiers. After the war, Tubman was active in the women's suffrage movement. She died in 1913.

1859: Joshua A. Norton declares himself Emperor. Norton, an English-born resident of San Francisco, declared himself "Norton I, Emperor of the United States" and later adopted the title "Protector of Mexico." He became a celebrated figure in San Francisco, and more than 10,000 people attended his funeral in 1880.

1862: The Battle of Antietam. The Army of the Potomac, led by General George B. McClellan, halted the advance of Robert E. Lee's Army of Northern Virginia into Maryland. The battle, fought near the Pennsylvania-Maryland border, was the worst single-day battle in the history of the United States, with more than 22,000 casualties. After the battle, Lee was able to withdraw to Virginia and McClellan, always overly cautious, declined to pursue. President Abraham Lincoln relived McClellan of his command as a result.

1900: Filipino revolutionaries defeat American troops at Mabitac. Filipino General Juan Kauppama Cailles's troops attacked advancing soldiers under Colonel Benjamin Cheatham during the Philippine-American War. Cailles's men were in entrenched positions, and the Americans, numbering about 300, were quickly pinned down in a flooded causeway. The Americans suffered between 40 and 180 casualties before the Filipinos withdrew.

1916: The "Red Baron" wins his first aerial battle during World War I. Manfred von Richthofen would win 80 more before he was shot down in April 1918.

1925: A bus accident seriously injures a young Frida Kahlo. As a result, she left medical school and took up art. Kahlo would go on to become one of the most celebrated artists of the 20th century.

1928: The Okeechobee hurricane strikes Florida. By the time it dissipated, the storm had killed more than 2,500 people in Florida and seriously damaged multiple cities and towns. .

1978: Israel and Egypt sign the Camp David Accords, which laid the foundation for a peace treaty between the two nations.

1983: Vanessa Williams becomes the first African-American Miss America winner.

September 18

1793: George Washington lays the first cornerstone of the US Capitol Building. The initial construction of the building took seven years to complete. During the War of 1812, British troops partially burned the Capitol building. It was later expanded, and its iconic dome added, in 1850, with much of the work being done by enslaved persons. It was expanded again in 1958.

1837: Tiffany and Co. is founded in New York City. Originally called "Tiffany, Young & Ellis" after its three proprietors, it advertised as a "stationary and fancy goods emporium." Tiffany was different from other stores in that it clearly priced its wares to prevent haggling and only accepted cash purchases. The store became famous for its jewelry.

1850: Congress passes the Fugitive Slave Act. The Act was part of the Compromise of 1850 between free states and slave-holding ones. It made it illegal for authorities to not arrest people accused of being runaway slaves, regardless of whether they were on free soil. Anyone helping runaways was subject to a fine and imprisonment. It made the cause of abolition a Northern one, where previously it had been concentrated on freeing slaves in the South. The Act was abandoned after the Emancipation Proclamation made it null.

1851: *The New-York Daily Times* is published for the first time. The newspaper would go on to become *The New York Times* in 1857 after its publishers decided to shorten the name. Called "The Gray Lady," the *Times* was initially sold for a penny.

1895: Booker T. Washington argues for the Atlanta Compromise. In it, he and other African-American leaders agreed that in the South, they would submit to political rule by whites, and Southern whites would guarantee due process and education to African Americans.

1919: Fritz Pollard plays a football game for the Akron Pros. A halfback for Brown University, Pollard was the first African American to play professional football. He led the team to the national championship in 1920. He went on to be the first African-American head coach in the National Football League (NFL), coaching the Pros while retaining his position as running back on the team. He also coached college football for Lincoln University in Pennsylvania. Pollard was removed from the NFL along with all nine other African Americans in the league in 1926. He reacted by organizing a barnstorming all-African-

American football league, and played on the Chicago Black Hawks and the Harlem Brown Bombers. Pollard went on to a career in music production and published New York City's first Black-owned newspaper, the *New York Independent News*, from 1935–1942. He died in 1986 at the age of 92.

1947: The Central Intelligence Agency is established with the signing of the National Security Act. The CIA replaced the Office of Strategic Services. Its goal was to coordinate intelligence efforts during the Cold War and prevent another surprise attack like the one that devastated Pearl Harbor.

1960: Fidel Castro visits New York City. While there, he attended the General Assembly of the United Nations, gave interviews to reporters, and met with Malcolm X and other activists. He also visited with Soviet Premier Nikita Khrushchev to discuss relations between the two allies.

1977: Voyager 1 photographs the Earth and Moon together.

2001: Letters containing anthrax are mailed from Trenton, New Jersey. The letters were sent to journalists and two US Senators. Five people were killed and 17 more became sick. No-one was ever charged in the attacks.

September 19

1356: The Black Prince, Edward II, is victorious in the Battle of Poitiers. Fought during the Hundred Years' War, the battle was significant in that the French King, John II, was captured along with his son and most of the French nobles. The resulting political chaos left France weak and unable to mount a counteroffensive against England. The French Dauphin entered into negotiations with Edward less than four years after the battle.

1676: Nathaniel Bacon's men burn Jamestown. Bacon and his followers in Virginia resented the policies of Governor William Berkeley. Berkeley had been unable to secure the frontier from attacks by Native Americans, and frontiersmen rebelled to oust him from power. The rebellion was put down by the English Royal Navy later that year.

1796: George Washington's Farewell Address is printed throughout the US. The address was written by Washington toward the end of his second term as President. Directed to his "friends and fellow-citizens," it argued for democratic republicanism and called for national unity. Washington asked the people to forgive any mistakes he may have made during his tenure, and closed by saying he was excited to become a private citizen after decades of public service.

1827: Jim Bowie fights a Louisiana banker in a duel. The fight soon turned into an all-out brawl, and Bowie killed the man with an early version of his infamous Bowie knife. The weapon was a foot long and invented by Bowie's brother Rezin.

1863: The Battle of Chickamauga begins. It would be the bloodiest two-day battle in the American Civil War, with over 34,000 casualties on both sides. General Braxton Bragg led the Confederate Army of Tennessee against Union positions commanded by General William Rosecrans in an effort to capture Chattanooga. Although James Longstreet's men were able to drive the Union soldiers from their position on the second day, the Union troops were able to set up a defensive position along a ridge. They defended it against repeated attacks before moving back into Chattanooga.

1865: George Pullman is granted a patent for the railway sleeping carriage.

1881: US President James Garfield dies. Charles Guiteau had shot Garfield in July, but the President held on for 79 days before succumbing to his wounds. Upon his death, Chester A. Arthur succeeded him.

1902: Orville and Wilbur Wright begin experimenting with gliders. The brothers performed more than 1,000 glider flights near Kitty Hawk in North Carolina in preparation for their first powered flight. A year later, they would take off in the world's first airplane.

1995: *The Washington Post* and *The New York Times* publish the Unabomber's manifesto. Ted Kaczynski, a mathematics professor turned terrorist, had been sending mail bombs to targets since 1978. He killed three people and wounded 23 others. In 1995 he sent a letter to the *Times* claiming he would cease his bombing campaign if they or the *Post* agreed to publish his manifesto. Both did. The manifesto was called *Industrial Society and Its Future* and was a critique of modern technology. Kaczynski's brother and sister-in-law already harbored suspicions that he was the Unabomber, and when they read the manifesto they realized it was highly likely he was the culprit. They turned him in to the FBI, who arrested him in April 1996.

September 20

1187: The Siege of Jerusalem Begins. A small force of knights led by the Crusader Balian of Ibelin was organized to defend the city when Saladin's forces surrounded it. Balian had 60 knights and several hundred archers and men-at-arms, while Saladin's army had more than 20,000 men. The defenders held out for two weeks before surrendering the city. The fall of Jerusalem sparked the Third Crusade in 1189, led by Richard the Lionheart.

1519: Ferdinand Magellan sets out on an expedition to sail around the world. With 270 men and five ships, Magellan departed from Spain and sailed south in the Atlantic Ocean to Patagonia, and around South America at the Strait of Magellan. The Strait is notoriously dangerous, and upon entering a calmer body of water he named it "the peaceful sea" or Pacific Ocean. Magellan's expedition eventually sailed through the Indian Ocean without him, as he had been killed in the Philippines in 1521. They completed the journey in 1522.

1737: The Walking Purchase is completed. William Penn's sons claimed they had a treaty with the Delaware Indians (also called the Lenape) that gave them all the land west of the Delaware River that someone could walk in one and a half days. While the document may have been a forgery, the Penns used it to force the Lenape to abandon their ancestral lands, ultimately claiming more than a million acres.

1893: Charles Duryea road-tests the first gas-powered automobile. Duryea, an engineer from Illinois, and his brother designed the machine by installing a gasoline engine on a horse buggy. The brothers founded the Duryea Motor Wagon Company in 1895.

1946: The first Cannes Film Festival is held. The French Minister of Education had conceived of the festival in 1932, and Cannes was selected as the location in 1939. The outbreak of World War II prevented the festival from being held, however. In the inaugural festival, filmmakers from 21 countries presented their films.

1973: Billie Jean King beats Bobby Riggs in the "Battle of the Sexes." King, at the time one of the top tennis players in the world, played Riggs, who had been the No. 1 player in the 1940s, at the Houston Astrodome. Riggs had previously beaten Margaret Court in two sets. He taunted King and other female tennis stars with sexist remarks, claiming no woman could beat a male player. King accepted the challenge. Thirty thousand fans packed the Astrodome to watch the match, and 50 million people watched on television. Before the match, Riggs presented King with a Sugar Daddy lollipop, and King responded by giving him a piglet, to symbolize male chauvinist "pigs." She then beat him in three straight sets, 6–4, 6–3, 6–3.

2001: President George W. Bush announces the War on Terror. Nine days prior, terrorist operatives of Al-Qaeda, an international Sunni Islamist organization, had attacked the United States. During an address to a joint session of Congress that was broadcast on television, Bush stated "the war on terror begins with al-Qaeda, but it does not end there."

2011: The US Military ends "don't ask, don't tell." The policy was the Clinton Administration's solution to the question of whether lesbian, gay, and bisexual (LGB) soldiers would be able to serve openly. It mandated that openly LGB people were prohibited from serving, but those who did not disclose their orientation could not be discriminated against.

2017: Hurricane Maria makes landfall at Puerto Rico. The Category 5 hurricane destroyed about 80 percent of the island's agriculture, wiped out its power grid, and killed upwards of 1,000 people. Puerto Rico's NEXRAD Doppler radar station was destroyed. Relief efforts were slow to arrive, exacerbating the situation for months afterward.

September 21

454: Valentinian III murders Flavius Aetius. Valentinian, the Roman emperor, had spent much of his time on the throne pursuing pleasure while Aetius handled the business of running the government. Valentinian killed Aetius with his own hands to prevent him from consolidating his power. Less than a year later, Marcus Avitus entered Rome at the head of an army and deposed Valentinian.

1348: Rioting against Jewish populations spreads through Switzerland, striking the cities of Chillon, Zurich, and Bern. Jewish people were blamed for an outbreak of plague after some "confessed" under torture that they had been given poison to put into wells in Venice.

1780: Benedict Arnold turns traitor. Arnold was a general in the Continental Army during the American Revolutionary War who had fought bravely in the Siege of Boston, Fort Ticonderoga, and the Battle of Saratoga. A brilliant battlefield commander, he won important strategic and military victories and greatly helped the revolutionary cause against the British. But he grew resentful of George Washington because he believed he was not sufficiently recognized for his exploits. Nevertheless,

Washington, who trusted Arnold completely, gave him command of the fortifications at West Point, New York. Arnold secretly plotted to surrender the fort to the British and defect. His plan was discovered, however, and he barely escaped. The British made Arnold a brigadier general, paid him 6,000 pounds, and gave him an annual salary. During the remainder of the war he led British troops in successful raids against Richmond and Petersburg, Virginia. After the war, he lived in exile in England.

1898: The Hundred Days of Reform in China come to an end.
The ultra-conservative Empress Dowager Cixi seized power from the emperor and put an end to his social programs.

1937: J.R.R. Tolkien publishes *The Hobbit*. The fantasy novel, which told the story of a mythical creature named Bilbo Baggins in his quest to steal treasure from the dragon Smaug, was immediately popular. Tolkien followed the book with the trilogy *The Lord of the Rings*. The novels radically changed fantasy literature and led to several film adaptations.

1972: Philippine President Ferdinand Marcos declares martial law. With the signing of Proclamation 1081, which was made public two days later, Marcos established himself as dictator, ruling with complete authority until 1986. In February of that year, the People Power Revolution deposed him peacefully, and Marcos and his wife Imelda fled the country.

1996: The US Congress passes the Defense of Marriage Act. The bill outlawed federal recognition of same-sex marriage. It was nullified by the Supreme Court in *Obergefell v. Hodges* in 2015.

September 22

1499: Holy Roman Emperor Maximilian I signs the Peace of Basel. The treaty, which came at the end of an unsuccessful conquest of Switzerland by Maximilian's forces, recognized the independence of the Swiss Confederation.

1554: Francisco Vázquez de Coronado dies in Mexico. The Spanish conquistador had searched fruitlessly for mythical cities of gold in the American Southwest and died in disgrace.

1789: The Office of the United States Postmaster General is created. Although the United States had already had three Postmasters General before the office was established—including Benjamin Franklin, who was Postmaster before the United States existed—the office was formally created by an act of Congress in 1789. George Washington appointed Samuel Osgood to the post on September 26 of that year.

1862: A preliminary draft of the Emancipation Proclamation is released. The order, which would free all enslaved persons held in states and territories loyal to the Confederacy, went into effect on January 1 of the next year.

1888: The first issue of *National Geographic Magazine* is published. The issue was sent to 165 charter members of the National Geographic Society. Since then its readership has grown to about 40 million people around the world. .

1896: Queen Victoria becomes the longest reigning monarch in British history. Victoria ruled as Queen of the United Kingdom of Great Britain and Ireland from 1837 until her death 1901. In 2015 her record was surpassed by Queen Elizabeth II.

1948: Gail Halvorsen starts dropping candy to children during the Berlin Airlift. Halvorsen was a US Air Force pilot who was assigned to the massive effort to bring supplies to the city. Earlier in the year, the Soviets had surrounded and blockaded Berlin in an attempt to break the Allied occupation of the city. When Halvorsen gave two pieces of Wrigley's Doublemint Gum to a group of German children who were near the runway where planes were landing, he was impressed by how thankful they were, as well as how well they shared the gum with one another. He agreed to drop more candy on his next run. Crowds of children gathered near the runway to wait for Halvorsen, who attached small parachutes to candy bars, which he dropped out of the plane as he approached the landing strip. When the story was reported in the news, General William Tunner decided to expand the effort, which was eventually called "Operation Little Vittles." Eventually more than 23 tons of candy were dropped on Berlin during the Airlift.

1975: Sara Jane Moore attempts to assassinate US President Gerald Ford. Moore was obsessed with Patty Hearst, who had been kidnapped by the Symbionese Liberation Army. Her interest in the case radicalized her politically, and she decided to kill Ford. She shot at him with a .38 caliber revolver but missed, and was tackled by Secret Service agent Oliver Sipple. Moore was sentenced to life imprisonment and released in 2007. She later expressed regret for her actions.

1981: The US Congress grants honorary citizenship to Raoul Wallenberg, a Swedish businessman and diplomat. Wallenberg had rescued hundreds of Hungarian Jews from the Holocaust during World War II.

September 23

1122: The Concordat of Worms is reached. The agreement between Pope Callixtus II and the Holy Roman Emperor Henry V resolved a power struggle between church and state. The conflict regarded who had the power to appoint church officials. It was the first step towards the recognition of national sovereignty that would culminate in the Peace of Westphalia in 1648.

1338: The first naval battle using artillery is fought. During the Hundred Years' War, the Battle of Arnemuiden was fought by the English and French fleets. One English ship, the *Christopher*, was armed with three cannons, the first time the weapons were deployed at sea.

1779: The Battle of Flamborough Head takes place during the American Revolution. In the battle, John Paul Jones, the commander of the American ship *Bonhomme Richard*, rammed the British ship *Serapis*. When the captain of the *Serapis* asked Jones "Has your ship struck," meaning "surrendered," Jones famously replied "I have not yet begun to fight!" Jones and the Americans won and captured both British ships.

1806: Lewis and Clark return to St. Louis. The explorers had left two years before. They traveled to the Pacific Coast in present-day Washington State, crossed the Continental Divide, and mapped much of the territory in the Louisiana Purchase.

1846: Neptune is discovered. Three astronomers, Jean Urbain Le Verrier, John C. Adams, and Johann Galle, collaborated on the discovery. Le Verrier predicted the planet's existence using celestial mechanics. Galle verified the calculations and found the planet using the Berlin Observatory's telescope. Unbeknownst to the pair, Adams had independently made the same prediction at about the same time.

1899: Nintendo is founded by Fusajiro Yamauchi. The company was originally a playing card manufacturer based in Kyoto, Japan. Beginning in the early 20th century, Nintendo mass-produced cards for bridge and *Hanafuda*, a Japanese game. In 1956 Yamauchi's grandson Hiroshi traveled to the United States to meet with the US Playing Card Company. He realized that there was a limit to the growth such a company could realize, and he shifted Nintendo's business ventures into games and toys. In 1974, the company entered the video game industry, securing the rights to distribute the Magnavox Odyssey, one of the world's first video game consoles, in Japan. Shigero Miyamoto joined the company as a console designer. In 1981, Miyamoto's game *Donkey Kong* revolutionized both Nintendo's fortunes and the video game industry forever. Over the next three decades, Nintendo became one of the most dominant video game and console producers in the world.

1942: The Mataniku Action begins as US Marines attack Japanese positions along the Mataniku river during the Battle of Guadalcanal. Over the next two days, 156 Marines and 750 Japanese soldiers were killed in brutal fighting.

1980: Reggae star Bob Marley plays his last concert in Pittsburgh, Pennsylvania. Marley had been diagnosed with a malignant form of melanoma earlier in the year. His health deteriorated rapidly after the concert, and the international hero died in 1981.

2002: The web browser Mozilla Firefox is released. The browser was the first free and open-source one released, meaning anyone could freely use, copy, and change the software of the program. The open-source design means that users are constantly improving the browser.

September 24

787: The Second Council of Nicaea is assembled at the Hagia Sophia. It was the final one of seven councils between representatives from the Eastern Orthodox Church and the Roman Catholic Church. The Council established rules governing holy images. It set the cross, iconography of Jesus, the Virgin Mary, angels and saints as those images that could be depicted in Christian churches, and decreed that every altar must contain a holy relic in it.

1398: Timur crosses the Indus river. Also known as Tamerlane, the conqueror of Central Asia was on his way to Delhi, which his forces would destroy in a ruthlessly bloody battle.

1789: The Office of the US Attorney General is created. The Congress passed the Judiciary Act of 1789, which established the federal judicial system. It created 13 judicial districts in the states, establishing a circuit court and district court in each. Edmund Randolph took office as the first Attorney General two days later. The Act also set the number of justices on the Supreme Court of the United States at six, including one Chief Justice and five Associate Justices.

1890: The Church of Jesus Christ of the Latter-Day Saints abandons its doctrine of polygamy. It had been a central tenet of the Church since Joseph Smith founded it sixty years before in western New York State. Polygamy had been a source of tension between Latter-Day followers (called Mormons) and non-Mormons. Brigham Young, who led the Church from 1844 until he died in 1877, led his followers in the Utah Mormon War of 1857–58 to retain control of the Utah Territory and maintain the practice of polygamy. The US Congress was so opposed to the doctrine that it ordered the Church's assets to be seized in order to put an end to it. As a result, Wilford Woodruff, the LDS Church President, issued a proclamation that officially ended the practice. Mormons who were found to be practicing polygamy were, and still are, excommunicated from the Church. Some fundamentalist factions continued the practice and split with the Church, forming their own Mormon sects.

1906: US President Teddy Roosevelt declares Devil's Tower a National Monument. The Tower, a 5,112-foot butte in north-eastern Wyoming, was the first natural area to be declared a National Monument. It attracts some 400,000 visitors each year.

1929: Jimmy Doolittle flies an airplane at Mitchel Field in New York. The feat was notable because Doolittle was flying entirely "blind." It demonstrated that flying an aircraft using only its instruments was possible.

1946: The Cold War policy of "containment" is introduced. Two military advisers to President Harry S. Truman recommended the strategy, which was geared toward stopping the spread of Communism and the influence of the Soviet Union, rather than confronting it. Containment guided US Cold War policy for the next forty years.

1957: President Dwight D. Eisenhower sends the 101st Airborne Division to Little Rock, Arkansas. The troops enforced the policy of desegregating public schools when the Governor of Arkansas refused to do so.

1964: The Warren Commission finishes its investigation. In a report that would be made public later that year, the Commission concluded that Lee Harvey Oswald had acted alone in assassinating President John F. Kennedy.

1968: *60 Minutes* debuts on CBS. The program was unique from other news formats in that it centered on investigative journalism rather than breaking reports.

September 25

1690: The first multipage American newspaper is published. Titled *Publick Occurrences Both Forreign and Domestick*, it was published in Cambridge. Before it debuted, newspapers were single-sheet affairs known as broadsides.

1513: Vasco Núnez de Balboa walks across the Isthmus of Panama. In doing so, Balboa became the first European to glimpse the Pacific Ocean.

1789: The US Congress passes the Bill of Rights. The Bill included ten amendments to the United States Constitution, which had been ratified earlier that year. It provided guarantees of personal liberty and rights, limited the power of the federal government, and specified that all powers not specifically given to Congress by the Constitution lay with the states or the people. Among the guarantees in the Bill of Rights are freedom of religion, the press, and public speech; the right to bear arms; and the right to due process and a speedy, public trial by jury for the accused. It prohibits quartering soldiers in private homes, a hated practice of the British Army during the Colonial Period. Finally, the Bill of Rights codifies the separation of powers and Federalism, the political doctrine of reserving powers among the states. James Madison brought the original nine amendments to Congress in June, and after two months of debate, the final draft was passed. The ten amendments were ratified by the states on December 15, 1791.

1818: James Blundell performs the world's first blood transfusion. The English obstetrician had previously argued that blood transfusions could treat severe postpartum hemorrhages, or bleeding after giving birth. After experimenting on animals, he extracted four ounces of blood from a patient's husband and transfused it in her bloodstream. Blood transfusions eventually became a mainstay of modern medicine and have saved countless lives over the decades.

1908: The Casablanca Affair begins. Six Germans who had deserted from the French Foreign Legion were jailed by French authorities, and the German Consulate was irate. The case went before a tribunal at the Hague, and while the court found the French had no right to seize them, they did not order the French to return the deserters.

1981: Sandra Day O'Connor is sworn in as the first female US Supreme Court justice. O'Connor had previously served on the Maricopa County Superior Court and the Arizona State Court of Appeals. She replaced Potter Stewart at the US Supreme Court. A Republican and federalist, she believed in very narrowly applicable rulings on cases before the court, and typically joined the conservative wing of the Court in her voting. She served for sixteen years before retiring in 2005 and being replaced by Samuel Alito.

September 26

1580: The English sailor Francis Drake completes his voyage around the world. He became the first British person to do so as he returns to Plymouth aboard his ship, the *Golden Hind*.

1687: The Parthenon in Athens explodes. While Turkish forces were occupying the city, they used the ancient temple as a munitions dump. Their powder magazine exploded, partially destroying the Parthenon, when a missile from a besieging Venetian force sparked the store.

1905: Albert Einstein publishes his first paper on special relativity. In it he argued that the laws of physics are unchanging and that the speed of light in a vacuum is the same for all observers, no matter how fast they or the light source are moving. The discovery solved the problems that Newtonian mechanics encountered when applied to electromagnetism. Special relativity revolutionized physical science and ushered in a new era of scientific discovery.

1933: "Don't shoot, G-men!" The gangster George Kelly Barnes, better known as "Machine Gun Kelly," famously shouted the line as he surrendered to agents of the Bureau of Investigation, the forerunner of the FBI. The nickname—short for "government man"—stuck, and FBI agents have been called "G-men" ever since.

1934: The *Queen Mary,* the largest ocean liner produced to date, is launched at Clydebank, Scotland. British Queen Mary herself was present for the launch. The ship completed its first transatlantic voyage the following Spring. In 1967 it was sold to an American company that turned it into a tourist museum and hotel in Long Beach, California.

1960: Richard Nixon debates John F. Kennedy on live television. It was the first debate between presidential candidates, as well as the first on TV. Nixon, who had recently been sick with the flu, appeared pale and sickly on television. He had turned down television make-up, and his five o'clock shadow was visible. He also sweated profusely under the hot studio lights. Kennedy, on the other hand, appeared young, relaxed, and tanned compared to Nixon. People who listened to the debate on the radio generally thought Nixon won the debate, while those who watched on TV believed Kennedy had won. Kennedy went onto win the election by just over 100,000 votes.

1969: The Beatles release *Abbey Road,* their last recorded album. The album, which featured the iconic cover image of the band walking across the titular road, reached Number 1 in the United Kingdom and United States and is viewed by critics as one of the band's best. The band broke up the following year.

1983: Stanislav Petrov, an officer in the Soviet Air Defense Forces, receives an early-warning indicator of a nuclear attack on the Soviet Union. Petrov correctly determined that the warning was a false alarm. He noted the warning system only reported five incoming missiles, and thought any real attack would be massive. His quick thinking prevented a Soviet retaliation and all-out nuclear war.

1996: American Astronaut Shannon Lucid returns to Earth after setting a new endurance record in space. She spent 188 days aboard the Russian space station *Mir*. Lucid's record was broken by Sunita Williams on June 16, 2007 while she was aboard the International Space Station.

September 27

1066: William the Conqueror assembles a massive army of heavy cavalry and infantry and sets sail from France, headed to Sussex, England. The Norman conquest was victorious, and William became the first Norman king of England, ruling until his death in 1087.

1540: Pope Paul III grants a charter to the Society of Jesus, a religious order that Ignatius of Loyola founded six years earlier in Spain. The Jesuits, as they came to be called, engage in evangelizing non-Christians and work in education.

1590: Only 13 days after being elected as pope, Urban VII dies of malaria. His tenure was the shortest papacy in history.

1825: The Stockton and Darlington Railway is opened in England. It was the first public railroad that used steam locomotives. The railway company operated until 1863.

1892: The Diamond Match company patents book matches. The company was one of the first matchmaking businesses. It made matches safer by reducing the amount of phosphorous in them, and with the invention of book matches, made them more portable and convenient.

1962: Rachel Carson publishes *Silent Spring*. In it, the marine biologist and conservationist showed that widespread pesticide use was causing significant harm to the environment. Its publication helped spark the environmental movement and eventually led to the creation of the US Environmental Protection Agency.

1968: The musical *Hair* opens at the Shaftesbury Theatre in London. It played 1,988 performances there and only closed when the theatre's roof collapsed in 1973.

1996: Taliban soldiers capture Kabul, the capital of Afghanistan. The Sunni Islamic fundamentalist army executed Mohammad Najibulla, the last Communist president of Afghanistan, and assumed power. They ruled until the US invasion of 2001, but continued a guerilla war for years after.

1998: The search engine Google is born. Stanford University PhD students Larry Page and Sergey Brin began working on the search engine as a research project in January 1996. The pair originally called the search engine "BackRub" because it ranked websites by how many other sites linked back to them. They changed the name to Google to reflect the search engine's ability to provide vast quantities of information. A Googol is the number written as 1 followed by 100 zeroes, and the pair misspelled it by mistake. Google rapidly grew to become the most widely used search engine in the world, and the company eventually expanded into other services including email and document storage. By the mid-aughts, Google had become one of the premier internet and technology companies in the world, with more than 80,000 employees. The company retroactively claimed September 27 as the search engine's birthday on its 15-year anniversary celebration in 2013.

2008: Zhai Zhigang, an astronaut with the China National Space Administration, becomes the first Chinese person to perform a spacewalk. Zhigang stepped out of the *Shenzhou* 7 while in orbit, and completed the spacewalk after 22 minutes outside the craft.

September 28

28 BC: The Roman General Pompey the Great is murdered. Pompey's army had been defeated by one led by Julius Caesar during the Roman Civil War, and he fled to Egypt to seek refuge. King Ptolemy of Egypt, seeking to win favor with Rome, ordered that he be killed as soon as he came ashore.

1634: A young John Milton debuts *Comus* before the Earl of Bridgewater. The masque—a kind of musical stage production popular in 17th-century Europe—was a cautionary tale about chastity and good and evil, themes he would later explore in his masterpiece epic poem *Paradise Lost*.

1781: The Siege of Yorktown begins. A combined force of Continental Army soldiers under the command of George Washington and French troops commanded by Comte d'Aboville trapped a British Army led by Lord Cornwallis in the city. The French Navy provided a blockade that completed the encirclement, preventing Cornwallis from escaping by sea. The Americans and French moved within cannon range of the city on September 29 and began bombarding the British defenses. The bombardment did not let up for two weeks, and French and American infantry made numerous attacks on the British lines. By October 14, the British defenses were sufficiently weakened, and the Continental Army attacked. The battle raged over two days of intense fighting, with Allied assaults breaking through

British lines and being repeatedly repulsed. By October 17, Lord Cornwallis realized his position was hopeless, and sent an officer to the American lines to negotiate a surrender. A British fleet sent to relieve Cornwallis arrived a week too late. The engagement was the last major battle of the American Revolution, and the victory over Cornwallis all but assured an American victory.

1850: The US Congress outlaws flogging as a punishment in the Navy. Proposals to abolish the practice had been put forward as early as 1820. Public opinion turned against flogging after the publication of *Two Years Before the Mast*, by Richard Dana, and *Cruise on the Frigate Columbia Around the World*, by William Murrell, both in 1840. In 1847 John Hale was elected to the Senate from New Hampshire, and he unsuccessfully introduced several bills to outlaw flogging in the Navy. Meanwhile, in March 1850 Herman Melville published *White-Jacket, or The World in a Man-of-War* in which he called for flogging to be eliminated. Hale introduced yet another anti-flogging bill, and it finally passed.

1865: Elizabeth Garrett Anderson is the first woman to obtain a license to practice medicine and surgery in Great Britain. Anderson would go on to be the first woman to become a mayor in England when she was elected the Mayor of Aldeburgh in 1908.

1918: The Fifth Battle of Ypres begins. Belgian troops led an attack on Ghent during what would begin the final Allied offensive of World War I. The battle lasted for four days with the Allies in control of Ypres and strategic points around the town. By November, the war would be over.

1939: Nazi Germany and the Soviet Union agree to divide Poland between them. Hitler and Stalin's foreign ministers signed the pact, which provided for joint invasion and occupation of Poland by the two powers.

September 29

48 BC: A Greek fleet composed of 370 triremes destroys and scatters a massive Persian fleet of more than 1,000 ships in the Battle of Salamis. It was the first major naval battle in recorded history, and turned the tide against Xerxes's planned invasion of Greece.

1227: Pope Gregory IX excommunicates the Holy Roman Emperor Frederick II for refusing to take part in the Crusades. Frederick had pledged to join the Crusade but later went back on his word, claiming illness. It was one of four times Frederick was excommunicated by Rome, with whom he had a hostile relationship. .

1789: The first session of the United States Congress adjourns after six months. During the session the Congress established the Department of War and the Department of the Treasury, passed the Judiciary Act to establish the Office of the Attorney General, and passed the first Tariff in United States history. Congress reconvened on January 4, 1790.

1829: London's Metropolitan Police Service is founded. Known informally as "The Met," the force was founded by English statesman Robert Peel, who is considered the father of modern British policing for his work. The first police officers, affectionately called "bobbies," patrolled the streets of London. The force eventually grew to more than 30,000 full time officers and an additional 2,700 special constables.

1923: Mandatory Palestine is created. After World War I, the victorious Allied nations carved up the territories that had been ruled by the Ottoman Empire among them. British Palestine was informally created in 1920 and officially recognized with the British Mandate. It existed until the creation of the state of Israel in 1948.

1975: WGPR goes on the air in Detroit, Michigan. It was the first television station that was owned and operated by African Americans. Ross Mulholland, a broadcaster who had worked for a number of other radio stations, founded the station.

2004: *Space Ship One* completes its first successful space flight. It was the first of two flights required to win the Ansari X Prize. The Prize was ten million dollars to the first non-governmental organization to do so.

2008: The Dow Jones Industrial Average falls 777.68 points. It was the largest single-day drop in history. The point loss followed the bankruptcies of Lehman Brothers and Washington Mutual, two major investment banks, as the sub-prime mortgage crisis enveloped Wall Street. The crisis precipitated the worldwide financial recession of 2008.

September 30

1189: After driving his own father from the throne with the aid of the French, Richard the Lionheart becomes the King of England.

1791: Wolfgang Amadeus Mozart debuts *The Magic Flute* in Vienna, Austria. It was the composer's last opera; he would die prematurely just two months later. The opera, which is a magical story about Prince Tamino's and the bird-catcher Papageno's quest to rescue Princess Pamina for the Queen of the Night. It remains one of the most frequently performed operas, an the Queen of the Night's aria is one of Mozart's best-known works.

1882: Thomas Edison opens the first hydroelectric plant in history at Appleton, Wisconsin. He called it the Appleton Edison Light Company. It provided electricity to light the Appleton Paper and Pulp Company building, a paper mill, and the home of H.J. Rogers, the president of the paper company.

1889: Wyoming grants the right to vote to women. It became the first state in the Union to do so. Women would not gain the vote nationwide until the passage of the Nineteenth Amendment was ratified in 1920.

1915: Radoje Ljutovac shoots down an enemy aircraft with a modified cannon during World War I. He was the first person in history to shoot down a plane from the ground.

1947: The World Series of Baseball is shown on television for the first time. The Series was played between the New York Yankees and the Brooklyn Dodgers; the Yankees won in seven games.

1954: The USS *Nautilus* is commissioned. The submarine was the first nuclear-powered vessel in the world.

1955: American actor James Dean is killed. Dean died in an automobile accident along with his mechanic Rolf Wutherich. The pair were driving Dean's Porsche 550 Spyder when they crashed head-on into another car. The young actor, who was only 24, was an icon of the malaise and disillusionment felt by many young Americans in the 1950s, and had cemented his reputation in his film *Rebel Without A Cause*, which had debuted earlier that year. Dean's other major films included *East of Eden*, in which he played a loner and drifter, and *Giant*, in which he played a disaffected ranch hand. Dean became the first actor to be nominated an Academy Award posthumously, for his work in *Rebel*. Dean took up auto racing as a hobby in 1954 and won several races. He was training for another race when the accident occurred. Dean was buried in Fairmount, Indiana, where he grew up.

1972: Roberto Clemente records his 3,000th and final hit in Major League Baseball. The Puerto Rican baseball star played right field for the Pittsburgh Pirates, where he was dominant offensively and defensively. During the 1972 off season, Clemente was in Nicaragua doing relief work for survivors of an earthquake when his chartered airplane crashed, killing everyone on board. He was 38 years old. The following year he was inducted into the National Baseball Hall of Fame.

Chapter 10

October

October 1

331 BC: Alexander the Great wins the decisive Battle of Gaugamela during his invasion of Persia. Alexander's army was heavily outnumbered by the Persians, who were commanded by Darius III. After the battle Darius was murdered by his own officers, who were led by a Persian noble named Bessus. When Alexander captured the Persian capital, he gave the dead king a full funeral and burial ceremony, and then had Bessus executed for betraying and killing his respected enemy.

1746: Bonnie Prince Charlie flees to France. Charles Edward Stuart was a claimant to the British throne whose enemies called "The Young Pretender." In 1745 he led a failed invasion of England by Scottish Jacobites and fled to France via Scotland and the Isle of Wight. He lived in France for the rest of his life.

1800: France acquires Louisiana from Spain. The Third Treaty of San Ildefonso outlined a deal in which Spain traded the Louisiana Territory, which it had claimed as part of the Viceroyalty of New Spain in 1762, to France. In exchange, Spain acquired parts of Tuscany, Italy.

1843: *News of the World* goes on sale in London. The newspaper was one of the first to report primarily on sensational stories about crime and vice in the United Kingdom. It was rebranded in 1984 as a celebrity tabloid paper, and remained in print until a scandal forced it to close in 2011.

1890: The US Congress establishes Yosemite National Park.
John Muir and Robert Underwood Johnson lobbied Congress
for years to make the California wilderness a protected Park.
Because the National Park Service was not created until 1916,
the Park was initially protected by the 4th Cavalry and Buffalo
Soldiers from the 9th Cavalry.

1903: The first baseball World Series begins. The American
League's Boston Americans played the National League's
Pittsburgh Pirates at Exposition Park in Allegheny City,
Pennsylvania, in a best-of-nine series. Deacon Phillippe pitched
five complete games for the Pirates, winning three. In his first
game, Phillipe struck out ten batters. The following day, Bill
Dinneen struck out eleven while pitching for the Americans,
evening the series at one game apiece. Pittsburgh won two more
to lead the series 3–1 before the Americans mounted a comeback. Cy Young—for whom baseball's annual award for pitching is named—won two more games for the Americans, who
went on to win four in a row and take the Series. Pittsburgh was
unable to mount a serious offense partly because Honus Wagner,
a dominant shortstop and powerful slugger, was hampered by
injuries. During the series he had just six hits in 27 at-bats and
committed six fielding errors. Wagner was so upset by his performance that he rejected an invitation to be inducted in a Hall
of Fame for hitters the following year. "I was a bum last year,"
he said.

**1931: George Washington Bridge opens in the Washington
Heights neighborhood of Manhattan, linking New Jersey
and New York.** More than 289,000 cars crossed the bridge
in 2016.

**1952: The National Aeronautics and Space Administration
(NASA) is created.**

**1975: Muhammad Ali defeats Joe Frazier in a boxing match
dubbed "The Thrilla in Manila."** The fight was held in the
Philippines and is considered one of the best in boxing history.

October 2

1528: William Tyndale publishes *The Obedience of a Christian Man*. Tyndale was a leader of the Protestant Reformation until his execution in 1536 for heresy. In addition to translating the Bible into English, he wrote *Obedience* to argue for the divine right of kings and the supremacy of a country's monarch as the head of its church. The book influenced King Henry VIII of England to declare himself Supreme Head of the Church of England in 1534.

1789: George Washington sends the proposed Constitutional amendments that would become the Bill of Rights to the States for ratification. The States took more than two years to do so, and ratification was completed on December 25, 1791.

1835: The Texas Revolution begins. After immigrating to Texas from the United States for several decades, white American arrivals began agitating for independence. Mexico responded by limiting immigration and banning slavery in the region, but Americans ignored the laws and continued to immigrate illegally, often bringing slaves with them. They began arming themselves, and a confrontation with authorities in Gonzalez, Texas, resulted. The rebels killed two soldiers before the Mexican troops withdrew. News of the victory spurred more Americans to travel to Mexico to join the rebels. After less than a year of fighting, the Mexican government signed a treaty allowing for the formation of the Republic of Texas.

1919: President Woodrow Wilson suffers a major stroke. The stroke partially paralyzed him and left him blind in one eye. Because of his incapacitation, Wilson was unable to complete his mission to convince the United States to join the League of Nations.

1950: Charles M. Schulz publishes the first *Peanuts* comic strip. The iconic strip featured Charlie Brown, his sister Sally, their dog Snoopy, and friends Franklin, Linus, Lucy, and Peppermint Patty, among others. It ran for 50 years and spawned a stage musical, a movie, and several television specials including *A Charlie Brown Christmas* and *It's the Great Pumpkin, Charlie Brown*. The final Sunday strip published on February 13, 2000.

1959: *The Twilight Zone* premieres on CBS. The television show was created and hosted by Rod Serling and ran for five seasons. It featured science fiction and psychological horror themes and often included a moralistic lesson.

1967: Thurgood Marshall becomes the first African-American Supreme Court justice. Marshall had studied law at Howard University and founded the NAACP Legal Defense and Education Fund. He was appointed to the US 2nd Circuit Court of Appeals by President John F. Kennedy, and appointed US Solicitor General by LBJ. Marshall served on the Supreme Court until October 1991.

1968: Soldiers open fire on student demonstrators in Mexico City at the Tlatelolco massacre. The students were protesting the government spending and repression in preparation for the Olympic Games. The soldiers shot students who had gathered in the Tlatelolco Plaza, killing between 300–400.

October 3

52 BC: Arverni King Vercingetorix, who led the Gauls in a revolt against Roman rule, surrenders to Julius Caesar. Vercingetorix had united several Gaul tribes in his fight against the Romans earlier in the year. They slaughtered the Romans living in their territory before raising an army to fight them. Vercingetorix won a few battles before the Romans besieged his army at Alesia. After surrendering, he was taken to Rome as prisoner and later executed.

1789: George Washington makes the celebration of Thanksgiving a national holiday. Since then the holiday has always fallen on the fourth Thursday of November. Washington proclaimed the holiday at the request of Congress.

1849: Edgar Allan Poe is found unconscious in a gutter in Baltimore. Poe died four days later, and the circumstances surrounding his death remain mysterious. Joseph W. Walker found Poe lying in the street, and later reported that he was delirious and "in great distress." Poe was taken to the hospital, but never regained consciousness.

1872: The first Bloomingdale's opens in New York City. Brothers Joseph and Lyman Bloomingdale opened the store in Midtown Manhattan. It pioneered the department store model, and the company eventually grew to have stores in many cities throughout the United States.

1952: The United Kingdom joins the nuclear age. In Operation Hurricane, the UK successfully tested a plutonium nuclear bomb in a lagoon near Western Australia. In doing so, it became the third country, after the United States and Russia, to acquire a nuclear weapon.

1957: The California State Superior Court finds that Allan Ginsberg's "Howl" is not obscene. The poem has references to illegal drug use and sexual acts, some of which were illegal at the time. In March, US Customs officials seized copies of the book that were being imported from London. In June, Lawrence Ferlinghetti was arrested for publishing the book. He was defended by the American Civil Liberties Union and found not guilty when the Superior Court found the book had "redeeming social importance."

1990: East and West Germany are reunified after 40 years as separate countries. With the dissolution of the Soviet Union and collapse of Communism in Eastern Europe, the German Democratic Republic ceased to exist on this day. Its territory joined the Federal Republic of Germany and its citizens became part of the European Community.

1993: Eighteen US soldiers are killed in the Battle of Mogadishu. The US invaded Somalia during the country's civil war. During an operation to capture leaders of one Somali faction, a detachment of US Army Rangers and Special Forces were pinned down in Mogadishu by militia and armed civilians. Two Black Hawk helicopters were shot down and one American soldier was captured. The 2001 film *Black Hawk* Down depicted the battle's events.

1995: A jury in Los Angeles, California, acquits O.J. Simpson of murder. Simpson, a celebrity and former NFL running back, was accused of stabbing his ex-wife Nicole Brown Simpson and her acquaintance Ron Goldman in the early morning hours of June 13, 1995. In a trial that gripped the attention of the nation for ten months, Simpson mounted a high-powered defense with some of the most expensive and well-known attorneys in the country. The daily proceedings of the trial were televised live on the newly created Court TV channel. Defense attorneys challenged the accuracy of DNA evidence presented by the prosecution and revealed that the case's lead investigator was a racist who had bragged about framing Black and Latino suspects. Simpson was found not guilty on all counts.

October 4

1582: Pope Gregory XIII implements the new Gregorian Calendar. The calendar introduced leap years that are spaced out every four years to make the average length of each year 365.2425 days. The change was made to stop the calendar from drifting away from the Spring and Fall equinoxes.

1883: The Orient Express makes its inaugural run. The long-distance passenger train originally ran between Paris, France, and Constantinople, in what was then the Ottoman Empire. The train has been frequently referenced in literature and film, most notably in Agatha Christie's 1934 novel *Murder on the Orient Express*. It continued running until service was discontinued in 2009.

1895: The first US Open Men's Golf Championship is held at Newport Country Club. The original tournament was a 36-hole competition played on a nine-hole course in a single day. Horace Rawlins, an Englishman, won the Championship. It has been played every year except during World Wars I and II.

1927: Gutzon Borglum begins sculpting Mount Rushmore. After working on Thomas Jefferson's face for two years, workers realized the rock was unsuitable. They dynamited it and began again. The entire sculpture, which includes the faces of Jefferson, George Washington, Abraham Lincoln, and Theodore Roosevelt, was finally completed in 1941.

1957: The Soviet Union launches *Sputnik 1*. The flight of the first artificial satellite to orbit the Earth signaled the beginning of the Space Race. *Sputnik* stayed in orbit and sent back radio broadcasts for three weeks until its battery died. It then continued to orbit for another two months before falling back to Earth. Just twelve years later, humans walked on the surface of the Moon.

1965: Pope Paul VI visits New York City, becoming the first Pope to travel to the Western Hemisphere. Paul VI was known for the many reforms he ushered into the Catholic Church during the Second Vatican Council, which closed the same year as his visit.

1993: Tanks bombard the Russian White House in Moscow during the Russian Constitutional Crisis. Communist hardliners had stormed the government building in an attempt to oust President Boris Yeltsin. He held onto power over ten days of conflict.

2006: Julian Assange launches WikiLeaks. The nonprofit organization publishes classified information, media, and government documents on its website of the same name. The website gained prominence in 2007 when it published footage of a US Apache helicopter firing on Iraqi journalists in Baghdad. In 2010, WikiLeaks released US State Department diplomatic communications, the US military logs of the War in Afghanistan, and field reports from the Iraq War. In 2010, Sweden issued an international arrest warrant for Assange based on accusations of sexual assault; Assange fled to the Embassy of Ecuador in London. The organization was later implicated in linking emails from the Democratic National Committee in an act that has been blamed for Hillary Clinton's loss in the 2016 US Presidential election.

October 5

1789: Ten thousand women march on the Palace of Versailles. During the French Revolution, the lack of grain led to scarcity and high prices on bread. After news broke that army officers had enjoyed a lavish banquet earlier in the week at Versailles, women in one of the marketplaces of Paris began marching toward the palace. The crowd grew as women from other markets joined in, and it numbered in the thousands by the time they reached Versailles. The crowd looted City Hall as it passed by, taking weapons and provisions with them. Arriving at the palace with cannons and other arms, they milled outside of it in an informal siege. The next morning, some of the marchers discovered an open gate and entered. They were shot by a guard, and the rest of the crowd angrily stormed into the palace. Only the intervention of the Marquis de Lafayette and Queen Marie Antoinette stopped the crowd from looting the palace.

1877: Chief Joseph of the Nez Perce surrenders to the US Army. Earlier that year, Joseph had led nearly 1,000 Nez Perce more than 1,170 miles in a long retreat from the US Army rather than submit to removal from their ancestral lands in Oregon. After the Army cornered the band in Montana, Joseph surrendered.

1914: The first air-to-air combat fatality in human history is recorded. French pilot Joseph Franz and his gunner Louis Quénalt shot down a German biplane during World War I.

1921: The World Series of Baseball is broadcast on the radio for the first time. The New York Giants defeated the New York Yankees, five games to three. Grantland Rice, a sportswriter from Pittsburgh, gave live play-by-play of the games on Pittsburgh's radio station KDKA.

1938: Nazi Germany declares Jewish citizens' passports to be invalid. As part of the Nuremberg Laws, which were initially passed in 1935, the Nazis seized passports belonging to Jews and had them stamped with the letter J.

1947: President Harry Truman delivers the first televised White House address. In the speech Truman asked Americans to conserve food to aid Europeans who were still recovering from the devastation of World War II. Among other requests, Truman suggested Americans stop eating meat on Tuesdays and forgo eggs and poultry on Thursdays.

1968: Police attack civil rights marchers in Northern Ireland, signaling the start of The Troubles. The Royal Ulster Constabulary beat demonstrators indiscriminately with batons, injuring more than 100. The attack provoked two days of riots in Derry, Northern Ireland.

1986: Mordechai Vanunu reveals that Israel secretly possesses nuclear weapons. Vanunu, a former nuclear technician from Israel, told the British *Sunday Times* about Israel's nuclear weapons program. The Mossad, Israel's intelligence service, abducted Vanunu from Rome and took him back to Israel, where he was imprisoned until 2004.

2001: Barry Bonds sets the record for most home runs in a single season. In a game against the L.A. Dodgers, Bonds hit two home runs for his 71st and 72nd that year, surpassing Mark McGwire's single-season record of 70.

October 6

1600: The opera *Euridice* is performed in Florence. The work by Jacopo Peri is the earliest surviving opera from the Baroque period of art and music in Europe.

1683: German immigrants found Germantown in Pennsylvania. Thirteen Quarter and Mennonite families from Krefeld, Germany, settled there. They later signed a charter with William Penn to incorporate the settlement, which was incorporated into the city of Philadelphia in 1854.

1723: Benjamin Franklin arrives in Philadelphia. Originally from Boston, Franklin got into trouble with authorities there for publishing a newspaper that was critical of the government when he was 17 years old. To avoid prosecution he traveled to Philadelphia. He found work as a typesetter in a printer's shop before becoming the publisher of *The Pennsylvania Gazette* in 1728. Franklin would go on to become one of the Founding Fathers of the United States of America as well as a respected scientist, statesman, and diplomat.

1876: The American Library Association is founded. During a meeting at the Centennial Exposition in Philadelphia, a group of eight librarians established the organization to support the budding public library movement. The organization is the oldest and largest library association in the world.

1927: *The Jazz Singer,* the first "talkie" movie, premieres. It was the first feature-length film to have a musical score and an audio track that was synchronized with the picture. Warner Brothers produced the movie, which starred Al Jolson and May McAvoy, who had become stars during the silent film era.

1973: Egypt and Syria attack Israel to begin the Yom Kippur War. Egyptian President Anwar Sadat declared the mission of the war was to recover land lost to Israel following the Six-Day War in 1967. In a surprise attack, Egyptian and Syrian forces took control of the Sinai Peninsula and the Golan Heights before Israeli forces stopped their advance after three days of fighting. The Israelis were able to repulse the Syrian forces altogether and invade Syria, penetrating far enough to threaten the capital city of Damascus. They then attacked the Egyptian forces and fought their way towards Suez in a week of intense combat. By October 25 a ceasefire was declared.

1981: Islamic extremists assassinate Egyptian President Anwar Sadat. In 1978 Sadat had shared the Nobel Peace Prize with Israeli Prime Minister Menachem Begin for signing the Camp Davide Accords. Sadat was killed by assassins led by Khalid Islambouli, an officer in Egyptian Army, during a military parade. The attackers threw hand grenades and sprayed the president's viewing stand with automatic gunfire.

2007: Jason Lewis finishes circumnavigating the globe using his own power. Lewis used a mountain bike, a pedal-powered boat, roller blades, a kayak, and his own feet to travel around the globe. He had started the journey in July of 1994.

2010: Kevin Systrom and Mike Krieger found Instagram. Two years after it was launched, Facebook purchased the photo and video-sharing app for $1 billion in cash and stocks.

October 7

1763: King George III of Great Britain issues a proclamation that closed lands west of the Allegheny Mountains to white settlement. Britain gained the territory from France after the French and Indian War. The King and Parliament wished to improve relations with Native Americans and stabilize trade and land appropriation on the frontier. American colonists were angered by the proclamation.

1864: In the Bahia Incident, a Union warship captures a Confederate gunboat off the coast of Brazil. The Union ship discovered the Confederates in the Bay of Bahia and attacked them, despite warnings from the Brazilian Navy not to. After attacking the gunboat, the Union sailors were able to outrun pursuing Brazilian ships, but their captain was later court-martialed after the Brazilian government complained to the United States.

1916: The Georgia Tech Engineers beat the Cumberland College Bulldogs in football, 222–0. The game was the largest blowout in the history of college football. Cumberland had recently cancelled its football program, so they were forced to field a team of unpracticed students. Georgia Tech's baseball team had recently been defeated 22–0 by Cumberland, and they ran up the score in the football game as payback.

1959: The Soviet space probe Luna 3 takes the first photographs from the far side of the Moon. The probe took a total of 29 pictures that included 70% of the far side and showed that its terrain was more mountainous than the near side. The images were received with great excitement around the world.

1993: The Great Mississippi and Missouri Rivers Flood ends after 103 days. The previous winter's heavy snowfall and severe rainstorms in the Spring saturated the ground across the Midwest. In May the Missouri and Mississippi Rivers rose above flood stage, inundating farmland and communities in an area of 30,000 square miles that covered parts of nine states. The flood caused about $15 billion in damages and killed 32 people before the waters receded.

1996: Fox News goes on air. Australian publisher Rupert Murdoch and American businessperson Marvin Davis founded the cable network as a 24-hour news programming platform. They expressly made the editorial bent of the channel right-wing, and hired Republican Party strategist Roger Ailes to oversee programming.

2001: US forces invade Afghanistan. After Al-Qaeda operatives attacked the World Trade Center and Pentagon with civilian airliners, the United States demanded Afghanistan's government turn over Osama bin Laden and other Al-Qaeda leaders. The Afghan rulers, called the Taliban, refused. US forces and their allies drove the Taliban from power by December, but remained in Afghanistan for years. Bin Laden was eventually captured and killed in Pakistan by US Special Forces in 2011.

2003: Arnold Schwarzenegger becomes Governor of California. Governor Gray Davis became widely unpopular after an energy crisis in the preceding years that caused electricity bills of many residents to triple. A campaign to recall him from office eventually gathered 1.6 million signatures, twice the necessary number to force a special election. The election both asked voters whether Davis should be recalled, and asked them to vote for a replacement. Democratic Lieutenant Governor Cruz Bustamante, Independent Arianna Huffington, a news publisher, and Republican Arnold Schwarzenegger, a former bodybuilder and movie star, were all on the ballot. In the vote to recall, 55.4 percent of voters responded "Yes." Schwarzenegger won the race to replace Davis with 48.6 percent of the vote.

October 8

1480: Russian Grand Duke Ivan III's forces stand up to Tartar Khan Akhmat's Great Horde. The Tartars were attempting to cross the Ugra river to link forces with Polish King Casimir's army. Ivan's soldiers held the banks of the river for four days, defending their position with firearms against the Tartars, who had none. After the battle Akhmat withdrew his forces from Russia, ending eight years conflict.

1645: The first hospital in North America opens. Jeanne Mance, a French nurse, and three sisters in the Order of St. Joseph, founded the Hôtel-Dieu de Montréal (Montreal House of God). The hospital burned down and was rebuilt three times over the next century. In 1901 the hospital opened a nursing school, and in 1996 it joined the University of Montreal health network.

1806: During the Napoleonic Wars, British troops bombard the port of Boulogne with newly-invented Congreve rockets. High winds and rough seas made many of the rockets miss their intended targets. The British only destroyed a single French ship, but the barrage had a devastating psychological effect and dissuaded Napoleon from attempting to invade England.

1829: The Rocket, George Stephenson's steam locomotive, wins the Rainhill Trials. The Trials were held in England to find the best design for a new railway, the first to run only on steam power. The Rocket was the basic design that was used for steam-driven locomotives for the next 150 years.

1871: The Great Chicago Fire breaks out. A barn owned by the O'Leary family on the city's near West side caught fire. The blaze quickly spread to surrounding buildings, which like most of Chicago's buildings were made of wood. There had been very little rain all summer, and the city was a tinderbox. By the time firefighters arrived in horse-drawn carriages, the fire was already moving out of the neighborhood and toward the central business district. A strong wind helped the conflagration jump across the Chicago River and being moving into the heart of the city and toward the North Side. Chicagoans fled into Lake Michigan to escape the flames, which burned until rain finally began to fall on October 10. Three hundred people were killed and most of the city was destroyed. Only five buildings were left standing in the burned area, which spread over 2,100 acres. In 1961 the Chicago Fire Department's training academy was built on the spot where the O'Learys' barn had originally stood.

1921: Pittsburgh radio station KDKA does the first live broadcast of a football game.

1956: New York Yankee Don Larsen pitches a perfect game in the World Series. In game five, Yogi Berra caught Larsen's pitches to the Brooklyn Dodgers over nine innings of flawless baseball. Larsen threw only 97 pitches, striking out seven batters and walking none. After the last out, Berra ran to Larsen and jumped into his arms; a photograph of the moment has become known as "the everlasting image."

1982: *Cats* opens on Broadway. The Andrew Lloyd Webber musical won seven Tony Awards, including Best Musical, Best Original Score, and Best Actress, which went to Betty Buckley for her performance as Grizabella.

2001: President George W. Bush announces the creation of the Office of Homeland Security. Former Pennsylvania Governor Tom Ridge headed the new anti-terrorism office.

October 9

768: Brothers Carloman and Charlemagne are crowned Kings of the Franks. Following the death of their father Pepin, the brothers jointly assumed the throne. Carloman died three years later. Charlemagne ruled until his death in 814. He was proclaimed the first Holy Roman Emperor in 800 after uniting much of central Europe under his rule.

1604: Kepler's supernova is first observed by astronomers on Earth. The supernova, which occurred in the constellation Ophiuchus in the Milky Way, was the most recent one to be seen from Earth. Although Kepler was not the first to observe it, he tracked it for a year and literally wrote the book on it, and it was subsequently named after him.

1635: Roger Williams is banished from the Massachusetts Bay Colony. The former Governor of the Colony, Williams had angered authorities when he questioned the colonists' right to settle land without compensating the Narragansett Indians who lived there. He was convicted of sedition and heresy for his views and banished from the Colony. He survived the winter when Wampanoag and Massasoit Indians gave him shelter in their camp. Williams later founded Providence, Rhode Island, on land he purchased from the Massasoit.

1701: The Collegiate School of Connecticut is founded in the Saybrook Colony. It was moved to New Haven and renamed Yale College after a benefactor. The Ivy League school awarded its first diploma to Nathaniel Chauncey in 1702.

1919: "Say it ain't so, Joe." The Cincinnati Reds beat the Chicago White Sox to win the World Series. Eight baseball players for Chicago were later accused of throwing games for pay in what became known as the Black Sox Scandal. Chick Gandil, the catcher for the White Sox, recruited seven other players who would split $100,000. At the time, Charles Comiskey paid his players just $4,000 per year. Pitcher Eddie Cicotte and out-fielder "Shoeless" Joe Jackson were two notable conspirators in the scheme. The following year, the team was dogged by rumors about the fix, and a grand jury was convened to investigate. Cicotte confessed his involvement to the grand jury, and the eight players were indicted but found not guilty by a jury. However, newly appointed Commissioner of Baseball Kenesaw Mountain Landis, determined to "clean up" baseball, banned them all from playing for the rest of their lives.

1967: Bolivian soldiers execute Ernesto "Che" Guevara. The Argentinian Marxist and Cuban revolutionary had been trying to mount a guerilla war to overthrow Bolivia's military government. His body was recovered and reburied to Cuba in 1997.

1986: *The Phantom of the Opera* opens at Her Majesty's Theatre in London's West End. The musical by Andrew Lloyd Webber eventually became the second longest-running musical in the world. It was performed for the ten thousandth time in October 2010.

2006: North Korea conducts its first test of a nuclear bomb. The North Korean military detonated an underground nuclear device after years of rumors that they were attempting to acquire atomic weapons. The government later issued an apology for conducting the test and agreed to return to international talks about disarmament.

October 10

680: Al-Husayn ibn Ali, the grandson of the Islamic Prophet Muhammad, is decapitated in the Battle of Karbala. Ali had refused to pledge allegiance to Yazid I, the Caliph of the Umayyad dynasty, and his small band of followers were overwhelmed by a much larger force of Umayyad fighters. His death is commemorated by Shia Muslims as the Day of Ashura.

1846: English astronomer William Lassell discovers Triton, Neptune's largest moon. Lassell was a brewer and amateur astronomer who made several improvements on the reflecting telescope. He began searching Neptune for moons soon after the planet was discovered, and found Triton after just 17 days.

1865: John Wesley Hyatt patents the first billiard ball made of celluloid. Hyatt was an earlier pioneer in the development and application of celluloid, the first plastic to be produced industrially. Before his discovery, billiard balls were made of ivory.

1903: Emmeline Pankhurst founds the Women's Social and Political Union in the United Kingdom. Pankhurst was active in the women's suffrage movement beginning when she was only 14 years old in 1879. She founded the WSPU as an all-women's organization dedicated to using civil disobedience and direct action to agitate for universal suffrage in England. The group was the first to be referred to as *suffragettes*. They held demonstrations in Parliament, broke windows in government buildings, and set fires. When imprisoned, they went on hunger strike. The government's orders to force-feed jailed WSPU activists won them broad sympathy in the public. By 1908 the organization was able to draw half a million supporters to a demonstration in London's Hyde Park. In 1914 Parkhurst was arrested outside of Buckingham Palace while attempting to deliver a petition to King George V. In 1918, nearly forty years after she began campaigning for women's suffrage, an Act of Parliament extended the vote to women.

1938: European powers sign the Munich Agreement with Nazi Germany. With the agreement, which British Prime Minister Neville Chamberlain said would mean "peace in our time," Czechoslovakia ceded territory to Nazi Germany. The appeasement did not work. The next year Germany invaded the rest of Czechoslovakia and Poland, beginning World War II.

1953: The United States and Republic of Korea sign a mutual defense treaty. The treaty, which was signed three months after the Korean Armistice Agreement ended fighting in the Korean war, provided for the US and Korea to provide mutual aid should either be attacked. As a result of the treaty the US stationed military forces in South Korea.

1967: The Outer Space Treaty goes into effect. The treaty was the basis for international space law. It forbids placing weapons of mass destruction in orbit, on the Moon, or any other celestial object as well as in outer space. It also prohibits any nation from claiming the Moon or any other celestial body as its territory.

1973: Spiro Agnew resigns. Agnew, Nixon's Vice President, had accepted bribes while he was still governor of Maryland and in the White House. He agreed to a plead guilty to federal income tax evasion and was sentenced to three years' probation.

October 11

1634: The Buchardi Flood kills 15,000 people in northern Germany. A strong storm combined with an incoming tide drove seawater over protective dikes and into low-lying coastal areas. The flood washed away whole towns and farmland. It also destroyed the island of Strand, creating three separate islands where the island had been.

1809: Meriwether Lewis dies in Tennessee under mysterious circumstances. Lewis, who had explored much of the Louisiana Purchase Territory with William Clark a few years prior, was staying at an inn in Natchez, Tennessee. The innkeeper heard gunshots from his room, and when she investigated, found him bleeding from wounds to his head and stomach. Although some contemporary accounts suspected Lewis was murdered, Clark concluded that the death was a suicide.

1890: The Daughters of the American Revolution is founded. The previous year was the 100th anniversary of George Washington's inauguration as the United States' first president. Centennial celebrations during that year rekindled patriotic interest in the history of the nation. Inspired by the story of the Revolutionary War patriot Hannah White Arnett, Mary Smith Lockwood published a newspaper article calling for the formation of a patriotic society. Arnett's great-grandson read the article and offered to help form the new society, which became the Daughters of the American Revolution. Its first General President was then-First Lady Caroline Scott Harrison, the wife of President Benjamin Harrison. The organization took the motto "God, Home, and Country." Its roughly 185,000 members are all directly descended from someone who was involved in the American Revolutionary War effort.

1899: The American League is created from the Western League. Originally a minor league, the American League became a major league after the American Association was broken up. The Boston Red Stockings, which became the Atlanta Braves, are the inaugural teams still in the AL.

1958: NASA launches Pioneer 1. The probe was the first spacecraft launched by the newly created agency. Pioneer 1 was designed to study the surface of the Moon. An equipment failure prevented the spacecraft from reaching the Moon, however. It remained in orbit for 43 hours before crash landing in the Pacific Ocean.

1984: Astronaut Kathryn Sullivan becomes the first American woman to perform a spacewalk. Sullivan engaged in extra-vehicular activity (the official term for a spacewalk) on the Space Shuttle Challenger for 3 hours with astronaut David

Leestma. The pair tested a system for refueling satellites during the spacewalk. Sullivan served on two more Shuttle missions, and later was the head of the National Oceanic and Atmospheric Administration.

1986: US President Ronald Reagan and Soviet Premier Mikhail Gorbachev meet at the Reykjavík Summit in Iceland. The Cold War adversaries discussed the elimination of nuclear weapons. The talks fell apart when Reagan refused to compromise on a missile defense system called the Strategic Defense Initiative. But the progress Reagan and Gorbachev made at Reykjavík led to a 1987 arms treaty between the two nations.

October 12

539 BC: Cyrus the Great's army captures Babylon (present-day Hilla, Iraq). The city was protected by the Euphrates river. Cyrus had his forces dig a canal to divert the river so they could wade across it in the middle of the night. Under his rule, the Persian empire eventually stretched from the Mediterranean Sea to the Indus River. It was the largest empire in history at its peak.

1492: Christopher Columbus lands in the Bahamas. The Italian explorer thought he was in Southeast Asia, or the "Indies." Columbus called the island he landed on "San Salvador," or "Holy Savior." The island's native Taino and Arawak people called it Guanahani. They met Columbus on the beaches and were friendly toward him and his men. Columbus noticed the Arawak's gold jewelry and took them prisoner to try to force them to turn over more gold. In his journal entry from that day, Columbus wrote "these people are very simple in war-like matters. I could conquer the whole of them with 50 men, and govern them as I pleased." On later voyages, Columbus and other Spanish colonists killed tens of thousands of Arawak and Taino people, whom they enslaved and forced to search for gold.

1773: The first insane asylum opens in North America. The Eastern State Hospital was opened in Williamsburg, Virginia, to house and care for mentally ill people. The House of Burgesses, Virginia's colonial legislature, established the asylum at the urging of the governor. Before the hospital was built, mentally ill people were put in jail after being "diagnosed" by a jury of twelve ordinary citizens. The hospital gave regular tours and parades of the patients.

1799: Jeanne Geneviève Labrosse is the first woman to jump from a balloon with a parachute. Labrosse studied ballooning with André-Jacques Garnerin, a French aeronautical inventor. Labrosse jumped 2,952 feet (900 meters) and parachuted safely to the ground. In 1802, she patented the first parachute design. She continued publicly demonstrating her own design, eventually parachuting from a height of 8,000 feet (2,438 meters).

1810: The first Oktoberfest celebration is held in Munich. The Crown Prince and Princess of Bavaria invited the people of Munich to their wedding celebration, which was held in fields outside the city. Horse races and a week-long fair were held. The festival has been held every year since, with Oktoberfest beer brewed for the occasion.

1823: Charles Macintosh invents the waterproof raincoat. Macintosh was a Scottish chemist who invented a new type of tarpaulin, or waterproof, material by sandwiching treated rubber between two pieces of fabric. Early versions of the Mackintosh raincoat tended to melt in hot water, but he kept working on the design. He eventually patented a method for vulcanizing the rubber, which greatly improved the coats' weatherproofing.

1901: President Theodore Roosevelt renames the Executive Mansion the White House. Roosevelt had the mansion expanded and renovated to modernize it. He also had the executive staff moved to the West Wing of the White House, where they have worked since.

1945: Desmond Doss is the first conscientious objector to be awarded the Medal of Honor. Doss became a conscientious objector because of his religious upbringing. He served as a combat medic with an infantry company. He was wounded in action four times while assisting wounded comrades during the battle of Okinawa.

1984: British Prime Minister Margaret Thatcher narrowly escapes a bomb planted by the Irish Republican Army. Thatcher's hotel was bombed, and five people were killed. She delivered a planned speech the next day despite the bombing, which made her widely popular in Britain.

October 13

54: Emperor Claudius is poisoned. Claudius's 17-year-old stepson Nero ascended the throne after his death. Many historians believe Claudius's wife Agrippina, Nero's mother, was behind the poisoning.

1269: Westminster Abbey is consecrated. King Henry III chose the location of the church as his own burial site in 1245. The original building was added to over the centuries. It has been the site of coronation and burial for English and British kings and queens for nearly 800 years.

1773: Charles Messier discovers the Whirlpool Galaxy. Messier, a French astronomer, was searching for celestial objects when he found the galaxy. In 1845 it was the first galaxy to be classified as a spiral galaxy after improved telescopes allowed William Parsons to get a better look at it.

1843: Henry Jones founds B'nai B'rith, a Jewish service organization. Jones and eleven other German Jewish immigrants in New York City founded the group to help other new arrivals in America. It eventually grew to be an international mutual aid and philanthropic organization, with hospitals and orphanages around the world.

1884: The International Meridian Conference sets the prime meridian for international use. US President Chester Arthur requested the conference in order to standardize time calculations globally. The recent spread of the railroad had made standardization necessary. The prime meridian was set as the longitude that passes through the Royal Observatory in Greenwich, London.

1928: The first iron lung is used. The contraption was used at Boston Children's Hospital to assist an eight-year-old girl who could not breathe on her own because of advanced polio. Within a minute of placing the patient in the lung, she improved significantly. The iron lung was widely used until polio vaccination programs nearly eradicated the disease in the US.

1967: In the first American Basketball Association game, the Oakland Oaks beat the Anaheim Amigos, 134–129. The ABA remained in existence until financial difficulties forced it to disband in 1976 and merge with the older National Basketball Association.

1983: Ameritech Mobile Communications establishes the first public cellular phone service. The service was rolled out in Chicago, Illinois. One subscriber, David Contorno of Lemont, Illinois, kept his Ameritech cell phone number since August 2, 1985, making it the longest continually used, according to the *Guinness Book of World Records*.

2010: The Copiapó mining accident concludes with all 33 miners surviving. A cave-in had trapped the miners underground in August. They were underground for a record 69 days while they awaited rescue.

October 14

1066: The Battle of Hastings is fought. Norman forces under William of Normandy defeated an Anglo-Saxon army commanded by Harold Godwinson, who was killed in the battle.

Following the victory, William assumed rule of England. The famed Bayeux Tapestry, which was commissioned by William's family, depicts the events leading up to the battle as well as the conflict itself.

1773: Patriots set several British East India Company ships ablaze in the port of Annapolis. The perpetrators were never caught. The episode occurred just before the American Revolutionary War broke out in earnest.

1888: Louis Le Prince films the first motion picture. *Roundhay Garden Scene* depicts the filmmaker's family and friends walking around the gardens of Oakwood Grange, in England. The film has a run time of just over two seconds.

1908: The Chicago Cubs win the World Series. The baseball club would not win another World Series for over a century. The alleged cause of the ensuing championship drought—the longest in sports history—eventually became known as the "Curse of the Billy Goat." While the Cubs reached the World Series in 1945, they failed to win it. During game four of the 1945 Series, William Sianis, owner of the local Billy Goat Tavern, was ejected from the ballpark. His goat, Murphy, whom he had brought to Wrigley Field with him, was attracting complaints from other fans because of its smell. Upon being thrown out, Sianis allegedly shouted "Them Cubs, they ain't gonna win no more!" In 2016, the Cubs finally broke the curse when they defeated Cleveland in seven games to win the World Series.

1912: John Schrank, a Milwaukee saloon keeper, shoots Theodore Roosevelt during a campaign stop. The bullet struck Roosevelt in the chest, but was slowed by the 50-page text of the former president's speech in his breast pocket. Although the shot wounded Roosevelt, he delivered his scheduled speech with the bullet still in his chest. Schrank, who claimed William McKinley had told him to shoot Roosevelt in a dream, was found to be mentally ill and placed in an institution until his death in 1943.

1926: A. A. Milne publishes _Winnie the Pooh_. The children's book describes the adventures of a teddy bear and his friends. It was widely popular and spawned three more books, as well as numerous television and film adaptations.

1947: Chuck Yeager breaks the sound barrier. The Air Force Captain flew a Bell X-1 rocket-powered plane named _Glamorous Glennis_ at Mach 1.06, or 700 mph, becoming the first pilot to do so in level flight.

1962: The Cuban Missile Crisis begins. A US Air Force U-2 spy plane flew over Cuba and took pictures of Soviet missile launchers being built on the island. The US established a naval blockade to prevent more missiles from being delivered, and President Kennedy demanded the USSR remove the missiles. After an 11-day standoff, Soviet Premier Khrushchev agreed to remove the missiles in exchange for the US dismantling nuclear missile launchers in Turkey.

October 15

1815: Napoleon begins his exile on Saint Helena. The deposed French emperor was sent to the island after surrendering to the British. He had attempted to reclaim his throne by force after returning from being exiled to Elba, off the Italian coast. Defeated at the Battle of Waterloo, Napoleon fled to Paris and considered escaping to the United States before Prussian and British troops closed in. Determined to prevent him from ever returning, the British sent him to Saint Helena—1,162 miles west of Africa in the Atlantic Ocean. He died there in 1821.

1878: The Edison Electric Light Company opens for business. Financed by industrial tycoon JP Morgan, Thomas Edison started the company to hold patents for his electric and lighting inventions. It went on to merge with another electric company in 1892 and form General Electric (GE).

1888: Jack the Ripper sends a letter "From Hell." Although the police investigating the serial killer received hundreds of letters that claimed to be from him, this was one of the few considered to possibly be real. The letter was sent, along with a human kidney, to George Lusk, the chairman of a volunteer citizens' patrol that operated in response to the killings. The letter taunted Lusk, and was signed "Catch me when you can Mishter Lusk." Although the letter and subsequent murders fueled a widespread manhunt that identified various suspects, no one was ever charged and the killer has never been identified.

1928: The *Graf Zeppelin*, the world's first passenger airship, makes its first flight across the Atlantic. The Zeppelin was hydrogen-filled and traveled from Germany to New Jersey in 111 hours. The airship traveled around the world the following year, visited the Arctic in 1931, and operated until 1937.

1951: The first episode of *I Love Lucy* premieres. The breakthrough television sitcom ran on CBS for 180 episodes, until May 6, 1957. It starred Lucille Ball, Desi Arnaz, Vivian Vance, and William Fawley as two couples living in New York City. It was the most-watched show on television for most of its run, and shaped the format and style of subsequent TV sitcoms.

1956: Fortran, the first modern computer language, is publicized. Engineers at IBM developed the language to replace hand-coding, which had been used until then.

1966: The Black Panther Party for Self-Defense is founded. Huey Newton and Bobby Seale started the organization in Oakland to monitor police activity in African-American communities. Their armed patrols and community social programs, such as the Free Breakfast for Children Program, soon attracted the attention of the FBI. Through an extensive and illegal domestic spying program called COINTELPRO, the government undermined, imprisoned, and assassinated leaders of the Black Panthers through 1972.

1989: Wayne Gretzky sets the record for the most career points scored in the National Hockey League. While playing for the LA Kings, Gretzky scored his 1,851st goal in a game against the Edmonton Oilers to break Gordie Howe's all-time points record. Gretzky went on to score a total of 2,857 points over the course of his career.

2001: NASA's *Galileo* spacecraft passes Jupiter's moon Io. The probe came within 112 miles of the moon and photographed volcanoes on its surface.

October 16

1384: Jadwiga (or Hedwig), daughter of Louis the Great, is crowned monarch of Poland. She was the first woman to rule the Kingdom of Poland, and reigned until her death in 1399. During her reign, Jadwiga converted many Polish pagans to Catholicism.

1793: French Revolutionaries execute Marie Antoinette. The Queen of France had become a target of the Revolution's ire for her lavish lifestyle and reported animosity toward social reforms to help the poor and working-class citizens. She was convicted of high treason by a Revolutionary Tribunal and guillotined at the Place de la Revolution in Paris.

1847: Charlotte Brontë publishes *Jane Eyre*. The novel revolutionized literature in that it was the first to feature the private thoughts of a protagonist in a first-person narrative style. Brontë used the plot to comment on class and social issues of the day.

1859: Abolitionist John Brown attacks Harpers Ferry, West Virginia. Brown led a group of 22 armed abolitionists in attacking the federal arsenal in the river town to try to spark a slave uprising. His party was quickly surrounded and trapped at the arsenal by US Marines. After four days they surrendered. Brown was tried and executed for treason. He was revered as a martyr by the abolitionist movement.

1869: The Cardiff Giant is "discovered." Workers digging a well on William Newell's farm in Cardiff, New York, uncovered the 10-foot-tall supposedly petrified remains of an ancient human. The specimen had in fact been buried in the spot by George Hull, Newell's cousin. Hull had had the stone man carved out of gypsum by a stonecutter in Chicago after hearing a preacher at a revival meeting claim that giants had once walked the Earth, as described in the Book of Genesis. Newell and Hull had the giant buried on Newell's land and waited a year before hiring workmen to dig on the spot and "accidentally" discover it. They then charged spectators to see it at 50 cents per viewing. Although archeologists were quick to discount the stone giant as a forgery, Hull sold it to an entertainment syndicate for $23,000.

1909: US President William Taft and Mexican President Porfirio Díaz narrowly escape assassination. The two were at the first meeting of American and Mexican presidents when an armed assassin was apprehended by Texas Rangers along a planned parade route.

1940: The Nazis establish the Warsaw Ghetto. They imprisoned over 400,000 Jewish civilians there until 1943. During that time, they systemically shipped thousands to concentration camps. Following an uprising the Nazis murdered or deported the remaining several thousand Jews and razed the Ghetto.

1950: C.S. Lewis publishes *The Lion, the Witch and the Wardrobe*. The fantasy novel was the first book in the author's acclaimed *Chronicles of Narnia* series, which comprised seven books when complete.

1998: Augusto Pinochet is arrested in London for murder. The former Chilean dictator, who overthrew the Allende administration in 1973 and ruled until 1990, was tried for the numerous human rights violations that were committed under his rule. Pinochet was found guilty and placed under house arrest. When he died in 2006, more than 300 charges were still pending against him.

October 17

1091: A tornado strikes London. The tornado destroyed London Bridge, over 600 houses, and several churches. Modern meteorologists have estimated the tornado as a T8 on the TORRO intensity scale, a rating system of T0–T11. Despite its power, only two people of a population of 18,000 were killed in the tornado's destruction.

1781: British General Cornwallis surrenders at the Siege of Yorktown. With the surrender, fighting in the colonies during the American Revolution came to an end. General George Washington's Continental Army had trapped Cornwallis' British troops in Yorktown in September. The British held out for nearly a month before surrendering nearly 8,000 men, 144 cannons, and over 40 ships. The following Spring peace negotiations began in Paris, and in 1783 the Treaty of Paris officially recognized the independence of the United States of America.

1814: The London Beer Flood. At the Meux an Company Brewery, a giant vat of more than 158,000 gallons (600,000 liters) of beer ruptured. Beer cascaded through the brewery, causing other vats to fail in a chain reaction. Eventually a wave of more than 300,000 gallons of beer flooded into the streets surrounding the brewery. The amber wave demolished a wall, crushing the first victim in the debris. It swept through several homes, destroying two, and washed through a funeral home where a wake was being held. Five mourners drowned. Continuing down a side street, the beer overwhelmed a family who were having tea. When the beer subsided, ten people were dead. Although survivors sued the brewery, a jury ruled the accident was an Act of God. The brewery operated until 1922. A pub located on the site now brews a special porter to commemorate the flood.

1860: The first British Open is held. The championship tournament is the oldest major event in professional golf. It is the only one of the four to be held outside the United States, and was first played at Prestwick Golf Club in Scotland.

1931: Gangster Al Capone is convicted of income tax evasion. Rather than attempt to jail him for the many murders he was suspected of ordering, the government used the new approach of charging him with tax-related crimes. Capone was sentenced to 11 years in a federal prison and was paroled in 1939 after he fell ill from syphilitic dementia. He died in 1947.

1933: Albert Einstein flees Nazi Germany. Because of his Jewish ancestry, Einstein fled the Nazi regime for the United States, where he was granted asylum. The physicist took a position at the Institute for Advanced Study in Princeton, New Jersey. He later contributed to the Manhattan Project to develop an atomic bomb, a program he had privately urged President Roosevelt to pursue. He remained at the Institute until his death in 1955.

1973: The OPEC Oil Crisis begins. Members of the Organization of Arab Petroleum Exporting Countries (OPEC) declared an oil embargo against countries supporting Israel during the Yom Kippur War. The embargo caused oil prices to rise from $3 to $12 per barrel, causing severe shortages in the US and other Western countries. The embargo lasted until March 1974.

October 18

1648: Shoemakers and barrel-makers form the first American trade union. At the time guilds did not exist in the colonies. The workers banded together to establish prices and quality standards for their products, and apprentice programs for newcomers.

1767: The Mason-Dixon line is established. The boundary between the then-colonies of Maryland and Pennsylvania settled a border dispute between them. Surveyors Charles Mason and

Jeremiah Dixon established the line. It later became the unofficial border between the North and South.

1775: Phillis Wheatley is freed from slavery. She had already become famous in England and the American colonies for her brilliant poetry. She had first published a volume of her work in 1773. After she was freed, she continued to write poetry in support of the American Revolution.

1851: Herman Melville publishes *The Whale* in England to tepid reviews and poor sales. The American edition of the book, renamed *Moby-Dick*, garnered more favorable reviews but did not sell very widely. It wasn't until the book was rediscovered in the 20th century that it became popular and was eventually regarded as a literary classic.

1867: Russia formally cedes Alaska to the United States for $7.2 million. At the time, there were no known natural resources in the Alaska territory. American opponents of the deal called it "Seward's Folly" after the Secretary of State, who negotiated the sale. In 1896 gold was discovered in Alaska, and oil and natural gas were found in the 20th century.

1898: The United States seizes Puerto Rico. The Caribbean island was claimed by the Spanish Empire when the United States invaded it during the Spanish-American War. In March 1898 the US Navy blockaded San Juan, the capital, and landed troops on the southern coast. American forces quickly defeated Spanish infantry. Meanwhile the US Army was also fighting Spanish forces in the Empire's colonies in the Philippines and Guam, where the Americans were also victorious. After signing an armistice, Spain relinquished control of its colonies to the United States. Puerto Rico was made an American territory and ruled by a military governor. In 1917 the government extended citizenship to Puerto Ricans and allowed a local legislature to be established. Movements for independence and for statehood were waged throughout the 20th century, but the island remained an American territory.

1945: Klaus Fuchs, a nuclear physicist at Los Alamos National Laboratory, New Mexico, sends plans for an atomic bomb to the USSR. Fuchs' spying was uncovered in 1950. He was sentenced to 14 years in prison but released after nine years. He emigrated to East Germany upon his release, where he was awarded the Order of Karl Marx. Fuchs remained in East Germany until his death in 1988.

1963: The French space program sends Félicette, a cat, into space. A black-and-white stray cat found on the streets of Paris, Félicette was the first cat to travel in space. She was launched into sub-orbital flight aboard an AGI-47 rocket, and remained aloft for 13 minutes. She survived the flight and parachuted safely to Earth. French scientists later euthanized her to study her brain.

October 19

1453: French troops capture Bordeaux, ending the 100 Years' War. The war between England and France had begun in 1337 and comprised dozens of battles punctuated by sporadic détente between the European powers. With growing unrest at home the English were unable to continue pursuing the war. Although they remained in a state of official hostilities, the belligerents did not fight any major battles after 1453.

1469: Ferdinand II and Isabella I unify Aragon and Castile to form the Kingdom of Spain. The monarchs ruled until 1504, during which time they notably commissioned Christopher Columbus' voyages to the Americas. Ferdinand and Isabella also oversaw the expulsion of the Muslim caliphate from the Iberian peninsula during their reign.

1789: John Jay becomes the first Chief Justice of the US Supreme Court. Jay, a believer in Federalism and a strong central government, made multiple landmark rulings during his tenure on the bench. He established judicial review, the process by which judges can rule on the constitutionality of executive and legislative actions.

1813: Napoleon's army is defeated by Coalition forces at the Battle of Leipzig. The battle, which involved more than 600,000 soldiers, was the largest in history until World War I. Napoleon's defeat forced him to retreat to France, where he was overthrown by the Coalition and exiled to the island of Elba.

1864: In the St. Albans Raid, Confederates attack Vermont by way of Canada. After being captured during the Battle of Salineville in 1863, Confederate officer Bennet Young escaped to Canada. While there he met with Confederate government agents and proposed a raid on the Union from the north. The plan was designed to seize funds for the Confederacy and divert Union troops away from the fighting further south. The agents agreed, and Bennet recruited a force of 21 Confederate volunteers for the raid. They slipped into Canada over a period of several days in October before crossing into St. Albans, Vermont, where they simultaneously robbed the city's three banks. They sized over $200,000 and forced bank tellers to pledge allegiance to the Confederacy at gunpoint. The raiders killed one townsperson and wounded two others in a skirmish before escaping back to Canada. Canadian authorities arrested them and returned about $90,000 to the Union, but refused to extradite the raiders.

1943: Albert Schatz discovers streptomycin. Schatz, a PhD student at Rutgers University, was involved in a project to discover multiple antibiotics when he isolated the compound. Streptomycin was initially used to treat tuberculosis, which was previously incurable, and later administered for a variety of bacterial infections.

1973: President Nixon refuses to turn over the Watergate tapes. A Federal Appeals Court had ordered him to relinquish the tapes, which contained recordings of his secret conversations about the burglary of the DNC headquarters at the Watergate hotel and subsequent coverup. Nixon was eventually forced to surrender the tapes, which implicated him and forced his resignation.

2005: Saddam Hussein is put on trial for crimes against humanity. The former dictator of Iraq, deposed in the US invasion of 2003, was found guilty and executed in 2006.

October 20

1720: The Royal Navy captures Calico Jack. The English pirate had operated in the Bahamas and Cuba for two years. He is famous for having been the first pirate to fly the Jolly Roger skull-and-crossbones flag, and for including two women, Mary Read and Anne Bonny, in his crew. Jack was tried and hanged in November in Port Royal, Jamaica.

1803: The US Senate ratifies the Louisiana Purchase. The Purchase transferred control of more than 828,000 square miles of territory from the French Empire to the United States. The US cancelled $3,750,000 of French debt and paid another $15 million in exchange for the territory.

1818: The United States and United Kingdom agree on the border between Canada and America. The Treaty of 1818 established the 49th longitudinal parallel as the boundary between the two nations. It has remained the border ever since.

1873: Four universities establish the rules of American football. Ivy league colleges Yale, Princeton, and Columbia, and Rutgers University, a public institution, agreed on the code of rules to govern their budding football league.

1947: The House Un-American Activities Committee (HUAC) begins its investigation. HUAC investigated hundreds of private citizens in an effort to root out Communist sympathizers in Hollywood. It eventually blacklisted more than 300 directors, actors, and screenwriters suspected of being "disloyal" to the United States.

1951: The Johnny Bright Incident. During a college football game in Stillwater, Oklahoma, a white player attacked Johnny Bright, an African-American quarterback. Bright, an offensive phenomenon and Heisman trophy candidate for Drake University, was the first African American to play football at Oklahoma A&M's stadium. Oklahoma players knocked Bright unconscious three times during the game. After one play was over, Oklahoma player Wilbanks Smith attacked Bright again, breaking his jaw. Although Bright completed a 61-yard touchdown pass later in that series, he was eventually forced to leave the game because of his injuries. Newspapers published photos of the assault. Several teams pulled out of the university league in protest. The incident later resulted in the NCAA changing rules and mandating better helmets with face guards to prevent such injuries in the future. In 2006, Drake renamed their football stadium in Bright's honor.

2011: Libyan rebels capture ousted dictator Muammar Gaddafi. The rebels captured him near his hometown, Sirte. They tortured and summarily executed Gaddafi, who had ruled Libya with an iron fist since he'd seized power in 1969.

October 21

1520: Ferdinand Magellan discovers the strait that would be named for him. The Strait of Magellan is the southernmost sea route around South America. Magellan was attempting to sail around the world when he found the passage around the tip of what is now Chile. After scouting the strait, his fleet passed through it on November 1.

1797: The USS *Constitution* is launched from Boston Harbor.
The 44-cannon, three-masted frigate is the oldest commissioned warship in the United States Navy. It saw action during the War of 1812, during which its crew captured multiple merchant vessels and five British warships. In 1907 the ship was turned into a floating museum.

1805: Admiral Nelson leads a British fleet to victory in the Battle of Trafalgar. Nelson fought a combined fleet of Spanish and French ships in the battle during the Napoleonic Wars. Twenty-seven British ships fought a force of 33 ships. Nelson beat the larger force by abandoning conventional battle practices. Normally two fleets would line up parallel to one another and exchange fire. Nelson sailed his fleet in a perpendicular maneuver toward the center of the combined fleet's line. Although he was mortally wounded in the battle, the tactic worked.

1824: Joseph Aspdin invents Portland cement. Aspdin was an English cement manufacturer. He used limestone from a quarry on the isle of Portland in his mixture, which improved upon existing designs. Portland cement remains the most widely used cement in the world today.

1892: The World's Columbian Exposition is dedicated in Chicago. The world's fair was held to commemorate the 400th anniversary of Columbus' arrival in the Americas. The fairground covered more than 600 acres and featured almost 200 temporarily-built structures surrounded by lagoons and canals near Chicago's Hyde Park neighborhood. Exhibition areas featured displays of the latest technological advances, including electric lights that illuminated many of the fair buildings. Although the fair's dedication and opening ceremonies were held in October, the exposition did not officially open to the public until the following May.

1921: President Warren G. Harding delivers an anti-lynching speech in Alabama. Following Reconstruction, white supremacists in the Deep South maintained the Jim Crow segregationist social order with violence and intimidation. Black men and women were murdered in illegal hangings known as lynchings. The practice became so widespread that by 1920 the NAACP alleged that two African Americans were lynched every week. Harding supported legislation outlawing lynching. The bill was ultimately filibustered in the Senate by Southern Democrats and never made it to a vote.

1959: The Guggenheim Museum opens in New York City. Architect Frank Lloyd Wright designed the art museum, which is located on Fifth Avenue in Manhattan. The Guggenheim attracts nearly a million visitors every year.

2005: Astronomers discover Eris, a dwarf planet. At the Paloma Observatory, Mike Brown and colleagues found the planet while examining images taken two years earlier by the Hubble telescope of the outer solar system. Eris is about a quarter of the Earth's mass and has one moon.

October 22

794: Emperor Kanmu moves the Japanese capital to present-day Kyoto. Originally called *Heian-ky*, meaning "tranquility and peace capital," Kyoto remained the official capital of Japan for more than a millennium. In 1868 the imperial capital was moved to Tokyo.

1790: Miami warriors defeat US Army troops in a battle in Indiana. Under the command of Chief Little Turtle, the Miami resisted American incursions into the Northwest Territory from 1785 until 1795, when they signed a treaty establishing Native lands in Indiana and Ohio. After the war, Little Turtle was received by George Washington, who presented him with a ceremonial saber. He died in 1812 and was buried in Indiana.

1844: Millerites await the end of the world in what became known as the "Great Anticipation." Devotees of William Miller, an American Baptist preacher, the Millerites believed this day was the date of the Second Coming of Christ. When the world did not in fact end on October 22, the following day came to be called the "Great Disappointment."

1879: Thomas Edison invents the incandescent light bulb. Edison used a filament of thread that he had treated with carbon in the light bulb's design. The design allowed the bulb to burn for 13 hours, a major advance at the time.

1964: Jean-Paul Sartre declines the Nobel Prize in Literature. The French existentialist had sent a letter to the Nobel jury to inform them of his intention to decline the prize if it were offered, but the letter arrived too late. Sartre turned down the Nobel anyway. In the letter, which the Swedish Academy kept secret until 2001, Sartre explained that he did not wish to be considered a cultural "institution," saying he always turned down "official honors."

1976: The US Food and Drug Administration bans Red Dye No. 4. The additive had been widely used in commercial food coloring. It was banned after experiments found that it caused bladder tumors in animal test subjects.

2001: The video game *Grand Theft Auto III* is released. It was the first game to use an open-world format. The format allowed players to explore a large playing area and find and use vehicles and other items. Its graphic violence spawned controversy, but the game was widely popular regardless and led to several sequels.

2014: Michael Zehaf-Bibeau attacks the Canadian Parliament. In the attack, he opened fire on a ceremonial guard, killing him. Parliament security then opened fire on Bibeau, killing him. The RCMP later classified the shooting as a terrorist attack, although there was evidence Bibeau was suffering from mental illness.

October 23

42 BC: Marcus Brutus commits suicide after his army is defeated at the Battle of Phillipi. Mark Antony and Octavian's forces won a decisive victory against an army led by Brutus and other leaders who had assassinated Julius Caesar two years earlier. Rather than be captured, Brutus ran into his own sword, held by two of his captains.

1850: The first National Women's Rights Convention in the US is held. Lucy Stone and Paulina Wright Davis presided over the convention, which was held in Worcester, Massachusetts. About 900 people attended. The convention stated its objective was "to secure for women political, legal, and social equality with man." Susan B. Anthony did not attend, but later said that reading a speech delivered at it by Stone convinced her to join the women's rights movement.

1861: President Abraham Lincoln suspends *habeas corpus*. At the beginning of the Civil War, rioting broke out in Baltimore. The unrest threatened the Union's ability to move troops by railroad to Washington, DC. Lincoln suspended the writ of *habeas corpus* (Latin for "Bring the body"), which allows prisoners to defend themselves in court, to combat the situation. With *habeas corpus* suspended, Union commanders were able to imprison rioters and Baltimore officials without trial. The Supreme Court, under Chief Justice Roger Taney, ruled that the suspension was unconstitutional. But Lincoln considered the ruling invalid in a time of armed resistance to the government and ignored it. Lincoln later suspended *habeas corpus* nationally for anyone suspected of interfering with the military draft or aiding the Confederacy. Congress officially suspended the writ with an Act passed in 1863. At the end of the Civil War the Act became null.

1911: The first military aviation takes place. During the Italo-Turkish War, an Italian pilot in Libya flew over Turkish army positions on a reconnaissance mission.

1917: In Russia, the Bolsheviks call for an armed revolution. Lenin, their leader, gave a speech in which he declared that Russia had waited long enough for an armed uprising and urged the Bolsheviks to seize power.

1935: The Chophouse Massacre: New York mobster Dutch Schultz and three associates are murdered in a saloon in New Jersey. Schultz had attempted to murder attorney Thomas Dewey, who was leading the prosecution of the gangster for tax evasion. But he'd done so without permission from the Mafia Commission, and they ordered him killed in retaliation.

1944: The Battle of Leyte Gulf begins. The naval battle was the largest in history, involving nearly 400 ships and about 2,000 aircraft. It was the first battle in which Japanese pilots engaged in kamikaze, or suicide, attacks. The battle lasted three days and resulted in a decisive Allied victory that gave them control of the Pacific for the remainder of the war.

October 24

1260: King Louis IX dedicates Chartres Cathedral. The cathedral took 26 years to complete. It was one of the first to use flying buttresses, which allowed it to have much larger windows than previously built structures.

1648: The peace of Westphalia is ratified. The peace treaty between the Westphalian cities of Osnabrück and Münster, in modern-day Germany, ended a period of religious conflict that had lasted more than a century. The treaty established the principle of international sovereignty, in which each nation state has exclusive control over its own territory, for the first time. It is considered the beginning of the modern system of statehood.

1851: William Lassell discovers moons orbiting Uranus. Lassell was an English astronomer who made improvements on the reflecting telescope. He named the moons Ariel and Umbriel.

1857: Sheffield F.C. is founded. The association football—or soccer—club in Sheffield, England, is the oldest still in operation. The club plays in the Northern Premier League Division. It originally played under a unique set of rules known as the Sheffield Rules, but adopted Football Association rules in 1878.

1901: Annie Edison Taylor is the first person to go over Niagara Falls in a barrel. Taylor, a 63-year-old unemployed widow, decided on the stunt as a financial venture. She hoped to use the experience to make her famous and secure work as a public speaker. Taylor had a barrel custom made from oak and iron with a mattress placed inside for padding. After she climbed in the barrel, her friends screwed the lid shut and compressed the air inside with a bicycle pump. They put the barrel in the Niagara River above the falls, and it soon went over. Rescuers found the barrel after about 20 minutes. Taylor emerged uninjured except for a cut on her head. Taylor was dubbed the "Queen of the Mist" for her stunt, and spent some time on the speaking circuit and selling souvenirs near the Falls. Her manager, Frank Russel, stole her barrel, and she spent much of her money on private detectives attempting to track him down. She died in 1921.

1926: Harry Houdini performs for the last time at Garrick Theater in Detroit. Two days earlier, a medical student named Jocelyn Whitehead had punched the illusionist in the stomach in a surprise attack. As a result of the blows and in severe pain, Houdini consulted a doctor, who diagnosed him with acute appendicitis. Against the doctor's orders and despite a fever of 102 degrees Fahrenheit, he went on with show. He passed out during the performance, was revived, and continued. Houdini died the following week.

1929: Wall Street crashes. In the worst stock market crash in US history, Wall Street lost 11 percent of its total value at the opening bell. The slide continued over the next few days, and by the end of the week, more than $30 billion had been wiped out. The crash marked the beginning of the Great Depression in the United States.

1945: The United Nations is founded. The organization was established with the goal of preventing another global conflict like World War II, which had just concluded. At its founding the UN had 50 member-states. This number eventually grew to more than 190.

1975: The Iceland National Women's Strike is held. About 90% of women in Iceland took part in the one-day work stoppage to protest gender inequality and unfair employment discrimination. The strike effectively shut down the country. The following year parliament passed a law guaranteeing equal rights to women.

1992: The Toronto Blue Jays win the World Series. In doing so, they became the first Major League Baseball team from a country other than the United States to win it.

October 25

1415: Henry V's army wins the Battle of Agincourt. Fought between English infantry and French cavalry during the Hundred Years' War, the battle was held on Saint Crispin's Day in northern France. Although the French troops outnumbered the English by more than two-to-one, Henry's troops were able to inflict heavy casualties with the use of the longbow. The battle marked the beginning of the dominance of longbowmen in medieval warfare and the decline of heavy cavalry. The battle was immortalized in William Shakespeare's play *Henry V*. In the play, the King urges his troops into battle with the famous St. Crispin's Day Speech. In it, the King calls them a "band of brothers" who will forever remember where they were that day.

1760: George III is crowned King of Great Britain and Ireland. During his reign the British fought the Seven Years' War (or French and Indian War) in the American colonies. They also notably lost the American Revolutionary War. As his mental health declined, he was replaced on the throne by his eldest son, George IV, who ruled as Prince Regent until his father's death.

1854: The Light Brigade charges at the Battle of Balaclava during the Crimean War. In the battle, a detachment of British light cavalry charged Russian positions. Due to a miscommunication, they attacked the wrong section of the Russian lines and were exposed to intense artillery fire. The attacking cavalry suffered heavy casualties and gained no ground. The attack was later immortalized by Alfred, Lord Tennyson, in his poem "The Charge of the Light Brigade."

1917: The Bolsheviks storm the Winter Palace. Lenin ordered Red Guards to assault the seat of Imperial power in Russia. Although the Bolsheviks later portrayed the attack as a ruthless battle, it was actually relatively bloodless. The palace guards and military cadets surrendered almost immediately. The attack started a civil war that ended in Bolshevik victory in 1921.

1940: Benjamin O. Davis Sr. becomes the first African-American general in the US Army. Davis enlisted in the Army in 1898 during the Spanish-American War as a first lieutenant. He was promoted to Brigadier General during World War II and retired from the military in 1948.

1973: The Yom Kippur War ends in a ceasefire. The war had begun when a coalition of Egyptian and Syrian troops attacked the Sinai and Golan Heights. It lasted just three weeks and resulted in a decisive Israeli victory.

1983: The United States invades Grenada. The Caribbean island was undergoing internal civil strife, and the US invaded ostensibly to rescue American medical students trapped there. After four days, American troops seized control and defeated Cuban military elements on the island.

October 26

1776: Benjamin Franklin sails to France to seek aid for the American Revolution. The diplomat was successful in recruiting the Marquis de Lafayette to the American cause, and soon won over the French King as well. France's support, particularly that of its Navy, was vital in helping the American Revolutionary War effort succeed.

1825: The Erie Canal opens. The canal ran 363 miles form Albany to Lake Erie and created a navigable water route from the Atlantic Ocean to the Great Lakes. Its establishment shortened the travel time from New York to Lake Erie by more than a week. The canal made New York City an important port and helped cement its civic and cultural dominance in the Northeast. With the spread of the railroad in the mid-19th century, the canal's importance declined.

1881: In Tombstone, Arizona, the Gunfight at the O.K. Corral takes place. The shootout was the result of an ongoing feud between sheriff's deputies and a gang of outlaws calling themselves the Cowboys. Wyatt Earp and his deputies had been subject to multiple death threats by the Cowboys because they interfered with their horse stealing and cattle rustling operations. To wrest control of Tombstone from lawless elements, Earp had the city council pass an ordinance outlawing the carrying of weapons in town limits. On the afternoon of October 26, Earp, his two brothers, and friend Doc Holliday confronted a group of Cowboys who were carrying arms. They tried to disarm them, but the men refused to surrender their pistols. Each side later claimed the other shot first. After the smoke cleared, three Cowboys lay dead. The shootout was the most famous gunfight in the Old West.

1892: Ida B. Wells publishes *Southern Horrors*. The book detailed accounts of lynchings in the Deep South. It alleged that lynch mobs used false charges of rape to hide their real motives, which were mainly the enforcement of African Americans' second-class status in the South. The book earned national support for the anti-lynching movement, a forerunner of the Civil Rights Movement of the 20th century.

1921: The Chicago Theatre opens. The theater was originally a motion picture house, but eventually shifted to mostly showing live performances. It is still in operation in Chicago's Loop. It was added to the National Register of Historic Places in 1979.

1977: The last natural case of smallpox is diagnosed. Ali Mao Maalin, a Somalian health worker, was the last person to be infected with the disease. He made a full recovery. The World Health Organization marks this date as the anniversary of the worldwide eradication of smallpox.

2001: President George W. Bush signs the Patriot Act into law. The legislation was passed in response to the September 11 attacks. It greatly expanded the government's ability to search telephone, email, and financial records without warrants. The act also authorized the indefinite detentions of immigrants and persons designated "enemy combatants."

October 27

312: Emperor Constantine has a vision of the Christian Cross. During the Battle of Milvian Bridge, Constantine, who worshipped the Sun at the time, had a dream in which he was commanded to place the sign of the Cross on his soldiers' shields. He did so, and his army was victorious. Constantine converted to Christianity as a result. As Roman Emperor, he helped spread the new religion across Europe.

1838: Missouri governor Lilburn Boggs orders all Mormons to leave the state. Boggs issued the infamous Extermination Order after a battle between Mormons and the Missouri State Militia during the Mormon War of 1838. The Order declared that Mormons had to leave the state or "be exterminated." As a result, Mormons were violently expelled from Missouri.

1904: The New York City subway opens for business. Before its construction, several aboveground public transit services and elevated trains had operated in Manhattan since 1868. A major blizzard in 1888 shut nearly all of them down and convinced civic leaders of the need for an underground transit line. Construction began in 1900. The original subway ran from City Hall to 145th Street. That branch was expanded two years later to run all the way to 225th Street. Over the following decades, the subway continued to be expanded until it eventually comprised 472 stations, more than any other metro transit system.

1914: The first British battleship is sunk in World War I.
The dreadnought battleship HMS *Audacious* weighed over 23,000 tons and had ten 13.5-inch artillery guns. It struck a German mine off the coast of Ireland at 8:45 A.M. and sank. There was one casualty as a result. The sinking was kept secret until the end of the war.

1936: Wallis Simpson obtains a divorce decree. It would allow her to marry King Edward VIII of the United Kingdom. Because she was an American commoner, Edward was forced to abdicate the throne in order to marry her. He was on the throne for less than a year.

1954: Benjamin O. Davis, Jr., becomes the first African-American general in the US Air Force. The son of the first African-American general in the US Army, Davis Jr. commanded the Tuskegee Airmen during World War II. In 1998 he was made a four-star general, the highest rank in the Air Force.

1967: Phillip Berrigan and accomplices occupy the Draft Offices in Baltimore to protest the Vietnam War. While there, Berrigan, a Catholic priest, poured blood on Selective Service records. He was sentenced to six years in prison for the protest.

1988: Soviet listening devices are discovered in the US Embassy in Moscow during construction. As a result of the discovery, President Ronald Reagan ordered construction to be halted. He also blocked Soviet diplomats from entering their embassy in Washington, DC. In 1994, the American embassy was partially dismantled and rebuilt to ensure it was not bugged.

October 28

306: Riots break out in Rome. Roman citizens were enraged by new taxes that had been ordered by Flavius Severus, the Augustus, or ruler, of the West. Severus marched on Rome to put down the growing rebellion. On the way there, however, most of his army deserted him. He was soon captured by forces loyal to his rival, Maximillian, and executed.

1349: Edward the Black Prince lands in Calais. Edward landed in France to put the Treaty of London into force. He had forced the French King John II to accept the treaty's terms after capturing him at the Battle of Poitiers in 1356. Upon landing, Edward besieged Calais in the opening hostilities of what would become known as the Hundred Years' War.

1636: Harvard is established. In a vote of the Massachusetts Bay Colony's Great and General Court, legislators approved an expenditure of 400 pounds to fund the new "schoale or college." The first classes were held two years later, making Harvard the first college established in the United States.

1726: Jonathan Swift publishes *Gulliver's Travels*. The satirical novel subtly mocked misogyny and misanthropy. Swift later said he wrote the novel "to vex the world" rather than to entertain. It was immediately popular and has come to be considered a piece of classic literature.

1776: British troops defeat Continental Army soldiers in the Battle of White Plains. Following a defeat in New York City, Washington was attempting to retreat northward when British General William Howe landed his army in upstate New York to cut them off. Washington was alerted to Howe's movements, however, and set up a defensive position on a hill in White Plains. British troops and Hessian mercenaries attacked the hilltop with artillery and infantry, and succeeded in driving the Continentals from their position. Despite the fierce onslaught, Washington's men were able to manage an orderly retreat from their positions into hills behind them. The following day a rainstorm prevented Howe from attacking, and Washington's army slipped away. The British chased the Continental Army through New Jersey and into Pennsylvania before Washington was able to secure a victory at the Battle of Trenton in December.

1886: President Grover Cleveland dedicates the Statue of Liberty. The Statue was a gift to the United States from France to commemorate the two countries' alliance during the American Revolution. Originally called Liberty Enlightening the World, the copper statue stands on Ellis Island, which in 1892 became the main entry point of European immigrants coming to the United States. For the next three decades, more than 12 million immigrants were welcomed to New York Harbor by the sight of the Statue.

1930: Saya San declares himself King of Burma. A monk and astrologer, San led an uprising of dispossessed peasants against British colonialists. His movement was ultimately crushed by superior British military might, and San and 125 of his followers were hanged.

1965: The Second Vatican Council declares the *Nostra aetate*. The "Declaration on the Relation of the Church with Non-Christian Religions" officially revoked the Church's position that had held Jewish people responsible for the death of Jesus. Pope Innocent III had originally blamed the Jews for Jesus' death 760 years before, which gave rise to centuries of anti-Semitic violence and prejudice.

October 29

53 BC: Cyrus the Great of Persia captures Babylon. Upon taking the city Cyrus allowed the Jews to return to their homeland. The capture of Babylon marked the beginning of the establishment of the Persian Empire, which would eventually stretch across much of the ancient world.

1390: The first trial for witchcraft is held. In Paris, Jehenne de Brigue was arrested and accused of witchcraft because she was accused of exchanging recipes for love potions with Macette de Ruilly. Allegedly de Ruilly was trying to find a spell to plague her husband, while de Brigue was trying to convince the man who had fathered her children to marry her. De Brigue was taken into custody and imprisoned for three months. She appealed their case to the Parliament of Paris, but the move ultimately backfired. The Parliament ordered her to be tortured to extract a confession. While being tortured, de Brigue named de Ruilly as her accomplice, and de Ruilly was also subjected to torture. The two women confessed to being witches under torture and were condemned to death. Both were burned at the stake in 1391.

1787: *Don Giovanni* premieres in Prague. The opera, written by Austrian composer Wolfgang Amadeus Mozart, was received to great acclaim. It tells the story of a nobleman who is plagued by the ghost of a man he killed in a duel. The ghost ultimately drags the nobleman, Don Giovanni, to the underworld in the opera's climax. The performance was widely acclaimed and the opera remains one of Mozart's best-known.

1811: The first steamboat begins operating on the Ohio River. The steamboat cruised downriver all the way from Pittsburgh, Pennsylvania, to New Orleans, Louisiana, on a journey that took nearly two weeks to complete.

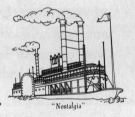
"Nostalgia"

1858: The first general store opens in Denver, Colorado. The general store sold goods to miners who arrived to search for gold deposits along the Cherry Creek and the South Platte River.

1863: Eighteen countries agree to form the International Red Cross. The organization is based in Geneva, Switzerland, and is charged with protecting the victims of war, including wounded, prisoners, refugees, and other noncombatants. It continues to operate around the world.

1942: British clergy and politicians condemn Nazi persecution of Jewish citizens. At a public meeting, a letter from Prime Minister Winston Churchill was read in which he registered his outrage over the "systematic cruelties" the Nazis were inflicting on the Jews of Europe.

1960: Cassius Clay wins his first professional boxing match in Louisville, Kentucky. The boxer would later convert to Islam and take the name Muhammad Ali. He went on to be one of the greatest athletes of the 20th century. His principled refusal to enlist during the Vietnam War resulted in his being banned from professional boxing for several years. The Supreme Court ultimately ruled in Ali's favor.

October 30

130: Roman emperor Hadrian founds the city of Antinoopolis on the banks of the Nile River in Egypt. The city was the western end of the Via Hadrianna, a road that Hadrian had built from the Red Sea to the Nile.

1817: Simón Bolívar establishes the independent republic of Venezuela. The South American nation had made repeated attempts at independence over the previous decade, but each one was crushed by Spanish Royalists. Bolívar defeated the Spanish in Venezuela and then went on to liberate Ecuador and Bolivia as well.

1862: Union General Ormsby Mitchell dies from his wounds in Beaufort, South Carolina. Mitchell had been a college professor before the war. He was notable for protecting enslaved Blacks who escaped to his lines. At the time, official Union policy was to return escaped slaves to the Confederacy.

1938: Orson Welles broadcasts *The War of the Worlds*. The radio play was performed as a Halloween episode for the series "Mercury Theatre on the Air." The program presented an adaptation of H.G. Wells' novel of the same name as a series of news bulletins about an invasion by Martians. In the purportedly live, on-location broadcast, a reporter excitedly related events at Grover's Mill, New Jersey. In the radio play, a strange object landed there, and Martians soon emerged from it, attacking onlookers with a heat ray. The radio broadcast sounded very realistic to listeners due to sophisticated sound effects and convincing acting. While it was not intended as a hoax, many listeners thought it was real newscast. The media condemned the broadcast in the following days as sparking a panic, although the degree to which people actually believed the radio play was real has been disputed. The incident cemented Orson Welles' notoriety as a skilled dramatist.

1925: John Logie Baird makes the first transmission of a moving image by television. The Scottish inventor built his transmitter with items that included a tea chest, piano wire, and knitting needles. He transmitted an image of his assistant, 15-year-old William Taynton. Taynton was thus the first person to appear on television.

1941: President Franklin D. Roosevelt approves $1 billion in loans to the Soviet Union. Roosevelt approved the lend-lease loans in an attempt to help the Allied cause while keeping the United States out of World War II. The attack on Pearl Harbor would ultimately draw the US into the war that December.

1953: President Dwight Eisenhower approves National Security Council Order No. 162/2. The secret document advised the expansion of the United States' nuclear weapons arsenal to counter the Soviet threat. It marked the beginning of the Cold War arms race that continued for the next thirty years and resulted in a major buildup of nuclear weapons by both nations.

1974: The Rumble in the Jungle: Muhammad Ali defeats George Foreman by knockout in the eighth round to claim the World Heavyweight Title. The fight was held in Kinshasa, Zaire (now the Democratic Republic of the Congo), and is considered one of the greatest sports events of the 20th century. Foreman was heavily favored coming into the fight. Ali used a novel "rope-a-dope" tactic throughout the fight to beat him. For seven rounds, he took Foreman's heavy blows while leaning on the ring's ropes for support. As Foreman ran out of energy in the eighth, Ali emerged to land a devastating series of blows that KO'd the champ and earned Ali the title.

October 31

475: Orestes, a Roman general, places his young son Romulus Augustulus on the throne after deposing Julius Nepos. Shortly after, Orestes' army abandoned him and switched their allegiance to the Italian barbarian king Odoacer. Upon overthrowing Romulus, Odoacer became the first king of Italy and ruled until 493. Odoacer's reign marks the fall of the Western Roman Empire.

1517: Martin Luther posts his *95 Theses* on the door of Castle Church in Wittenberg. In them, Luther outlined his grievances with the Catholic Church. Chief among these was his opposition to the clergy's sale of plenary indulgences. These certificates were sold to lay persons to reduce their punishment in the afterlife for their sins. Luther argued that people could repent directly to Jesus Christ, thus circumventing the authority of the Catholic clergy. Luther's posting of his *Theses* marked the start of the Protestant Reformation and is commemorated as Reformation Day. The *Theses* were widely reprinted and made Luther famous. He was ultimately excommunicated from the Catholic Church in 1521, even though he believed he was not in disagreement with the Pope in his views.

1822: Just five months into his reign, Emperor Agustín de Iturbide dismisses the Mexican Congress. In its place he established a junta of loyalists to govern. As a result, his rivals organized a coup, and he was overthrown shortly thereafter by a rebellion led by General Santa Anna.

1861: Union General Winfield Scott retires. Scott had gained fame during the Mexican-American War, but cited failing health upon retiring even as the Civil War was breaking out. Scott was replaced by General George McClellan, who had great success in building and training the Army of the Potomac, but whose caution made him an ineffective battlefield general.

1870: The Blanquist Uprising begins in France. The Blanquists were followers of the French socialist revolutionary Auguste Blanqui. Napoleon III had recently surrendered to the Germans, and the revolutionaries saw an opportunity to seize the Third Republic. They ultimately failed in overthrowing the government.

1888: John Dunlop patents the pneumatic bicycle tire. The Scottish veterinarian and bicyclist ultimately had his patent revoked when it was discovered that Robert William Thompson had already patented the idea in 1845. Thompson's invention, which was designed to be used on horse-drawn carriages, had been forgotten in the forty years between them.

1917: In the Battle of Beersheba, Egyptian cavalry attack Ottoman forces in Palestine. The attack is considered the last time a cavalry charge was successful in modern war.

1956: During the Suez Crisis, French and British warplanes begin bombing Egypt. Egyptian President Gamal Abdel Nasser nationalized the Suez Canal. When Israel invaded the Egyptian Sinai, the Egyptians withdrew but blocked the canal to all international shipping. The Allied forces eventually were forced to withdraw, and the canal was reopened following international negotiations.

1984: Indian Prime Minister Indira Gandhi's own bodyguards assassinate her. The bodyguards were ethnic Sikhs and they killed Gandhi after she ordered the Indian Army to assault a Sikh temple. The assassination led to widespread rioting in which thousands of Sikh civilians were murdered in retaliation.

Chapter 11

November

November 1

1254: Hayton of Corycus leaves the Mongol capital city of Karakorum after visiting for six weeks. Hayton was an Armenian nobleman and historian from modern-day Turkey. His description of the journey and visit became one of the volumes in his historiography *Flower of the Histories of the East*. The four-volume work was one of the first detailed accounts of the history of the Mongol Empire's conquest of Asia.

1512: The ceiling of the Sistine Chapel goes on public view. Painted by Michelangelo, the ceiling is one of the most famous Renaissance works in the Vatican City.

1604: Shakespeare's tragedy *Othello* is produced in London for the first time. The performance was mentioned in an account that says on "Hallamas Day" (now called All Hallow's Day), a play called "The Moor of Venis" by one "Shaxberd" was shown at Whitehall Palace.

1800: President John Adams moves into the Executive Mansion, later renamed the White House. Adams was the first President to live there. Construction of the mansion had begun in 1792. It was designed by James Hoban and built of Virginia sandstone by a workforce that included Scottish stonemasons and enslaved African-American laborers.

1848: The Boston Female Medical School opens. The institution was the first medical school for women in North America.

1870: The Unites States Weather Bureau makes its first weather forecast. The Bureau, which was renamed the National Weather Service in 1970, was established by President Ulysses S. Grant. Grant had the first weather stations established at military outposts. They relayed information about storms moving through the Great Plains to cities back East via telegraph.

1911: During the Italo-Turkish War, Italian pilot Giulio Gavotti drops the first bombs from an airplane on an enemy position. Gavotti served as a Second Lieutenant in combat in Libya. He flew a monoplane over Turkish lines and dropped four grenades. There were no casualties. The Ottoman Empire formally protested the bombing, citing the Hague Convention of 1899, which outlawed dropping bombs from balloons. Italy replied that the law did not apply to heavier-than-air planes.

1938: The "Match of the Century." Seabiscuit defeated War Admiral at Pimlico Race Course in an upset. War Admiral was given 4–1 odds to win the race by bookmakers. About 40,000 fans from around the United States attended the event. Seabiscuit led for the first half of the $1\,3/16$ race, but War Admiral pulled level and took the lead. Over the closing stretch, however, Seabiscuit again took the front position and won by four lengths.

1950: Puerto Rican nationalists attempt to assassinate US President Harry Truman. Griselio Torresola and Oscar Collazo tried to shoot their way into Blair House, Truman's temporary residence while the White House was being refurbished. Secret Service agents and police repelled the attack in a gun battle that left Torresola and a policeman mortally wounded. Truman was napping on the building's second floor, but was not harmed.

1957: Mackinac Bridge opens. The five-mile-long suspension bridge connects Michigan's Upper and Lower Peninsula at the Straits of Mackinac. Its main span, at 3,800 feet, is one of the longest in the world.

November 2

1783: General George Washington issues a Farewell Order.
It officially disbanded the Continental Army following the
American Revolutionary War. Washington thanked his soldiers
for "eight long years" of service, which he wrote was "little short
of a standing miracle." The order was distributed by newspapers
throughout the newly liberated United States.

1795: The French Directorate takes power in Paris. The
five-man revolutionary government replaced the Committee
of Public Safety following the Reign of Terror. The Directorate
governed France until 1799, when it was overthrown by
Napoleon Bonaparte.

**1840: Afghan rebels led by Dost Mohammad defeat British
forces at Parwan.** The victory was the only one the partisans
had during the campaign to drive the British from Afghanistan.
They surrendered the following day.

**1861: President Abraham Lincoln recalls Union General
John C. Fremont from the Western Department of War during the Civil War.** Fremont, a controversial general, had declared
martial law and emancipated enslaved people in Missouri, causing a delicate political situation in the border state.

1917: The Balfour Declaration is proclaimed. British Foreign
Secretary Arthur Balfour sent the declaration to Lord Lionel
Rothschild, a prominent British leader of the nascent Zionist
movement for a Jewish homeland. In the letter, Balfour declared
the British government's support of establishing a national home
for Jewish people worldwide in Palestine. At the time, Palestine
was controlled by the Ottoman Empire, which Great Britain
fighting in World War I. Balfour added that British support of
the Jewish homeland came with the clear understanding that
nothing would be done to infringe on the rights of existing
non-Jewish communities in the region. After the War, Britain
established Mandatory Palestine as part of the partition of the

Ottoman Empire by the Allies. Israel was officially established in 1948.

1920: KDKA begins broadcasting in Pittsburgh. It was the first commercial radio to go on air and broadcast the results of the 1920 presidential election, in which Warren G. Harding was elected. The station was first in several other radio milestones, including the first broadcasts of a live report from an on-site location, a speech by an American president, and a professional baseball game.

1959: Charles Van Doren admits to Congress that he cheated on the TV quiz show *Twenty-One*. Producers had provided Van Doren with questions and answers during his three-month winning streak on the game show. The resulting scandal led Congress to pass legislation prohibiting the fixing of TV game shows.

2016: The Chicago Cubs finally win the World Series. After a 108-year drought—the longest in modern professional sports history—the Cubs beat the Cleveland Indians in seven games to clinch the championship. Two days later, over five million fans crowded a parade route from Wrigley Field to Grant Park to cheer their team, no longer "loveable losers."

November 3

1295: Mahmud Ghazan, a direct descendant of Genghis Khan, seizes the throne of Il-Khan during the Mongols' occupation of the Il-Khanate, modern-day Iran. Ghazan converted to Islam when he took the throne. He ruled until his death in 1304.

1534: The English Parliament passes the first Act of Supremacy. The Act made King Henry VIII the supreme head of the Anglican Church. Henry split with the Catholic Church after the Pope refused to grant him a divorce.

1793: Olympe de Gouges is executed. De Gouges, born Marie Gouze, was a French feminist, playwright, and journalist who published pamphlets during the French Revolution that demanded women be granted the same rights as men. She criticized the Revolutionary Government during the Reign of Terror and was executed by guillotine.

1838: *The Times of India* is founded. Originally called *The Bombay Times* and *Journal of Commerce*, it is the highest selling English-language daily newspaper in the world, as well as the oldest English-language newspaper in India still in circulation. It reaches an estimated 2.7 million readers.

1863: J.T. Alden patents a method for preserving yeast. Alden's "Improvement in the Preparation of Yeast" was the first to produce a granular form of dried yeast. Earlier methods had prepared a thick, dry cake of yeast that bakers had to crush manually. Alden's form reduced damage and made it easily dissolved.

1883: Charles Earl Boles, aka "Black Bart," robs his last stagecoach. The notorious robber operated in the Pacific Northwest and gained a reputation as a gentleman bandit. In his last robbery, he was wounded and fled, but left behind clues that led to his capture. He spent four years in San Quentin Prison, and vowed never to return to a life of crime.

1911: William C. Durant founds Chevrolet. The automobile manufacturer was the first to compete with the Ford Motor Company. In 1929, Chevrolet surpassed Ford as the best-selling car in the United States.

1957: Laika, a stray dog from Moscow, becomes the first animal launched into space. Laika was enlisted in the Soviet space program and went into orbit aboard *Sputnik 2*. At the time, the effects of space travel on animals was poorly understood. Laika was provided with a sophisticated—for the time—life support system and attached to sensors that would monitor her bodily functions. After liftoff, she survived for only a few hours and ultimately died from overheating. The Soviets claimed the dog had survived for six days, and said she was euthanized when her oxygen began to run out. The spacecraft disintegrated on re-entry in April 1958. After the collapse of the Soviet Union, Oleg Gazenko, a scientist involved in the experiment, expressed regret at the mission's outcome.

1969: President Richard Nixon appeals to "the great Silent Majority" during a televised address. As public opposition to the Vietnam War led to massive peaceful demonstrations and violent resistance in American cities, Nixon asked for calm. He claimed he was attempting to work for "peace with honor" in Vietnam. In reality, he was secretly ordering bombing of Laos and Cambodia.

November 4

1429: Joan of Arc captures Saint-Pierre-le-Moûtier during the Lancastrian War. Joan's army was repulsed in their first attack on the city. She rallied her troops and led them in a second assault, which overwhelmed the defenders. King Charles VII granted her noble status as a result of the victory.

1520: Christian II of Denmark and Norway is crowned King of Sweden. Upon taking the throne, Christian immediately had more than 100 opponents executed in what became known as the Stockholm Bloodbath. The massacre sparked a rebellion that ultimately led to Swedish independence.

1677: Prince William of Orange and the future Mary II of England are married. The pair would later be called William and Mary. Both Protestants, they became King and Queen of England following the Glorious Revolution of 1688. Mary's father James II and VII, a Roman Catholic, was deposed in the Revolution. William and Mary established the English Bill of Rights during their reign. It established civil rights for com-moners, set free elections and freedom of speech in Parliament, and limited the powers of the throne. The Bill also allowed

Protestants to bear arms for their own defense and prohibited cruel and unusual punishment for crimes. Many of the features of William and Mary's Bill of Rights were later adopted by the United States following the American Revolution.

1839: The Newport Rising begins. Extremists in the Chartist parliamentary reform movement took up arms in Newport, Wales. The revolt was put down by British and its leader, John Frost, was found guilty of treason.

1879: John Beers patents an improved dental crown. Beers, a California dentist, developed a gold crown that could be glued into place with enamel and cement. Before his invention, gold crowns were hammered into place. Beers called that procedure "costly, difficult, and tedious" in his patent application.

1880: James and John Ritty invent the first cash register. They named it "Ritty's Incorruptible Cashier." The brothers were saloon owners in Dayton. James had seen a machine for counting the revolutions of a ship's propeller, and applied the concept to keeping track of sales. The brothers' register did not have a drawer, but recorded the number of sales and amount of each.

1942: Erwin Rommel's Afrika Corps is defeated by the Allies in the Battle of El Alamein. Hitler had ordered Rommel to fight to the death, but he disobeyed the order. Rommel proceeded to engage in a protracted five-month retreat across North Africa.

1952: UNIVAC, the world's first commercial electronic computer, predicts the results of the upcoming US presidential election. UNIVAC correctly called the outcome a landslide victory for Dwight D. Eisenhower over Adlai Stevenson.

1995: While attending a peace rally in Tel Aviv, Israeli Prime Minister Yitzhak Rabin is assassinated. He was killed by an Israeli Jewish law student who opposed Rabin's recent efforts to reach a peace agreement with the Palestinians.

November 5

1605: Guy Fawkes is arrested. Fawkes, an English Catholic, was the leader of a conspiracy against King James I that became known as The Gunpowder Plot. James, a Protestant, had made no efforts to stop the ongoing persecution of Catholics in England, and the conspirators decided to depose him. The plotters intended to blow up the House of Lords during the opening session of Parliament, killing the King. They would then install James' nine-year-old daughter Elizabeth as the head of state. An anonymous informer sent a letter to William Parker that detailed the plot. Parker led a detachment of soldiers to search the House of Lords and discovered Guy Fawkes guarding a massive store of gunpowder. Fawkes was tortured into confessing his role in the plot. He was later hanged, drawn and quartered—the typical method of execution for traitors. November 5 has since become known as Guy Fawkes Day, and bonfires and fireworks mark the holiday.

1688: William of Orange lands with a Dutch fleet in England. Upon landing, William declared "the liberties of England and the Protestant religion I will maintain." His army and navy vastly outnumbered James the Pretender's forces, and the latter quickly surrendered. William was proclaimed King of England the following year.

1831: Nat Turner is convicted of insurrection. Turner led an uprising of enslaved and free African Americans in Southampton County, Virginia. His rebellion rampaged from one plantation to another over two days and killed some 60 white men, women, and children. Turner was captured after two months as a fugitive. He was hanged, posthumously beheaded, and buried in an unmarked grave. Thomas Ruffin Gray published *The Confessions of Nat Turner* shortly after his execution. The book was based on Gray's research and interviews he conducted with Turner before the rebel's trial.

1862: Abraham Lincoln fires General George McClellan. The West Point graduate had organized the Army of the Potomac to great success, but his overly cautious battlefield manner prevented him from using it effectively. Despite considerably outnumbering the Army of Northern Virginia, McClellan's Union army failed to destroy the Confederates at the Battle of Antietam. McClellan later ran for the nomination of the Democratic Party but lost. He later served as the Governor of New Jersey.

1872: Susan B. Anthony votes for the first time. The suffragist organizer voted in her hometown of Rochester, New York. She was arrested and convicted of illegally voting in a widely-publicized trial. Anthony refused to pay the fine levied against her, and authorities quietly dropped the matter.

1891: Marie Curie enrolls at the Sorbonne, France. The future Nobel Prize winner, who made discoveries about radiation and nuclear physics, began teaching classes the same day. She was the first female physics instructor in the university's history.

1916: The Everett Massacre takes place. During a labor strike in the Pacific Northwest, sheriffs' deputies and vigilantes attacked a group of Industrial Workers of the World, an international labor union. The attackers killed as many as 12 IWW men and wounded another twenty.

2007: Google unveils the Android mobile operating system. Android became one of the two premier operating systems for touchscreen phones, along with the iPhone OS.

November 6

963: Holy Roman Emperor Otto I calls a papal council. The council replaced the 25-year-old Pope John XII with Leo VIII. John would later depose Leo in a papal coup. Three months after retaking the papacy, John died while in bed with his mistress.

1528: Spanish conquistador Alvar Nuñez Cabeza de Vaca is shipwrecked on a low, sandy island off the coast of Texas. The dehydrated, desperate explorer was the first European to set foot on what would become the Lone Star State.

1860: Abraham Lincoln is elected the 16th President of the United States. He was the first Republican candidate to hold the office. Southern slaveowners, wary of his opposition to expanding slavery further west, had threatened to leave the Union if he was elected. By the time Lincoln was inaugurated in March 1861, seven Southern states had rebelled against the United States. They would fire the first shots that April, attacking the Federal garrison at Fort Sumter and sparking the American Civil War. Lincoln went on to be reelected while the Civil War was raging. Following the defeat of the Confederacy, he urged the nation to come together "with malice toward none, and charity toward all." He was assassinated shortly after by John Wilkes Booth.

1861: Jefferson Davis is elected the first (and only) president of the Confederacy. Davis, formerly a lieutenant in the United States Army during the Mexican-American War, had served as a Senator and Representative of Mississippi in the United States Congress. He resigned from Congress on January 21, 1861, and quickly gained support from Southerners who wanted him to serve as their president. After the war, he was imprisoned and indicted for treason. Released on bail, he fled to Canada. President Andrew Johnson's 1868 pardon of every Southerner who rebelled included Davis, and he returned to the United States the following year.

1899: The first Packard automobile is test-driven in Warren, Ohio. The Model A Packard had just one seat and featured a single-cylinder engine and chain drive. It sold for $1,250.

1947: *Meet the Press* debuts on NBC. The weekly news and interview program is the longest-running show in the history of television. Its creator Martha Rountree was its first host.

2012: A suite of firsts. Barack Obama became the first African-American to be elected President of the United States, and Tammy Baldwin became the first openly gay person to be elected to the United States Senate.

November 7

1492: The Ensisheim meteorite crashes into a wheat field outside of the village of Ensisheim, Alsace, in France. It is the oldest meteorite whose date of impact is known.

1497: Vasco da Gama arrives in the Santa Helena Bay near the Cape of Good Hope on the southern tip of Africa. Da Gama, a Portuguese explorer, was searching for a sea route from Europe to India. He reached India the following May, proving it could be reached by sea.

1775: John Murray, Lord of Dunmore, issues the Offer of Emancipation. The proclamation, which the royal governor of the then-British Colony of Virginia issued, was intended to undermine the efforts of American revolutionaries. Dunmore's Offer, as it was known, promised freedom to any enslaved persons who left their plantations and enlisted in the Royal forces. The Offer enraged Virginia's slave owners. They reacted by declaring that all fugitive slaves who did not immediately surrender to colonial commanders would be executed. The Continental Congress directed Virginian colonists to resist the Dunmore Proclamation. Slave patrols were organized to find and capture fleeing captives. Between 800 and 2,000 escapees eventually reached Dunmore and enlisted in a unit called "Dunmore's Ethiopian Regiment." The unit fought in the Battle of Great Bridge. In 1779, British General Clinton issued the Phillipsburg Proclamation, which freed all slaves owned by American rebels. During the course of the War, some 100,000 enslaved Blacks escaped to British lines.

1811: The Battle of Tippecanoe. Shawnee warriors led by Laulewasikau the Prophet, the brother of Tecumseh, were defeated by US Army troops led by General William Henry Harrison. Harrison would go on to be elected the ninth President of the United States with the slogan "Tippecanoe and Tyler Too," after Harrison's running mate John Tyler.

1869: James Moore wins the first multi-city bicycle race. The English cyclist rode from Paris to Rouen, France, a distance of 84 miles, in 10 hours and 25 minutes.

1875: Verney Lovett Cameron reaches the coast of Angola. In doing so, the British explorer was the first person known to have crossed the African continent, traveling from the Indian Ocean to the Atlantic.

1908: Butch Cassidy and the Sundance Kid are killed in Bolivia. The pair of outlaws had embarked on a crime spree in the American Old West in the late 19th century, robbing banks and trains, before being pursued by the Pinkerton detective agency. They fled to South America, first traveling to Argentina and then Bolivia, where they subsequently met their end in a shootout with police.

1938: Herschel Grynszpan shoots German diplomat Ernst vom Rath in Paris, France. Grynszpan was a 17-year-old Polish-Jewish refugee who declared he was acting in the name of persecuted Jews in Nazi Germany. The Nazis used the assassination as a pretext to attack Jews in the pogrom known as Kristallnacht.

1944: President Franklin D. Roosevelt is elected to a fourth term. FDR carried 432 of 531 electoral votes in a landslide defeat of Thomas Dewey. He died in office the following April, and was succeeded by Harry Truman. In 1947 the 22nd Amendment was ratified, limiting Presidents to just two terms.

1989: David Dinkins is elected Mayor of New York City. He was the first African American to hold the office. During his one-term tenure, crime in New York decreased more quickly than ever before.

November 8

1047: Theophylactus of Tusculum becomes Pope Benedict IX for the third time. He was initially elected to the papacy after his father bribed electors. He sold the position once, was deposed due to "immoral" behavior once, and was driven out of Rome the third time after only a year as Pontiff.

1520: The Stockholm Bloodbath begins. Danish King Christian II seized the throne in Sweden and ordered the execution of more than 80 Swedish nobles over two days. As a result, outraged Swedes joined together to expel Christian from their country.

1861: A Union warship seizes the British naval vessel HMS *Trent*. Federal soldiers took two Confederate diplomats prisoner, sparking an international incident. The United States was forced to apologize to Great Britain, narrowly avoiding war.

1867: David Livingstone arrives at Lake Mweru. The Scottish missionary and explorer was searching for the source of the Nile river. He was the first European to see Lake Mweru, part of the Congo River drainage.

1923: Adolf Hitler leads the Beer Hall Putsch. It was his first attempt at seizing power in Germany. Hitler, backed by a group of Nazi stormtroopers, barged into a meeting of Bavarian government officials in Munich and fired a pistol into the air, shouting that "the national revolution has begun." His uprising failed, and he was imprisoned. While in prison, he dictated *Mein Kampf*, which laid out his anti-Semitic ideology.

1935: John L. Lewis, leader of the United Mine Workers, announces the Committee for Industrial Organization. The CIO was one of the first "big" unions that merged labor organizations from various industries into a single group.

1987: The IRA commits the Remembrance Day Bombing. At a ceremony honoring British military war dead in Enniskillen,

Northern Ireland, the IRA set off a bomb during a parade of Ulster Defense Regiment soldiers. The explosion killed 11 people, including civilians and children, and injured more than 60.

1994: The Republican Revolution in Congress. Also known as the "Gingrich Revolution" after its chief architect, Newt Gingrich, it was the largest midterm electoral success by Congressional Republicans in American history. Republicans picked up 54 seats in the House of Representatives and eight seats in the Senate. Under Gingrich's direction, Republicans opted for a midterm strategy of campaigning nationwide under a single, unified message, rather than campaigning independently in each Congressional district. The message was known as the Contract with America. It was presented as an alternative to the policies of the Clinton administration. Republicans gained control of both houses of Congress. They had not been the majority in the House since 1952. Republicans clashed with Clinton, who vetoed several of their bills, ultimately resulting in a government shutdown in 1995. Thousands of federal employees were furloughed, including hundreds who worked at the White House. During the shutdown, the White House was largely staffed by unpaid interns, including Monica Lewinsky.

November 9

694: Egica, King of the Visigoths of Spain, accuses Jews of helping Muslims at the 17th Council of Toledo. The anti-Semitic ruling decreed that Christians could enslave Jewish Spaniards and confiscate their property.

1620: Aboard the *Mayflower*, Pilgrims sight land at Cape Cod. They spent the next several days attempting to sail south to Virginia, but were turned back by strong winds. The Pilgrims established a settlement on Cape Cod. They later relocated to Plymouth, Massachusetts, after they angered Nauset Indians by stealing their food stores.

1799: The French revolutionary government known as the Directory overthrows the revolutionary Consulate. The resulting chaos opened the way for Napoleon Bonaparte to seize power and establish a dictatorship.

1862: General Ambrose Burnside takes command of the Union Army. Burnside replaced General George McClellan, who had been ineffective on the battlefield. Burnside's unique facial hair became known as "sideburns." Within months, he was also removed from command.

1938: The Nazis launch the *Kristallnacht*. The "Night of Broken Glass" was a pogrom against Jews in Nazi Germany. It was carried out by stormtroopers aided by gangs of anti-Semitic civilians. They used the recent assassination of a Nazi diplomat in Paris as an excuse to begin the attacks. The perpetrators ransacked Jewish-owned businesses, as well as Jewish homes, schools, and hospitals. They also destroyed 267 synagogues throughout Germany and Austria. More than 90 Jews were murdered in the violence. After the attacks, about 30,000 Jewish men were arrested and sent to concentration camps. The *Kristallnacht* was the beginning of the Holocaust against European Jews and other minorities the Nazis deemed "undesirable."

1960: Robert McNamara becomes president of the Ford Motor Company. He was the first person to serve in that position who was not a member of the Ford family. Within a month, he left the post to join the new administration of John F. Kennedy as Secretary of Defense. In that position, he oversaw the United States' Cold War strategy, and eventually became one of the chief architects of the Vietnam War.

1965: Roger Allen LaPorte immolates himself in protest of the Vietnam War. LaPorte, a 22-year-old member of the Catholic Worker movement, committed the act of protest in front of the United Nations in New York City. He declared "I'm against all wars, all wars. I did this as a religious act," before fatally setting himself on fire.

1967: The first issue of *Rolling Stone* magazine is published. The issue's cover story featured the Monterey Pop Festival, a three-day concert held in California in June. The magazine was named after the song "Rollin' Stone," by Muddy Waters. It became one of the primary cultural magazines in the United States, eventually expanding to political reporting as well.

1989: East Germany opens checkpoints along the Berlin Wall. It had divided the city into Allied and Communist sectors since 1961. The next day, jubilant crowds began using hand tools to tear down the concrete symbol of the Cold War.

November 10

1202: During the Fourth Crusade, a French army besieges the occupied Venetian city of Zara, Croatia. They eventually captured the city on behalf to the Venetians, who returned the favor by providing them with ships to transport them to the Middle East.

1241: Just two weeks after being elected Pope, Celestine IV dies. He was the first Pope to be elected by a Papal Conclave of cardinals, rather than appointed to the position or seizing it by force.

1775: Samuel Nicholas forms the United States Marine Corps. Two battalions of Marines were organized as an amphibious landing force for the Continental Navy. The force was disbanded after the Revolutionary War. It was reestablished in 1798 when the United States was engaged in naval hostilities with French warships in the Quasi-War. The Marine Corps gained widespread fame as an expeditionary force when it was put into action against the Barbary Pirates in North Africa. During the American Civil War, the Marines supported blockade actions against Confederate ports. About one-third of the force deserted to join the Confederacy during the war. The Marines have fought in every American war since, and played a central role in the fighting in the Pacific during World War II.

1808: Osage Indians are sent to a reservation in Oklahoma.
The tribes were forced to abandon their ancestral lands in
Missouri and Arkansas. The lands they were relocated to were
later found to be the site of oil and gas deposits. In the 1920s,
whites murdered several prominent Osage people in an attempt
to seize the wealth from the oil wells.

1865: Henry Wirz is hanged for war crimes. Wirz, a Swiss
immigrant who joined the Confederate Army during the
American Civil War, was the commander of Andersonville
Prison in Georgia. The prison's inhumane conditions caused the
deaths of nearly one-third of its 46,000 captives.

1871: "Dr. Livingstone, I presume?" On the shore of Lake
Tanganyika in Tanzania, the American journalist Henry Stanley
finally found the Scottish missionary and explorer David
Livingstone. Stanley had been sent to find Livingstone, who had
disappeared several years before and was presumed to be dead,
by the *New York Herald*.

1888: Jack the Ripper commits his last known murder. The
serial killer had stalked the streets of London's East End for over
a year, killing seven women. He taunted police with letters, but
was never caught or even identified.

1928: Michinomiya Hirohito becomes Emperor of Japan.
He was the 124th monarch in an unbroken line that dated back
to 660 BCE. Hirohito oversaw the expansion of Japan's military
and imperialist conquests in Asia and the Pacific. Following
Japan's defeat in World War II in 1945, he was forced to abdicate
the throne.

**1995: Nigerian playwright and environmental activist Ken
Saro-Wiwa is hanged.** Saro-Wiwa led a nonviolent campaign
against the destruction of land and water by the Royal Dutch
Shell company. He was framed for the murders of several Ogoni
chiefs and executed along with eight other activists, despite a
worldwide protest.

November 11

1215: The Fourth Lateran Council declares the Catholic dogma of transubstantiation. The doctrine declared that the bread and wine served during the Catholic Mass turned into the actual body and blood of Jesus Christ.

1572: Tycho Brahe observes the supernova B Cassiopeiae (SN 1572). The appearance of this "new star" prompted Brahe, a Danish astronomer, to publish a work titled "Concerning the Star" about it. He challenged existing ideas about the cosmos that had assumed the stars were fixed and unchangeable. The discovery prompted a revolution in the science of astronomy.

1620: The Pilgrims sign the Mayflower Compact. The Compact was the first governing document of Plymouth Colony, and thus the first establishment in writing of a colonial government in the Americas. At the time, the Pilgrims were divided among themselves into conflicting factions. Because they were in an area that had no government, a minority faction declared they were free to do what they wished. To prevent them from doing so, the Pilgrims established a government aboard the *Mayflower* itself while it was anchored in Provincetown Harbor, off the northern tip of Cape Cod. Of the 101 aboard the ship, 41 passengers, all men, signed it. It declared the establishment of "a civil Body Politick, for our better Ordering and Preservation."

1675: Gottfried Leibniz discovers integral calculus. The discovery established one of the two branches of calculus. The other branch, differential calculus, was discovered by Sir Isaac Newton at the same time. Calculus revolutionized mathematics and has led to applications in engineering, science, and economics.

1838: Four hundred members of the secret organization called the Hunters' Lodges mount a rebellion in Canada. The rebels invaded Ontario in an attempt to inspire other Canadians to rise up against British rule. They were ultimately unsuccessful.

1853: David Livingstone departs from the Linyanti River in present-day Angola. The Scottish explorer and missionary set out to cross the African continent. He declared that he intended to "open up a path into the interior, or perish." Two years later, he arrived at the mouth of the Zambezi River at the Indian Ocean.

1887: The Haymarket Martyrs are hanged. August Spies, Alpert Parsons, Adolph Fischer, and George Engel were convicted of conspiracy after a bomb killed police while the anarchists were rallying for an eight-hour workday in Chicago.

1921: President Warren G. Harding dedicates the Tomb of the Unknown Soldier. The Tomb, located in Arlington Cemetery, contains the remains of an unidentified American soldier who died in World War I.

1968: The United States Air Force begins bombing the Ho Chi Minh Trail. The Trail was an overland route through Laos and Cambodia that the People's Army of Vietnam used to send soldiers and supplies south. The US dropped more than three million tons of bombs on Laos during the course of the war.

2004: The Palestinian Liberation Organization announces the death of their leader, Yasser Arafat. Arafat had led the PLO in its campaign against Israeli occupation of Palestine since 1969.

November 12

1660: John Bunyan is arrested for teaching "nonconformist" theology. The English Puritan minister was tried in January and sentenced to three months in prison. At the end of his sentence, he refused to disavow his preaching and his period of imprisonment was extended. He ultimately remained in prison for 12 years. While he was in prison, he wrote *Grace Abounding* and began working on *Pilgrim's Progress*. The book, a Christian allegory, was published in 1678 and eventually became one of the most significant works of religious literature in the English language. It has since been translated into more than 200 languages. The book was distributed by Protestant missionaries throughout Asia and Africa. Bunyan wrote nearly 60 books in all, although *Pilgrim's Progress* is his best-known work. He died in 1688 in London and remains a major figure in the Church of England.

1864: General William Tecumseh Sherman orders Atlanta to be burned. His troops had captured the Confederate city in September. They prepared to march to the Atlantic coast, and Sherman did not want to leave the city intact as a base of Confederate operations.

1912: Robert Scott's body is discovered. Members of the British explorer's shore party began searching for him after he failed to return from his expedition to the South Pole seven months earlier. They found the bodies of Scott and two of his companions, along with diaries detailing their ordeal, in a windbattered tent on the vast plains of Antarctica.

1927: The Soviet Communist Party expels Leon Trotsky. As a result, Joseph Stalin was able to assume sole control of the Party. Trotsky, one of the leaders of the October Revolution that swept the Bolsheviks into power, fled the country. He was later murdered by Stalinist agents while he was a fugitive in Mexico.

1941: Winter comes early to the Moscow front during World War II. Soviet troops deployed ski troops to attack the frostbitten German troops encircling the city. Two months later, the siege was broken, and the Nazi troops were forced to begin retreating west.

1956: The Rafah massacre is committed. During the Suez Crisis, Israeli troops occupied the Gaza strip. In Rafah, they rounded up hundreds of Palestinian Arab civilian men. They machine-gunned more than 100 of the men, whom they suspected were soldiers.

1980: *Voyager I* passes within 77,000 miles of Saturn. The interplanetary space probe had been launched three years before. It sent pictures of Saturn's hundreds of rings back to Earth, some 950 million miles away.

1997: Ramzi Yousef is convicted of masterminding the 1993 World Trade Center bombing. The bombing was intended to destroy the towers. While it failed to do so, it killed six people and injured more than 1,000. After the attack, Yousef fled the United States, but was later arrested in Pakistan. He was sentenced to life in prison in the U.S. for his role in the bombing.

November 13

1002: Æthelred II orders the massacre of all Danes in England. Denmark had invaded England multiple times over the preceding years, and the king—who was known mostly for his poor judgment and the assassination of his own brother—was seeking revenge. The Saint Brice's Day massacre, as it became known, only provoked new invasions by the Danes.

1775: Continental Army troops led by General Richard Montgomery capture Montreal, Canada. The Americans occupied the city for only a few weeks. When they attempted to march on Quebec, they were turned back and forced to retreat all the way back to New York State.

1790: William Herschel, a German-born English astronomer, observes a strange spectacle in the night sky. He described it as a star in the center of a glowing cloud. He had discovered a nebula, or massive cloud of interstellar dust and gas. The discovery led him to propose a theory of nebulae as stars forming from gasses condensing under the force of gravity.

1841: James Braid observes a demonstration of "animal magnetism." The Scottish surgeon and "gentleman scientist" attended a demonstration by Charles Lafontaine where he first observed the phenomenon. Lafontaine "magnetized" two men in the demonstration. The men became drowsy and unresponsive during the performance. Braid began to study the subject intensely. He eventually determined that magnetism was not causing the effect. Via a series of experiments on himself, Braid discovered that he could place himself in a trance. The outcome was the discovery of hypnotism. He began lecturing and giving demonstrations of hypnotism to audiences in Manchester, England. The topic was controversial, and ministers denounced it as a form of witchcraft. Braid persevered, reporting his findings in scientific journals. He is considered the "Father of Modern Hypnotism" for his work.

1916: The First Battle of the Somme concludes during World War I. The battle had lasted for more than four months. In the fierce fighting, whole battalions charged entrenched positions repeatedly and were cut down by machine gun fire. More than a million casualties were suffered by the two armies. After all the carnage, no ground had been gained by either side.

1940: Walt Disney releases *Fantasia*. The ambitious animated film had no plot, but paired imaginative scenes with orchestral scores. It was well-received by critics but failed to make a profit in its first run. It later became one of the best-known animated films in history.

1947: Mikhail Kalashnikov completes the development of the AK-47. The assault rifle became an iconic symbol of the Cold War. Sturdy and dependable, it was widely used by Warsaw Pact armies and anticolonial rebels worldwide.

1953: Mrs. Thomas J. White, a member of the Indiana Textbook Commission, calls for all references to Robin Hood to be removed from school books. She said that Robin Hood's practice of robbing from the rich and giving to the poor was "the Communist line." White later condemned Quakers for their antiwar stance.

2001: President George W. Bush orders military tribunals to try persons suspected of participating in terrorist attacks on the United States. It was the first time military tribunals had been enacted in the United States since World War II.

November 14

1552: Lady Jane Grey is arrested. The 16-year-old great-granddaughter of Henry VII had been the Queen of England for just nine days when she was accused of high treason and beheaded on the orders of Mary Tudor.

1587: While sailing around the globe for the third time, English navigator Thomas Cavendish captures the Spanish galleon *Santa Ana* off the coast of California. The 600-ton treasure ship was returning from a trade mission in the Philippines. It was carrying so much cargo that Cavendish could not bring all of it aboard his two ships. His men took all the gold the galleon was carrying, as well as silks, spices, and wine worth millions of pesos.

1770: James Bruce reaches Lake Tana, the source of the Blue Nile in Ethiopia. Bruce, a Scottish explorer, wrote *Travels to Discover the Source of the Nile*. It is a classic of travel writing.

1799: Following the fall of the Directory in Revolutionary France, Napoleon Bonaparte becomes the first Consul of France. The move was the first step in the general's road to complete dictatorship.

1862: President Abraham Lincoln approves General Burnside's plan to capture the Confederate capital at Richmond, Virginia. The decision would prove to be disastrous. It led to the Battle of Fredericksburg, which was one of the worst defeats the Union suffered during the Civil War.

1882: Buckskin Leslie shoots Billy Claiborne in a gunfight in Tombstone, Arizona. Claiborne was one of the survivors of the gunfight at the OK Corral the previous year. He confronted Leslie while drunk and was shot down. In his last words he referred to the OK Corral shootout, saying "Frank Leslie killed Johnny Ringo, I saw him do it."

1885: During a trade dispute, Great Britain invades Upper Burma. British troops captured Mandalay Bay and deposed Thibaw, the last King of Burma.

1940: The German Luftwaffe bombs Coventry, England. The air raid involved 500 German bombers, killed more than 1,000 civilians, and destroyed the city's medieval cathedral, which had stood for more than 900 years. Adolf Hitler ordered the ruthless attack in response to Allied bombing of Berlin, the German capital. Historians believe that British Prime Minister Winston Churchill had advance warning of the bombing raid, but decided not to tell Coventry's officials. Churchill made the difficult decision because he knew that an evacuation of the city in preparation of the raid could tip off the Nazis that the Allies had cracked the German war codes. The shell of the destroyed cathedral still stands in Coventry. A modern cathedral was built next to it after the war. The new cathedral was consecrated in 1962.

1945: Tony Hulman buys the dilapidated Indiana Motor Speedway. The racetrack had fallen into disuse during World War II. Hulman extensively renovated the track, which allowed for races to resume. The Speedway became one of the most famous auto racetracks in the world.

1982: Lech Walesa, the leader of Poland's Solidarity movement, is released from prison after eleven months. Solidarity was the first trade union in the Soviet bloc that was not controlled by the state. Walesa won the 1983 Nobel Peace Prize for his efforts.

November 15

1315: The Battle of Morgarten. An army of peasants known as the Swiss Confederation sprang an ambush on the Hapsburg army of Leopold I, and routed them. The victory was the first in a long struggle to establish the independent nation of Switzerland.

1532: Francisco Pizzaro meets Atahualpa, the Incan emperor, outside of Cajamarca. Pizzaro, a Spanish conquistador, invited Atahualpa to a feast in his honor the next day. At the feast, Pizzaro's men massacred thousands of Incan nobles and statesmen and took the emperor prisoner. They ordered Atahualpa to provide them with vast amounts of gold in exchange for releasing him. They then forced him to convert to Christianity before murdering him.

1805: The Corps of Discovery arrives at the mouth of the Columbia River at the Pacific Ocean. Led by Meriwether Lewis and William Clark, the party decided to settle in at the site for the winter. They remained there until March, when they began making their way back East.

1864: General William Tecumseh Sherman begins the March to the Sea. During the Civil War, Sherman was tasked with destroying the Confederate's economic and strategic assets in the Deep South. After occupying and burning Atlanta, he led his army across Georgia to the Atlantic Ocean. Sherman left his supply lines behind. His army foraged for supplies and food along the way, raiding Southern plantations, burning cotton fields, and destroying mills. They wrecked railroad rails to prevent the Confederates from using them to move troops and supplies. To do so, they heated the rails over fires and wrapped them around tree trunks. As they moved through Georgia, Sherman's soldiers also freed thousands of enslaved persons, who followed the army to the sea. The march was a decisive blow to the Confederacy. On December 20, Sherman captured Savannah, Georgia. He telegraphed the news to President Lincoln in a message that read "I beg to present you as a Christmas gift the City of Savannah, with one hundred and fifty guns and plenty of ammunition, also about twenty-five thousand bales of cotton."

1914: Harry Turner dies from injuries sustained in an Ohio League football game. The Ohio League was the predecessor of the modern NFL. Turner fractured his back and severed his spinal cord while tackling another player. He was the first person to die from an injury in an American football game.

1926: The National Broadcast Company (NBC) goes on air. NBC was originally launched on 24 television stations across the United States. The first program was aired from the Waldorf-Astoria Hotel in New York City and carried across the nation.

1943: German SS leader Heinrich Himmler orders the Holocaust be extended to include Romani. In the order, he declared that Romani (known then as Gypsies), a migratory ethnic group in Europe, be put "on the same level as Jews and placed in concentration camps." Death squads often murdered Romani on sight. Between 220,000 and 1.5 million Romani were murdered during the Holocaust. Himmler was captured by the British and committed suicide in 1945.

1996: The Stone of Destiny is returned to Scotland. According to legend, the Stone was the pillow on which Jacob had his vision of angels climbing a ladder to Heaven. It was stolen by Edward I in the 13th century.

November 16

534: Justinian I publishes the Codex Justinianus. In doing so, the Eastern Roman Emperor updated and codified the Roman legal system. After the fall of the Empire, the Codex was lost for centuries. It was partially rediscovered in the 12th century. In 1877, Paul Krüger, a German professor of Roman law, recreated and published a modern edition of the Codex.

1632: King Gustavus Adolphus is killed during the Battle of Lützen. The Swedish "Lion of the North" led his army against that of the Holy Roman Empire in the important battle. His troops were victorious. The Thirty Years' War would drag on until 1648, when the Peace of Westphalia established the right of nation-states to govern themselves.

1776: During the American Revolutionary War, a force of 3,000 Hessian mercenaries captures Fort Washington. Patriot riflemen protecting the fort, located on Long Island, were overwhelmed by the Hessians' superior numbers. The attackers were aided by detailed knowledge of the fort's defenses. William Demont, an American officer who deserted and turned traitor, had provided the plans to the British.

1793: Ninety Roman Catholic priests are drowned during the French Revolution's Reign of Terror. The Drownings at Nantes, as they became known, were carried out against the priests for suspected allegiance to the Royal family. Jean-Baptiste Carrier, the French Revolution's emissary to Nantes, ordered the executions. The drownings continued for weeks. As many as 4,000 suspected Royalist sympathizers were systematically drowned in the river. In 1794 Carrier was tried and executed for his role in the massacres.

1821: Trader William Becknell arrives in Santa Fe in present-day New Mexico. Becknell brought goods and supplies that he sold to Native Americans at an enormous profit. On his return to Independence, Missouri, he took a shortcut that became known as the Santa Fe Trail. The Trail became an important overland route for traders and settlers. During the Mexican-American War in 1846, the American Army used the Trail in their invasion of Mexico.

1849: Fyodor Dostoyevsky is sentenced to death by a Russian court. The young novelist had been found guilty of association with a radical literary discussion group that opposed the Tsar. The group was accused of reading banned books and distributing copies of them. Dostoyevsky's sentence was later reduced to a mock execution, which was carried out in Semonovsky Square in Saint Petersburg. After the mock execution was carried out, Dostoyevsky was taken to a gulag in Siberia, where he spent the next six years in exile. His hands and feet were constantly shackled for the entire time. After his release in 1854, he began working on a novel based on his experience. It became *Crime and Punishment*, which is widely regarded as a classic of Russian literature.

1871: The National Rifle Association is founded in New York. The organization was started by Civil War veterans. General Ambrose Burnside was elected the first president. Its original goal was to train riflemen in marksmanship. It focused on target shooters and hunters until the 1970s, when it began forming coalitions with conservative politics.

1945: In "Operation Paperclip," the United States imports about 90 German scientists and engineers to assist in developing missile technology. Many of the scientists were former Nazis. The government justified the move as necessary to counter the growing Soviet threat.

1959: *The Sound of Music,* by Rodgers and Hammerstein, opens on Broadway. It would later be adapted to a hit movie.

November 17

1603: Sir Walter Raleigh is tried for treason. Raleigh, an English explorer and courtier, played a major role in the English colonization of North America on behalf of Queen Elizabeth I. After her death in 1603, he became involved in a plot to overthrow her successor, James I. He was found guilty, but James spared his life. He was imprisoned until 1616. In 1618, while exploring South America, he attacked a Spanish settlement against the King's orders. For that offense he was brought back to London, tried, and beheaded.

1777: The United States Congress submits the Articles of Confederation to the states for ratification. The Articles were the first constitution of the newly formed nation. They provided the federal government with extremely limited powers. This weakness became problematic, particularly after Shay's Rebellion in 1786. Ultimately the Articles were replaced by the Constitution in 1789.

1837: Giuseppe Verdi's first opera debuts in Milan. The impresario—or artistic director—of La Scala Theater was extremely impressed by the young composer's opera, which was called *Oberto, Conte di San Bonifacio*. The next day he hired Verdi to write three more. One of them, *Nabucco*, was a hit. He went on to write *Aida*, his best-known opera, in 1871.

1858: Denver, Colorado is founded. A group of prospectors from Lawrence, Kansas first settled there during the Pikes Peak Gold Rush. The city was named for James W. Denver, the Territorial Governor of Kansas. At an elevation of 5,690 feet, Denver became known as the "mile-high city."

1863: Abraham Lincoln begins writing the first draft of the Gettysburg Address. Over the next two days he rewrote the address several times. The final version was just 272 words long.

1894: H. H. Holmes is arrested in Boston. Holmes was one of the first modern serial killers. Holmes had studied anatomy and dissection at the University of Michigan. While he was a medical student there, he was suspected of using cadavers to defraud life insurance companies, but never charged. He worked in a drugstore in Philadelphia after completing his studies. After a boy died from taking medicine Holmes had prescribed him, he moved to Chicago to escape arrest. While there, he built a mansion with false doors and hidden passages and began using it to carry out his grisly murders. The World's Columbian Exposition was being held at the time, and young women traveled to Chicago from all over the country to find work. Holmes lured several to his "Murder Castle," where he killed and buried them.

Police eventually found the bodies in the basement. He confessed to 27 murders, but only nine could be confirmed.

1903: The Russian Social Democratic Labor Party splits into two factions. Nearly fifteen years later, the Bolshevik (Russian for "majority") faction would seize power and drive out the Menshevik ("minority") faction during the October Revolution.

1993: The US House of Representatives ratifies the North American Free Trade Agreement. The regional trade bloc included the United States, Mexico, and Canada. It was a controversial subject in the 1992 elections, when independent candidate Ross Perot campaigned against it.

November 18

1307: William Tell shoots an apple off his son's head. According to legend, Tell refused to remove his cap in deference to the Austrian governor of Switzerland, Albrect Gessler. In punishment, the governor gave him the choice of being executed or attempting to shoot the apple from his son's head. Tell, an expert marksman, succeeded in the attempt. He then told the governor that had he failed, his next arrow would have killed Gessler. Enraged, the governor had him sent to prison, but he escaped while being taken there. He then returned and assassinated Gessler with his crossbow. The incident sparked the William Tell Rebellion. In the rebellion, Swiss communities formed a confederation that eventually succeeded in throwing the Austrian occupiers out of the region.

1421: The St. Elizabeth's Day Flood inundates the Netherlands. A seawall at the Zuider Zee dike broke, and the North Sea flooded 72 villages. Some 10,000 people drowned.

1493: Christopher Columbus sights the island of Puerto Rico. The native Taino people called the island *Boriken*. Columbus later reported his finding to the Spanish king, who sent Juan Ponce de Leon to invade and pillage Puerto Rico.

1626: Pope Urban VIII consecrates the new St. Peter's Basilica in the Vatican.

1803: The Battle of Vertières is fought. An army of formerly enslaved people defeated a French army that was tasked with capturing and returning them to bondage. Led by Jean Jacques Dessalines, the Haitians had repeatedly overwhelmed the French despite being outnumbered. Vertières marked the final stage of the Haitian Revolution. By 1804 the Haitians had won their independence.

1865: Mark Twain publishes *The Celebrated Jumping Frog of Calaveras County.* It was the American writer's first short story to be published. Two years later, Twain published a collection of short stories under the same title. He became one of the best-known writers and humorists of the 19th century in America.

1883: American and Canadian railroads begin using four continental time zones at noon. The time zones were designed to end the confusion of managing hundreds of different local times determined independently from one city to another.

1928: *Steamboat Willie* is released. The animated short film is considered to be the unveiling of the characters Mickey and Minnie Mouse, although it was the third time the pair had appeared in a cartoon. It was the first use of synchronized sound in an animated film. It was first shown in the Broadway Theater in New York City as an introduction to the film *Gang War*.

1978: Jim Jones' followers commit mass suicide in Jonestown, Guyana. Jones founded the Peoples Temple in the 1950s and attracted followers in San Francisco during the following decades. They relocated to rural Guyana and established a village. Jones became addicted to drugs while there, and Congress began investigating the commune in 1977. Jones' guards shot and killed Congressman Leo Ryan as he was departing after a fact-finding mission. The same day, Jones ordered his 909 followers to drink cyanide-laced Kool Aid.

2002: A United Nations delegation arrives in Iraq to search for weapons of mass destruction. Led by Hans Blix, the investigators found no weapons. The United States government insisted that its intelligence agencies knew of the existence of weapons anyway. Under the leadership of George W. Bush, it used this as a pretext to invade Iraq the following March.

November 19

636: The Battle of al-Q disiyyah. The Rashidun Caliphate, the first Islamic empire established after the death of the Prophet Muhammad, defeated the Sasanian Empire of Persia. The four-day battle was a turning point in the control of Iraq, which had been controlled by Persia until then. The Caliphate lasted until 661, when it was overthrown by the Umayyad Caliphate.

1703: A masked prisoner dies in the Bastille, Paris. The so-called "Man in the Iron Mask" actually wore a mask of black velvet cloth. He was known only as "Marchioly." The first records of his imprisonment date to 1669. A letter describing the prisoner called him "Eustache Dauger," but no previous records of him exist. He was held at various French prisons for three decades, mostly in isolation. He was guarded by two musketeers whenever he left his cell, and they were said to have orders to kill him if he ever removed his mask. While his true identity has never been revealed, many theories were developed about him beginning almost as soon as he had died. One argued that he was the King's own twin brother, while others said he was Oliver Cromwell's son. In 1850, French author Alexandre Dumas wrote *The Man in the Iron Mask*, which was loosely based on the historical figure.

1794: The United States and Great Britain sign the Jay Treaty. The treaty resolved lingering issues from the American Revolutionary War, including compensation for losses, regulating commerce and naval control of the waterways, and provisions for extraditing criminals.

1863: Abraham Lincoln delivers the Gettysburg Address.
The speech was one of several delivered at the dedication of the Soldiers' National Cemetery at Gettysburg, the site of one of the bloodiest battles of the Civil War. It is the best remembered, however. In it, Lincoln referenced the founding principles of the United States and urged the Union to continue its cause.

1916: Samuel Goldfish and Edgar Selwyn found the Goldwyn Pictures company. The movie studio went on to become one of the most successful ones of the 20th century. It started the careers of stars such as Lucille Ball and Gary Cooper as well as writers including Sinclair Lewis.

1943: After a mass uprising at the Janowska concentration camp in the Ukraine, the Nazis murder all of its prisoners. At least 6,000 Jewish prisoners were slaughtered and the camp was closed. The Nazis then retreated ahead of the advancing Soviet Red Army.

1969: Brazilian soccer star Pelè scores his 1,000th goal. The striker played for Santos Football Club from 1956–1974 and the New York Cosmos from 1975–1977, as well as the Brazilian national team. He is widely considered the greatest soccer player.

1985: US President Ronald Reagan and Soviet General Secretary Mikhail Gorbachev meet for the first time. The Geneva Summit was a series of talks on nuclear disarmament aimed at easing relations between the two powers.

1998: *Portrait of the Artist Without Beard,* by Vincent Van Gogh, sells for $71.5 million. At the time it was the third-most expensive painting ever sold. It is believed by some art historians to be the impressionist's last self-portrait.

2004: The Malice at the Palace. During an NBA game between the Indiana Pacers and Detroit Pistons, a brawl broke out in the fourth quarter. Stephen Jackson was suspended for 30 games, and Ron Artest (who later changed his name to Metta World Peace) for the remainder of the 2004 season.

November 20

284: Diocletian is proclaimed the Roman emperor. During his reign, he appointed Maximian co-emperor and divided the Empire. Diocletian ruled the Eastern Empire, and Maximian ruled the Western Empire. Together they secured the Roman Empire and led a successful campaign against Persia. Diocletian ruled until 305, when he became the first emperor to voluntarily give up the throne. He died in 312.

1272: Edward I becomes the King of England. During his reign, he established legal reforms including the reduction of Church authority to strictly ecclesiastical, or spiritual, matters.

1820: A sperm whale attacks the whaling ship *Essex* off the coast of South America. The whale sank the ship, stranding its crew thousands of miles from shore. The men were forced to use the ship's surviving whaleboats to make their way back to land. Accounts of the ordeal later inspired Herman Melville to write *Moby Dick.*

1805: The opera *Fidelo* premiers in Vienna. It was the first and only opera written by Beethoven.

1866: Pierre Lallemont invents the rotary-crank bicycle. The invention was the first to use pedals to turn the bicycle's wheels. Previous models, called hobby-horses, required the rider to propel the cycles along the ground with their feet.

1910: The Mexican Revolution begins. Francisco Madero issued the Plan de San Luis Potosì, in which he denounced the government of President Porfirio Diaz after a rigged election. He called for armed revolution to oust Diaz from power. Thousands took up arms and rallied to the cause. The following year a new election was held, and Madero was elected president. The war dragged on for eight more years, however.

1945: The Nuremberg Trials begin. Following World War II, the Allied forces tried 24 Nazi leaders for their roles in the

Holocaust and other war crimes. Defendants included high-level officials in the Nazi Party, officers in the German military command, and lawyers and doctors who had aided the persecution of Jews and other minorities before and during the war. The Nuremberg Trials were the first instance of an international court trying persons accused of crimes against humanity and war crimes. Twelve of the defendants were sentenced to death by hanging. The trials established a set of guidelines for determining what a war crime is. These guidelines were later codified as the Nuremberg Principles by the United Nations. They remain in use in the International Criminal Court at the Hague.

1947: Elizabeth II marries Phillip Mountbatten at Westminster Abbey in London. In 1952, Elizabeth ascended the throne on the death of her father. She went on to become the longest-ruling British monarch in history.

1985: Microsoft Windows 1.0 is released. The operating environment was the first mass-produced graphical user interface (GUI) that used icons and a point-and-click model. It influenced all operating environments that came after.

1998: The first module of the International Space Station is launched. The ISS eventually grew to include 16 pressurized modules. Five are Russian, eight are American, two are Japanese, and one is European.

November 21

1676: Ole Rømer demonstrates the speed of light. Rømer, a Danish astronomer, was working at the Royal Observatory in Paris when he made the discovery. He used the eclipses of Jupiter's moon Io to determine the speed of light. He calculated that light took 22 minutes to travel a distance equal to the diameter of the Earth's orbit around the Sun. From that, he was able to estimate the speed of light as 220,000 kilometers per second. The estimate was pretty close: the true speed of light is 299,792.458 km per second, as measured in 1975.

1783: Jean-François Pilâtre de Rozier and François Laurent d'Arlandes make the first untethered flight in a hot-air balloon. The pair flew 5.5 miles over Paris, France, in a half hour. They used a cloth balloon made by brothers Jacques-Ètienne and Joseph-Michael Montgolfier.

1806: The French Navy begins a blockade of Great Britain on the orders of Napoleon I. The British refused to negotiate with Napoleon. Instead, the Fourth Coalition, which included Britain, Prussia, Sweden, and Russia, attacked France. The Coalition ultimately lost the war.

1843: Thomas Hancock patents vulcanized rubber in England. The process made rubber more durable and pliable. Eight weeks later, American inventor Charles Goodyear patented the discovery in the United States. The two became involved in a patent dispute in British courts, which Hancock ultimately won. Goodyear, however, is remembered as the inventor of the process.

1905: $E=mc^2$. Physicist Albert Einstein published his most famous research paper, titled "Does the Inertia of an Object Depend Upon Its Energy Content?" on this day in the journal *Annalen der Physik* ("Annals of Physics"). It was one of several papers he published during his *Annus Mirabilis*, or "Miraculous Year" of discovery, and described the discovery of mass-energy equivalency, which means that objects with mass can be transformed into energy at a measurable rate. The amount of energy, E, in any object is equal to that object's mass, m, times the speed of light squared, c^2. The discovery revolutionized physics and made Einstein the most famous scientist in the world. One of the implications of the discovery was the Manhattan Project, which created the atomic bomb. The project used the transformation of a relatively small amount of uranium to create an explosion of energy that was equal to 20,000 tons of dynamite. Einstein would go on to win the Nobel Prize for other scientific discoveries, but "$E=mc^2$" remains his best-known.

1922: Rebecca Latimer Felton becomes the first female United States Senator. Felton had long campaigned for women's suffrage. She was also a white supremacist and had given speeches in favor of lynching. She was appointed Senator from Georgia by the state governor, who was trying to secure the women's vote following the passage of the 19th Amendment. Felton served for one day, and was replaced on the 22nd after a special election.

1953: The Natural History Museum of London announces that the Piltdown Skull is a hoax. The skull had been presented as an early human skull by Charles Dawson in 1912. Anthropologists questioned its accuracy almost from the start, but could not prove it was a forgery for decades.

1969: ARPANET goes online. The Advanced Research Projects Agency Network was an early predecessor of the modern Internet. ARPANET allowed "packets" of data to be sent from one computer server to another. In 1982, the Internet Protocol Suite (IP) was designed as the standard protocol for use on the ARPANET. The system was decommissioned in 1990.

November 22

1220: After promising the Pope that he would support the Fifth Crusade to capture Jerusalem, Frederick II is crowned emperor of the Holy Roman Empire. The crusade ended in failure, and the Pope blamed Frederick for the defeat.

1307: Pope Clement V issues orders for the arrest of all the Knights Templar. The Catholic military order had existed for nearly two centuries and fought several crusades, acquiring vast wealth from their conquests of the Middle East. King Phillip IV of France had fallen deeply into debt to the order, and he pressured Clement to have them arrested. The knights were tortured into giving false confessions of witchcraft and idolatry and burned at the stake. In 1312 the order was officially disbanded.

1699: Russia creates an alliance with Denmark and Poland to counter Swedish power in Europe. The alliance started the Great Northern War in 1700. Although the Swedish army was initially able to win several victories, the alliance won the Battle of Poltava in 1709. Swedish power declined after the battle, and Russia became the dominant force in Northern Europe.

1718: The pirate Blackbeard is killed. Edward Teach had operated in the Caribbean for years. He was killed in a fierce battle off the coast of North Carolina by the British Navy.

1935: Airmail service begins over the Pacific. The *China Clipper*, a "flying boat" operated by Pan American Airways, carried mail from San Francisco to Manila until 1945, when it crashed on landing.

1963: John F Kennedy is assassinated in Dallas, Texas. While riding in a presidential motorcade through Dealey Plaza in an open-top limousine, Kennedy was shot in the back of the head by Lee Harvey Oswald. Oswald, firing an Italian Carcano bolt-action rifle from the 6th floor of the Texas School Book Depository, struck Kennedy and Texas Governor John Connally. The motorcade rushed to the hospital, where Kennedy was pronounced dead. Oswald was arrested by Dallas Police less than two hours after the shooting. Lyndon Johnson was sworn in as the 36th President later the same day. Oswald was shot and killed two days later by Jack Ruby as he was escorted from the Dallas Jail to a waiting car. Ruby died in prison in 1967. The Warren Commission investigated the assassination, and found that Oswald had acted alone in killing Kennedy, and Ruby had acted alone in shooting Oswald. The assassination sparked numerous conspiracy theories that implicated multiple additional shooters as well as the Mafia, the CIA, Lyndon Johnson, and the Soviet KGB.

1967: The UN Security Council passes Resolution 242. The Resolution set out the principles for peace between Arab states and Israel. It was the foundation for later negotiations between

Israel, Egypt, and Jordan, as well as the Palestinian Liberation Organization.

1995: *Toy Story* is released. It was the first feature-length film that only used computer-generated images. The film was a major hit, grossing $373 million worldwide, winning three Academy Awards, and spawning multiple sequels.

2005: Angela Merkel becomes the first female Chancellor of Germany. The former physicist and leader of the Christian Democratic Union became the *de facto* leader of the European Union and one of the most powerful women in the world.

November 23

1165: Pope Alexander III returns to Rome after exile. Alexander was chosen to succeed Pope Adrian IV. He was opposed by Victor IV and forced to leave Italy until Victor's death. When he returned, Alexander established the procedure for inaugurating new Popes by a vote of two-thirds of the Catholic cardinals. The practice persists to this day.

1644: John Milton publishes *Areopagitica*. The pamphlet argued against censorship and defended the right to freedom of speech and expression. It greatly influenced other Enlightenment thinkers, including the authors of the American Declaration of Independence.

1733: A slave revolt breaks out on the island of St. John. About 150 enslaved persons revolted against plantation owners in the Dutch West Indies (present day US Virgin Islands). The rebellion lasted until August of 1734, making it one of the largest and longest-lasting slave revolts in the Americas. The rebels took control of most of the island before being defeated by French and Swiss troops.

1863: The Battle of Chattanooga is fought. General Ulysses S. Grant's Union army attacked the Confederate Army of Tennessee under Braxton Bragg. The Confederates retreated to

Lookout Mountain, where Union Forces defeated them the next day.

1876: Boss Tweed, the corrupt leader of New York's Tammany Hall, is delivered to authorities. William Magear Tweed controlled the Democratic Party's political machine in New York City beginning in 1863. He ruled the machine from Tammany Hall, the Party's seat of power, where he controlled who ran for office, getting friends and allies elected. He also used the position to engage in considerable corruption and bribery. He eventually stole between 20 and 40 million in taxpayer money. Tweed was not subtle about his graft. He took to wearing a huge diamond on his shirt and became one of the largest landowners in the City. Following a scandal in 1871, Tammany's power declined, and he was arrested on the orders of the city Comptroller. During his trial, he fled the country and went to Spain. Spanish authorities captured him and returned him to the United States to stand trial. He remained in prison until his death in 1878.

1897: Andrew J. Beard invents the "jenny coupler." The device, which is still used to this day, allows railroad cars to hook together automatically simply by bumping into one another.

1889: The first jukebox is installed at the Palais Royale Saloon in San Francisco. Louis Glass and William S. Arnold invented the device, which was coin-operated and contained an Edison cylinder phonograph. Listeners could choose from four songs for a nickel a play.

1948: Frank Back patents the first zoom camera lens. The invention replaced the use of multiple focal-length lenses and made long-distance photography much more accessible.

1964: The Vatican discontinues the use of Latin as the official language of the Roman Catholic Church. The move was part of the Vatican's liberalization and modernization campaign that was called "throwing open the windows of the Church."

1981: President Ronald Reagan signs National Security Decision Directive 17. The top-secret executive order authorized the CIA to train Contra rebels in Nicaragua. To acquire funds to arm the rebels, the Reagan administration secretly sold arms to the Iranian regime. The resulting scandal was known as the Iran-Contra affair.

2005: Ellen Johnson Sirleaf is elected president of Liberia. Sirleaf was the first woman to become a head of state on the African continent.

November 24

1105: Rabbi Nathan ben Yehiel finishes translating the Talmudic dictionary. The Roman rabbi traced the root of each word in the Talmud, a text of the Jewish Oral Law, and cited the section in which it appeared.

1642: Abel Tasman sights an island south of Australia. The Dutch navigator named the island Van Diemen's Land, after his ship's captain. It was later named Tasmania in his honor.

1859: Charles Darwin publishes *On the Origin of Species*. The book, in which he laid out his theory of evolution by natural selection, revolutionized the life sciences. The work was the foundation for all studies of evolutionary biology that came after.

1877: Anna Sewell publishes *Black Beauty*. The novel told the story of a horse named Black Beauty from his perspective, and advocated better treatment of animals. It became an instant bestseller and is considered a classic of animal welfare.

1947: The Hollywood Blacklist is created. The US House of Representatives voted to approve charges of contempt against ten writers and directors who refused to cooperate with an investigation into communist influences in the movie industry. The next day, Hollywood began compiling a list of dozens of artists who were suspected of having communist sympathies. They were prevented from working in the industry for over a decade.

1971: D.B. Cooper hijacks a plane in Portland, Oregon.
On Thanksgiving Eve, Cooper purchased a one-way ticket on Northwest Orient Airlines to Seattle, Washington. He boarded the plane carrying only a black attaché case, took a seat, and ordered a scotch and soda. Shortly after the plane took off, Cooper handed a note to a flight attendant. The note said that Cooper had a bomb in his briefcase. He then showed the attendant his briefcase, which appeared to contain several sticks of dynamite. He demanded $200,000 and four parachutes. The plane landed in Seattle, where FBI agents turned over the ransom money and parachutes. Cooper allowed the plane's passengers to depart the plane. The plane then took off again with just Cooper and the flight crew aboard. Cooper then parachuted out of the back of the plane as it flew through a severe thunderstorm. In 1980, some of the ransom money was discovered on a riverbank in Washington state. Despite a massive manhunt, Cooper was never found.

1974: Donald Johanson and Tom Gray discover the remains of Lucy. An incomplete skeleton of an *Australopithecus afarensis* female that lived in present-day Ethiopia some 3.2 million years ago, Lucy is one of the earliest known ancestors of modern humans. She was named for the Beatles song "Lucy in the Sky with Diamonds."

2016: The Revolutionary Armed Forces of Colombia (FARC) sign a peace deal with the government. The peace process brought an end to a more than 50-year-long civil war between the FARC and the Colombian government that claimed thousands of lives, most of whom were civilians.

November 25

885: The Vikings attack Paris. A force of 30,000 Viking warriors sailed several hundred ships up the Seine river to take the city. They demanded tribute, but the Parisian Count Odo refused. They were unable to break through the city's walls and set in for a siege. King Charles the Fat arrived in October of 886 with an army and negotiated with the Vikings. He paid 700 pounds of silver to convince them to leave. When Charles died in 888, the French elected Odo to be King.

1491: The Siege of Granada ends. Granada was the last Moorish stronghold in Spain. Spanish forces besieged the city beginning in April. After the defenders ran out of food and supplies, they signed a conditional surrender. The Treaty of Granada marked the end of Moorish rule in Spain.

1783: The last British soldiers withdraw from New York City almost three months after the Treaty of Paris ended the American Revolutionary War. Following their departure, General George Washington led a column of Continental Army soldiers into the city to the cheers of its residents.

1864: Confederate operatives attempt to burn Manhattan. A group of Southern spies calling themselves the Confederate Army of Manhattan entered the Union through Canada and made their way to New York City. Once there, they attempted to start multiple simultaneous fires throughout the city. The plot was made in revenge for General William Tecumseh Sherman's burning of Atlanta. Targets included 19 hotels, a theater, and PT Barnum's American Museum. They failed in their attempt. The fires either failed to spread or were quickly extinguished. All of the spies but one, Robert Cobb Kennedy, escaped. Kennedy was captured while attempting to return to Richmond, the Confederate capital. Kennedy had attended West Point before the Civil War and joined the Confederacy when hostilities broke out. His defense lawyer at his trial was General Edwin Stoughton, a Union officer and former West Point classmate of his. Kennedy was found guilty, and hanged for his role in the plot on March 25, 1865. He was the last Confederate soldier executed by the United States government during the Civil War, which ended less than a month later with the surrender of the Confederate armies.

1884: John B. Meÿenberg files a patent for evaporated milk.
He founded the Helvetia Milk Condensing Company, and later assisted in the creation of the Carnation Evaporated Milk company. Carnation became a major food brand in the 20th century and was acquired by Nestle in 1985.

1936: Japan and Germany sign the Anti-Comintern Pact.
The alliance stated their shared animosity to the Communist International organization. Italy and Spain later joined the pact, forming an alliance that became known as the Axis Powers.

1952: Agatha Christie's play *The Mousetrap* opens in London. The murder-mystery became the longest-running play in history at 27,174 performances.

1999: Elian Gonzalez, a five-year-old Cuban boy, is rescued by fisherman off the Florida coast. Gonzalez became the center of an international custody battle between his Cuban father and the Cuban American community in Miami. He was ultimately returned to Cuba after the US Supreme Court declined to hear the case.

November 26

1789: President George Washington proclaims Thanksgiving Day a national holiday.
The holiday originally celebrated the signing of the US Constitution. It was not officially recognized by the government again until 1863, when President Abraham Lincoln proclaimed it a national holiday to be celebrated every year on the third Thursday of November.

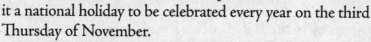

1812: Starving and decimated, Napoleon Bonaparte's army begins crossing the Berezina River after suffering a terrible defeat in Russia. The army hastily built two temporary bridges to cross the river. They were pursued by the Russian army in a rout that drove them all the way back to Poland.

1842: The University of Notre Dame du Lac is founded.
Known simply as "Notre Dame," the Catholic research university gained fame in the 20th century for its football program. The team, called "the Fighting Irish," is considered one of the top NCAA Division I squads. It was named after the Irish Brigade, which fought for the Union during the Civil War. The University's third president, William Corby, had been the Brigade's chaplain and served at the Battle of Gettysburg, where the unit gained fame for its bravery.

1867: JB Sutherland patents the refrigerated railroad car.
The invention was an insulated boxcar with two large bunkers of ice on either end. Cool air from the bunkers was circulated through the car by the movement of the train. Sutherland's invention allowed perishable goods to be transported over long distances.

1917: The National Hockey League is formed. The league originally consisted of just five teams: The Montreal Canadiens, Montreal Wanderers, Ottawa Senators, Quebec Bulldogs, and Toronto Arenas. Only the Canadiens still exist, but the NHL grew to include more than 30 teams.

1942: *Casablanca* premieres in New York City. The movie starred Humphrey Bogart and Ingrid Bergman. It was set in Casablanca, Morocco, during World War II. Bogart's character, an American expat, falls in love with Bergman's, who is involved with the Czech resistance against the Nazis. Ultimately the two are forced to part, with Bogart uttering the now-iconic line, "Here's looking at you, kid."

1976: Anarchy in the U.K. The Sex Pistols, a band from London, released an album that will usher in the era of punk rock music. Among their early fans were Souxsie Sioux and Billy Idol, who would go on to become punk rock icons in their own right. The Sex Pistols' music was characterized by fast beats, stripped-down instrumentations and heavy feedback, and anti-authoritarian, often profanity-laced lyrics. The band lasted just two and a half years, but their cultural impact was immense. In 1979, their bassist, Sid Vicious, died of a heroin overdose, and the Sex Pistols broke up. The musical movement they helped spawn included bands such as the Clash, the Buzzcocks, and the Ramones. Multiple "waves" of punk rock followed as the genre spread to the United States, where the club CBGB became a mainstay of punk music in New York City.

1983: Thieves steal 25 million pounds of gold bars from the Brinks-Mat security warehouse in London's Heathrow Airport. Several of the robbers were tried and convicted for the robbery in the 1980s, but most of the gold was never recovered.

November 27

1095: Pope Urban II makes an appeal to Christians in Europe to seize Jerusalem from the control of Muslims. The appeal led to the first Crusade. Ultimately, several Crusades would be waged. The conflicts occurred over several centuries as the Holy Land was captured and recaptured by Christian and Muslim armies.

1798: Rabbi Shneur Zalman is released from a Russian prison. Zalman wrote the *Tanya*. The book is considered sacred by members of the orthodox Hasidic movement. Russian authorities had accused Zalman of high treason because he sent funds to the poor of Jerusalem, which was controlled by the Turkish Sultan at the time.

1810: Theodore Hook perpetrates the Berners Street hoax. Hook bet his friend Samuel Beazley that he could make any house in London the most famous address in a week. Beazley took the bet, and the pair settled on 54 Berners Street. Hook sent out thousands of letters that requested assistance, visitors, and deliveries. Beginning at nine o'clock in the morning, the hoax began to play out when a chimney sweep arrived at the door. The maid answered, and told him no sweep had been ordered by anyone in the house. While she was speaking with him, eleven more chimney sweeps arrived at the door. As they were leaving, a series of deliveries began to arrive. Carts full of coal, wedding cakes, doctors, lawyers, and priests all began showing up at the door. A dozen pianos and an organ were delivered. The Governor of the Bank of England, the Archbishop of Canterbury, and the Lord Mayor of London also came to the address that day. Traffic became so congested that part of London came to a halt. Hook and Beazley spent the day across the street from the house, watching the commotion. Hook left London after the hoax. He later received the world's first postcard, which he likely sent to himself.

1834: Thomas Davenport invents the first successful electric motor. The motor operated using a direct, rather than an alternating, current.

1924: Macy's Department Store holds its first Thanksgiving Day parade. The parade was started by the store to boost holiday shopping. It drew 250,000 spectators and was made an annual event that now draws millions of onlookers to Broadway Avenue each year.

1978: Dan White assassinates San Francisco mayor George Moscone and city supervisor Harvey Milk. Milk was the first openly gay elected official in the history of California. He was elected to office in 1977. White, a former city supervisor, was angry at Moscone and White for opposing his reappointment to the Board of Supervisors. White served five years for manslaughter and committed suicide after his release.

2006: The Québécois are recognized as an internal nation in Canada. The Canadian House of Commons passed a resolution that recognized the province of Quebec and its French-speaking population as a nation-within-a-nation. The Québécois had engaged in various struggles for autonomy and recognition, some violent, for the better part of 50 years.

November 28

1520: Ferdinand Magellan enters the Pacific Ocean with a fleet of three ships. In doing so, he became the first European person in recorded history to sail from the Atlantic Ocean to the Pacific, which he did by sailing below South America through the Straits that bear his name.

1582: William Shakespeare and Anne Hathaway pay £40 for a marriage license in Stratford-upon-Avon. At the time Hathaway was pregnant with the couple's first child. Little else is known about Hathaway other than that she outlived Shakespeare by seven years.

1660: Christopher Wren and eleven other scholars found the Royal Society. The organization was chartered at Gresham College to promote the study of the sciences. It included natural philosophers and physicians. Famous scientists from Sir Isaac Newton to Stephen Hawking have been elected Fellows of the Royal Society over the years.

1678: King Charles II of England accuses his wife, Catherine of Braganza, of treason. The reason he cited for the accusation was that she had suffered three miscarriages. The House of Commons passed an order to remove her from the Palace and have her stand trial. The King later intervened, preventing the trial. He took multiple mistresses and fathered children with them.

1720: Anne Bonny and Mary Read are tried for piracy. The Irish pirates had operated in the Caribbean for two years before they were captured. Bonny became one of the most famous female pirates in history despite the short period in which she sailed the high seas. She worked with Calico Jack Rackam to attack coastal shipping lanes. Early one morning in October 1720, a British Navy ship attacked Rackam's ship. Bonny and Read were the only two who resisted the attack, but they were quickly overpowered by the British, who outnumbered them.

They were found guilty in a court in Jamaica and sentenced to death. Because they were pregnant, they appealed to English common law, which provided a stay of execution for women who "pleaded their belly." Read died during childbirth. Bonny was released after giving birth.

1895: The first American auto race is held. Frank Duryea won the race, traveling over 54 miles on a course that wound from Jackson Park in Chicago to Evanston, Illinois, in about ten hours.

1905: Arthur Griffith founds Sinn Féin, an Irish nationalist party. The organization became a vital part of the early 20th-century movement for Irish independence along with the Irish Republican Brotherhood. It remained active after Ireland won its independence and was partitioned, and continued to campaign for a united Irish republic. During the Troubles of the 20th century, as the civil strife and guerilla campaigns waged in Northern Ireland were called, the party became the political wing of the Irish Republican Army.

1907: Louis Mayer opens a movie theater in Massachusetts. The company grew as movies became popular, and Mayer began producing his own films. After a series of mergers with other film companies, the company became known as Metro-Goldwyn-Mayer, one of the major film studios in the United States.

November 29

1787: King Louis XVI of France issues an Edict of Tolerance. The decree extended civil status to Protestants living in the majority Catholic country for the first time.

1864: Colonel John Chivington leads a detachment of Volunteers against a peaceful village of Cheyenne near Sand Creek in the Colorado Territory. The Volunteers slaughtered men, women, and children mercilessly. As a result, Chivington was widely despised by Americans for the rest of his life.

1887: The King of Hawaii grants the United States the right to establish a coaling and repair station at Pearl Harbor. The United States would use the foothold to encroach on the islands over the following years, eventually occupying them and over-throwing the monarchy.

1899: FC Barcelona is established. The Spanish football club, founded by a group of Spanish, Catalan, and English players, would go on to become one of the premiere athletic clubs in the world, and the third most valuable sports team (behind the Dallas Cowboys and the Real Madrid soccer club). Superstars such as Lionel Messi and Ronaldinho rose to prominence in FC Barcelona.

1911: In the first championship of the National Football League, the Pittsburgh Steelers defeat the Philadelphia Athletics by a score of 11 to 0.

1963: President Lyndon Johnson establishes the Warren Commission to investigate the assassination of JFK. The Commission was chaired by Chief Justice Earl Warren. It spent nearly a year investigating the assassination. On September 24, 1964, the Commission's 888-page report was presented to Johnson. The Commission found that Lee Harvey Oswald was solely responsible for the assassination. The American public was widely skeptical of the Commission's findings, as were four of its own seven members. The report led the Secret Service to modify the methods it used to protect the President.

1972: Atari releases Pong. The table tennis sports game was one of the first arcade video games. It was designed by Allan Alcorn, a computer scientist at Atari. He was given the task by Nolan Bushnell, one of the company's co-founders, who intended it to be a training exercise. Alcorn's creation became quickly pop-ular and led to the establishment of the home console and video game industry. In the game, two players engage in an electronic game of ping-pong on a two-dimensional, interactive screen. The game was very similar to one developed by Magnavox Odyssey, and the latter sued Atari. The lawsuit ultimately ended in an out-of-court settlement. Pong spawned several sequels and made Atari a household name in video gaming. Alcorn later hired Steve Jobs to work at Atari. Jobs went on to found Apple, one of the most successful computer companies in history.

1993: The British government reveals it has secretly been negotiating with the Irish Republican Army. Several members of Parliament denounced the talks because John Major had publicly refused to negotiate with the outlawed paramilitary group. But the negotiations eventually led to the IRA agreeing to a cease-fire.

November 30

1215: Pope Innocent III ends the Fourth Lateran Council. It was the first Papal Council to use the word "transubstantiation" to describe the Eucharistic ceremony in the Catholic mass.

1609: Galileo Galilei observes the moon with a telescope for the first time. The Italian astronomer made drawings of his observations, and later published them in his pamphlet *Starry Messenger*. It was the first time the public saw images of the Moon with a heavily cratered surface.

1774: Thomas Paine arrives in Philadelphia. The patriotic activist wrote a number of influential pamphlets in support of the American cause of independence, including *Common Sense* in 1776 and *The American Crisis*, published from 1776 to 1783. His work *The Rights of Man* in 1791 defended the French Revolution and argued that political revolution is permissible in some circumstances.

1803: France acquires the Louisiana Territory from the Empire of Spain. Spain had fallen deeply into debt with the French, and the Spanish King did not wish to have to defend the Territory against the new United States of America. In less than a month, France sold it to the Americans.

1864: The Battle of Franklin. Union troops under General John Schofield defeated the Army of Tennessee, commanded by John Bell Hood. As a result of this battle and the Battle of Nashville two weeks later, the Army of Tennessee was effectively destroyed.

1887: The first softball game is held in Chicago. The game was played indoors at the Farragut Boat Club on Thanksgiving Day. A group of sports fans had gathered to hear the outcome of a football game between Yale and Harvard. After the game was over, one fan tossed a boxing glove at another, who swung at it with a stick. The game soon developed into a way to play baseball indoors. It later became popular throughout Chicago, where amateur leagues play outdoor softball all summer. In the 20th century the game caught on as a college sport, with slow-pitch and fast-pitch variations being played.

1947: The United Nations proposes a Partition of Palestine. The partition recommended the creation of two independent states, one Arab and one Jewish. As a result of the partition, a brutal civil war characterized by reprisals and civilian killings broke out. It lasted until 1949, but hostilities between the two communities continued for decades.

1979: Pink Floyd releases the rock opera The Wall.

1999: The Battle of Seattle. Thousands of demonstrators converged on Seattle, Washington, to protest the Ministerial Conference of the World Trade Organization. The protesters used an array of tactics to shut the city down and bring the conference's proceedings to a halt. Their grievances included environmental damage caused by multinational corporations, labor violations against sweatshop workers, and the concentration of wealth at the highest tiers of society. The Seattle police, unprepared for the massive demonstrations, responded with heavy-handed and brutal tactics, beating and pepper-spraying nonviolent protestors. The demonstrations were a high point in the anti-globalization movement that had begun in Berlin in 1988.

Chapter 12

December

December 1

1640: Portugal revolts against Spain after a 60-year occupation. Spain's military power was concentrated on its ongoing involvement in the Thirty Year's War. The Portuguese Restoration War lasted until 1668, when the Treaty of Lisbon established its independence.

1824: Congress declares that no one has won a clear majority in the presidential election of that year. As directed by the 12th Amendment to the Constitution, the House of Representatives then decided the winner. The House chose John Quincy Adams, the son of former President John Adams, as President. He served one term and was succeeded by Andrew Jackson. Adams went on to serve in the House of Representatives.

1835: Hans Christian Andersen publishes his first collection of fairy tales. *Fairy Tales Told for Children* was the Danish author's third book. It included classics such as "The Princess and the Pea" and was followed by a second collection, which included "Thumbelina." Andersen's third collection of fairy tales, published in 1837, featured "The Little Mermaid" and "The Emperor's New Clothes." His fairy tales were later adapted into plays and films.

1838: Mexico declares war on France. France had accused Mexico of mistreating French citizens residing there and demanded reparations be paid after a French pastry chef claimed Mexican army officers had looted his shop. When Mexico refused, the French established a naval blockade and begun bombarding the port of Veracruz. The resulting conflict, called the Pastry War, lasted until Britain negotiated a treaty between the two countries in 1839.

1862: President Abraham Lincoln delivers his State of the Union Address. In it, he asserted that enslaved persons freed by the Emancipation Proclamation would remain forever free. At the same time, he attempted to ease fears of conservative Senators by promising that there would be a gradual emancipation of enslaved persons still in bondage in border states.

1865: Shaw University is founded in Raleigh, North Carolina. It was the first historically black college or university (HBCU) founded in the American South. Nearly a century later, the Student Nonviolent Coordinating Committee (SNCC), a leading organization in the Civil Rights Movement, was started in a meeting at Shaw University organized by Ella Baker.

1913: Ford Motor Company invents the moving assembly line. Factory workers assembled Ford automobiles as they moved past them, with each worker assigned to perform the same task on every car.

1955: Rosa Parks refuses to give up her bus seat in Montgomery, Alabama. Parks, the secretary of the Montgomery chapter of the National Association for the Advancement of Colored People (NAACP), decided to resist the segregation of city buses. She sat near the "whites only" section of a bus upon boarding. When the bus driver told her to move to the back of the bus to make room for white passengers, Parks refused. The driver called the police, who arrested Parks. Her arrest sparked the Montgomery Bus Boycott. African Americans in Montgomery refused to use city buses for over a year, with private citizens organizing car pools and taxi drivers charging the equivalent of bus fare for rides. The city gave in and agreed to stop segregating buses. Parks' action and the boycott were catalysts in the Civil Rights Movement.

December 2

1697: St. Paul's Cathedral in London is consecrated. The cathedral was designed by Sir Christopher Wren following the destruction of the Old St. Paul's Cathedral in the Great Fire of London.

1804: Napoleon Bonaparte crowns himself Emperor Napoleon I at Notre Dame Cathedral in Paris. Napoleon had designed the entire coronation ceremony, which included customs from various ancient and contemporary traditions. Pope Pius VII handed the crown to Napoleon, who then placed it on his own head.

1823: US President James Monroe declares American neutrality in European conflicts, and warns European nations not to meddle in the Americas. The policy, which became known as the Monroe Doctrine, established the United States as a world power. It was a defining moment in the history of the hemisphere, and remained official US foreign policy through the 20th century.

1848: Franz Josef I is crowned emperor of Austria. During his reign, the Austrian Empire's power slowly and steadily declined. In 1914, his nephew Archduke Franz Ferdinand was assassinated in Sarajevo, triggering a series of events that led to World War I.

1939: LaGuardia Airport opens in New York City. The airport, located in the Queens borough, grew to become the 20th busiest in the United States, serving about 30 million passengers a year.

1942: Enrico Fermi oversees the first controlled nuclear chain reaction. The experiment took place in a secret laboratory beneath the bleachers of Stagg Field at the University of Chicago. The Manhattan Project, as the experiment was called, ushered in the nuclear age.

1954: The United States Senate votes to censure Senator Joseph McCarthy. The Senator from Wisconsin had been the public face of anti-communist efforts during the Cold War for four years, claiming he had a secret list of hundreds of communist sympathizers working in the US government. He did not.

1956: The *Granma* lands at Cuba's Oriente Province. Aboard yacht are Fidel Castro, Che Guevara, and 80 members of the 26th of July Movement. Castro had left Cuba the previous year after his imprisonment for anti-government activities on the island. He returned with a guerilla force that was determined to spark a revolution against the Cuban dictator, Fulgencio Batista. The guerillas were repeatedly attacked by government forces as they made their way inland through a mangrove swamp. Once they regrouped, only 19 remained. They established a base of operations in the Sierra Maestra mountains and began launching hit-and-run attacks on Batista's army. Peasants joined the guerillas by the hundreds. Within three years, they had gained control of most of the island's rural territory. On January 1, 1959, the revolutionaries triumphantly entered Havana, the capital, and Batista fled the country. They established a communist government.

2001: Enron files for bankruptcy. The energy and commodities company engaged in one of the largest instances of accounting fraud in history under its CEO Kenneth Lay. Days before Enron filed for bankruptcy, it paid $55 million in bonuses to about 500 of its top executives.

December 3

1775: John Paul Jones raises the Grand Union Flag aboard the USS *Alfred*. The Continental Congress purchased the ship so it could be used to capture British arms shipments coming from England. The *Alfred* was the first ship to fly the Grand Union Flag, a precursor of the Stars and Stripes. It first saw action in the Battle of Nassau.

1800: The US presidential election ends in a tie. Before the 12th Amendment was passed in 1804, the Electoral College members cast two votes each. They did not distinguish whether these votes were for president or vice-president. Thomas Jefferson, Aaron Burr, John Adams, and Charles Pinckney ran. Jefferson and Burr won 73 votes each, Adams won 65, and Pinckney won 65 votes. The tie resulted in 35 rounds of subsequent votes, of which neither Burr nor Jefferson won a majority. On the 36th round, Jefferson won after Alexander Hamilton threw his support behind him and convinced other electors to do the same. Burr was elected vice president. Hamilton was highly critical of Burr's politics, and publicly feuded with him. As a result of the feud, during the last year of his term as Vice President, Burr killed Hamilton in a duel.

1910: Georges Claude demonstrates neon lights at the Paris Motor Show. Claude was considered the "Thomas Edison of France" for his inventions. Neon quickly became popular, particularly among auto dealerships in the United States. During World War II Claude collaborated with the Nazis, and was later imprisoned for it.

1927: Laurel and Hardy's first film debuts. The comedy duo starred in *Putting Pants on Phillip*, a short silent film. The pair would make more than a hundred films.

1964: During the Free Speech Movement on the campus of the University of California, Berkeley, police attack a sit-in at the university administration building. The students were protesting a decision to ban demonstrations on campus. Over 800 were arrested.

1967: Christiaan Barnard performs the first human heart transplant at Groote Schuur Hospital in Cape Town, South Africa. The patient, 53-year-old Louis Washkansky, lived for eighteen days following the transplant.

1976: Gunmen attempt to assassinate reggae superstar Bob Marley two days before the Smile Jamaica concert. The concert had been organized by Prime Minister Michael Manley to cool tensions between rival political factions. Marley's wife and manager were seriously injured but later recovered. Marley was slightly wounded. He went ahead with the concert in spite of the shooting.

1984: The Bhopal disaster occurs in India. A gas leak at a pesticide plant exposed more than half a million residents in nearby towns to toxic methyl isocyanate. Over 16,000 people died as a result. The Union Carbide Corporation paid $470 million in damages and seven employees were imprisoned as a result of the disaster.

1992: A computer engineer sends the first text message. The engineer, an employee of Sema Group, sent a text message from a personal computer to a colleague's phone using the Vodafone network.

December 4

1791: *The Observer,* the world's first newspaper published on Sundays, goes on sale. W.S. Bourne started the paper believing its novelty would make him rich. He quickly fell into debt. He tried to sell *The Observer* to the British government. The government refused to buy the paper outright, but agreed to subsidize it on the condition it could oversee the paper's editorial content. As a result, *The Observer* regularly ran editorials denouncing political activists such as Thomas Paine and Joseph Priestly. Ownership changed hands several times over the centuries, and the paper's editorial bent changed with it.

1881: The first edition of the *Los Angeles Times* is published by Nathan Cole Jr. and Thomas Gardiner. In the early 20th century, the paper took an editorial position against trade unions, which resulted in its offices being bombed in 1910. Twenty-one people were killed in the attack.

1905: Alpha Phi Alpha, the first intercollegiate African-American Greek fraternity, is founded. The fraternity's first chapter was established at Cornell University in Ithaca, New York, by 15 students. Notable members include Dr. Martin Luther King, Jr, W.E.B. DuBois, Duke Ellington, Thurgood Marshall, and Andrew Young.

1953: The first Burger King opens in Jacksonville, Florida. Originally called Insta-Burger King, the fast food restaurant became one of the largest in the world, with more than 15,000 franchises serving Whoppers around the globe.

1956: The "Million Dollar Quartet" meets at Sun Record Studios in Memphis, Tennessee. The group included Elvis Presley, Jerry Lee Lewis, Johnny Cash, and Carl Perkins in an impromptu jam session. Jack Clement, the studio's sound engineer, decided to record the session on a whim. At the time, Lewis, Presley, and Cash were not yet the superstars they would go on to become. Perkins recorded a set while the others listened from the control room. They then joined him in the studio and played 17 tracks together, most of which were impromptu. After the group played a few songs together, Presley quietly left while Lewis was playing the piano. A local reporter came to the studio and wrote an article about the session called "The Million-Dollar Quartet." The recording was not released until 1981.

1965: The Grateful Dead perform their first concert as the Grateful Dead. They were previously called the Warlocks, and before that had been Mother McCree's Uptown Jug Champions. The concert was part of Ken Kesey's Acid Tests—a series of shows in which Kesey passed out LSD to the audience. The Grateful Dead were associated with the hippie counterculture and psychedelia for the remainder of their career, which lasted for most of the 20th century.

1969: Chicago Police assassinate Black Panther leaders Fred Hampton and Mark Clark. Hampton was a captivating organizer and the chairman of the Chicago chapter of the Black

Panther Party for Self-Defense. He had organized Free Breakfast and community education programs on Chicago's West Side. Under his leadership, the Panthers established a "Rainbow Coalition" of revolutionaries from different communities, including the white Young Patriots and the Latino Brown Berets. Hampton was also negotiating a truce between local gang leaders. Police attacked the Panthers' house before dawn, firing hundreds of rounds into the bedroom where Hampton was sleeping next to his pregnant wife.

December 5

1484: Pope Innocent VIII issues a papal bull authorizing German inquisitors to prosecute suspected witches. The bull, titled *Summis desiderantes affectibus* ("Desiring with supreme devotion"), ruled that persons found guilty of witchcraft could be excommunicated from the Catholic Church. It opened the door for later, more severe witch hunts.

1766: James Christie opens an auction house at Pall Mall in London. It would grow to become one of the premiere auction houses in the world, with fine art, treasures, and precious jewels selling for small and large fortunes. It is still in operation.

1775: Henry Knox sets out with the Noble Train of Artillery from Fort Ticonderoga. During the American Revolutionary War, a Continental Army detachment led by Benedict Arnold captured Fort Ticonderoga in May 1775. The Americans had targeted the fort because it was known to have a large store of artillery pieces, and the Continental Army needed the cannons to support its siege of Boston. George Washington gave the task of transporting the weaponry from the fort to Henry Knox, who was serving in the Massachusetts militia during the siege. Knox organized an oxen-sled train to carry 59 cannons to Boston. The journey, which had been planned for two weeks, took ten. Knox later became the chief artillery officer of the Continental Army and the United States' first Secretary of War.

1933: The 21st Amendment to the US Constitution is ratified. The Amendment repealed the 18th Amendment, which had outlawed the sale of alcohol in 1919. It was the only one that was ratified by special state conventions rather than state legislatures, as well as the only to repeal a previous amendment. News of the ratification was greeted by enthusiastic crowds across the nation.

1941: General Georgy Zhukov of the Soviet Red Army orders a huge assault on German positions besieging Moscow. More than 18 infantry divisions supported by 1,700 tanks and 1,500 aircraft were transferred to Moscow for the attack. The battle to break the siege lasted for weeks and the Soviets drove the Germans between 60 and 155 miles back along the Eastern Front. The Germans were finally repulsed from positions that threatened Moscow in 1943.

1952: The Great Smog of London kills more than 4,000 people. A combination of cold weather and lack of wind allowed a thick layer of airborne pollution to settle over the city for four days. At least 100,000 people fell ill as a result. In the months after the incident about 6,000 more people died from complications related to the smog.

1983: The military junta that has ruled Argentina for eight years disbands. During its tenure, the military dictatorship engaged in a dirty war against its own people. At least 9,000 Argentine civilians were "disappeared" by the state—arrested, tortured, and murdered without trial. The dictatorship was supported by the American government. Almost every military officer who led the junta has since been convicted of crimes against humanity and genocide as a result of the dirty war.

2014: The *Orion* makes its first test flight. The uncrewed spacecraft launched from Cape Canaveral at 7:05 Eastern time and made two orbits of the Earth. NASA developed the interplanetary spacecraft to carry four astronauts to the Moon and beyond.

December 6

1704: A Mughal army defeats a Sikh Khalsa force at the Battle of Chamkaur Sahib. The battle, fought in present-day Punjab, India, pitted a much larger force against just 42 Khalsa soldiers. The Mughals had demanded the surrender of the Khalsa's leader, Guru Gobind Singh. Rather than allow him to be captured, the Khalsa soldiers—which included Singh's sons—held off the army while he escaped.

1768: The first edition of *Encyclopædia Britannica* is published. The Encyclopedia's original title was "Encyclopædia Britannica, or, A Dictionary of Arts and Sciences, compiled upon a New Plan." It was published during the Scottish Enlightenment, a period of about two centuries in which many great thinkers came out of that country.

1877: *The Washington Post* is founded. The newspaper won 47 Pulitzer Prizes over the course of the 20th and early 21st century. In the 1970s reporters Bob Woodward and Carl Bernstein published a series of investigative pieces that exposed the Watergate Scandal. Partly as a result of their reporting, the Nixon coverup of the burglary of the Democratic National Committee's headquarters at the Watergate Hotel in Washington, D.C. became public. The scandal led to the first resignation of a sitting US President.

1884: The Washington Monument is completed. The monument to the nation's first president was finished 101 years after its location was approved by George Washington. It stands to the east of the Lincoln Memorial and the Reflecting Pool on the National Mall. Construction on the 555-foot-tall granite obelisk began on July 4, 1848. The project was halted in 1854 because Congress could not afford to keep funding it. In 1876, President Ulysses S. Grant ordered construction to begin again. When the capstone was put in place, the monument was the tallest artificial structure in the world, and held that distinction until

1889. In 2011, the monument was damaged by the 5.8 magnitude Virginia earthquake. National Park Service employees and engineering consultants climbed the monument to inspect the damage, and it underwent extensive restorations. It was reopened in 2014 following repairs.

1912: Ludwig Borchardt discovers the bust of Nefertiti in Amarna, Egypt. The bust of the ancient Egyptian Queen was sculpted by the royal sculptor, named Thutmose, in 1345 BC.

1921: The Irish War of Independence concludes with the signing of the Anglo-Irish Treaty. The treaty established the Irish Free State and partitioned Ireland, with the six counties of Northern Ireland remaining part of Great Britain. Its signing almost immediately sparked a civil war in Ireland, with pro-treaty Free State troops battling anti-treaty Irish Republicans. Michael Collins, the leader of the Irish Volunteers during the struggle for independence, was one of the signers of the treaty. He was assassinated by anti-treaty Republican guerillas within a year.

1945: While working on a prototype radar device, Percy Spencer accidentally invents the microwave oven. Spencer, an engineer at Raytheon, noticed that a chocolate bar in his pocket melted when he stood in front of a magnetron tube. He sold the first commercial microwave oven in 1947.

1984: Martina Navratilova loses a tennis match to Helena Sukova. The loss ended the longest winning streak in women's singles tennis, 74 matches played over 11 months.

December 7

1703: An extratropical cyclone hits London. The Great Storm of 1703 destroyed 2,000 large chimneys and piled over 700 ships together in the Thames river. Thousands of ancient trees were destroyed in the New Forest. At sea, hundreds of ships were blown off course. As many as 15,000 people lost their lives.

1776: The Marquis de Lafayette takes a commission in the Continental Army during the American Revolution. Lafayette commanded American troops in a number of pivotal battles, including the Siege of Yorktown and the Battle of Brandywine. At Brandywine, which was a British victory, Lafayette was wounded but still commanded his troops in an orderly retreat, saving many lives. Halfway through the war, he briefly returned to France to convince the French King to support the Revolution. He became known as The Hero of the Two Worlds for his acts in America and Europe.

1869: Jesse and Frank James rob the Daviess County Savings Association in Gallatin, Missouri. It was Jesse's first bank heist. The pair didn't get much money from the robbery. Jesse shot the cashier and escaped with his brother through a posse. The James Gang went on to rob trains and banks throughout the west. Jesse was later assassinated by a member of his own gang named Robert Ford, who was seeking reward money.

1932: Albert Einstein is granted an American visa. Einstein was a German with Jewish ancestry, and while he was visiting the United States, Adolf Hitler seized power in Germany. Einstein remained in the US and became a citizen in 1940. He helped develop the atomic bomb that the US used to win World War II.

1941: A date which will live in infamy. The Imperial Japanese Navy attacked the United States' naval base at Pearl Harbor, Hawaii. An attack wave of 183 planes were launched from aircraft carriers and island airstrips in the Pacific and approached their targets at 7:53 am. A second wave of attackers struck an hour later. The surprise attack caught the base's defenders completely off guard, and most of the US aircraft at Pearl Harbor were destroyed on the ground. Dive bombers sank or damaged all eight of the US battleships in the Pacific fleet and nine other ships. Nearly 200 aircraft were destroyed, 2,403 Americans were killed and 1,178 were wounded. The attack prompted the United States to enter World War II in both the Pacific and Europe.

1963: Instant replay makes its television debut. During the Army-Navy football game in Philadelphia, CBS Sports used a videotape machine to produce instant replays of the live broadcast. It was used only once during the game, to show a replay of a touchdown play. The technology made American football far more popular than it had previously been.

1972: Apollo 17 launches from the Kennedy Space Center. The mission was the last in the Apollo program and the final time humans would land on the surface of the moon during the 20th century. Commander Eugene Cernan and Lunar Module Pilot Harrison Schmitt touched down on the Moon on December 11 while Command Module Pilot Ronald Evans monitored them from orbit. The astronauts returned to Earth on December 14.

December 8

1660: A woman appears on stage in a theater production England for the first time. She was either Margaret Hughes or Anne Marshall, and played Desdemona in the play *Othello*.

1854: Pope Pius IX declares the dogma of the Immaculate Conception. The Catholic belief holds that Mary, the mother of Jesus, was conceived without Original Sin.

1927: The Brookings Institution is founded. The think tank was created by philanthropist Robert S. Brookings. Its research is mainly focused on economics, government, foreign policy, and development.

1931: Lloyd Espenchied and Herman Affel patent coaxial cable. The technology can transmit a wide range of frequencies at the same time. It is widely used for television, computing, and—until it was replaced by fiber-optic cable—high-speed internet connections.

1941: The United States formally declares war on Japan. President Franklin D. Roosevelt called the attack on Pearl

Harbor "dastardly," and vowed in a speech to a joint session of Congress that "the American people in their righteous might will win through to absolute victory." Four years later, the United States and its allies prevailed in the war.

1952: On an episode of *I Love Lucy,* pregnancy is acknowledged on television for the first time. The word "pregnant" was never actually mentioned during the episode due to censorship concerns. Instead the episode was called "Lucy Is Enceinte."

1953: "Atoms for peace." US President Dwight D. Eisenhower ushered in an era of nuclear technology in a speech before the UN General Assembly. Citing the use of atomic weapons, he urged world leaders to use the new technology to instead power schools, hospitals, and research institutions. The United States then started an international aid program known as Atoms for Peace in which it supplied nuclear technology to civilian projects around the world.

1962: The New York City Newspaper Strike begins. Employees of four newspapers in New York City, the *Daily News, New York Journal American, The New York Times,* and *New York World-Telegram & Sun* walked off the job. Five other newspapers voluntarily ceased printing in support of the labor action. The strike lasted for 114 days. During it, *The New York Review of Books* was created, and its circulation quickly ballooned to 75,000 subscribers. Local radio station WABC-FM started an all-news format to fill the void left by the lack of newspapers. An all-news radio station had never existed before, and the format became a prototype for many other stations in the following decades. After extensive negotiations with the publishers, the strikers agreed to conditions that included a pay raise from eight dollars to $12.63 a week.

1980: Mark David Chapman murders John Lennon. Chapman, who was mentally ill, believed that killing Lennon would attract the attention of actor Jodie Foster. After shooting the former Beatle, Chapman sat down on a curb and began reading *The Catcher in the Rye*. He later told police the book was his statement. He plead guilty to murder and was sentenced to 20 years to life in prison.

1991: The Soviet Union is officially dissolved. Leaders of three member states—Russia, Belarus, and Ukraine—signed an agreement that abolishes the Union of Soviet Socialist Republics, which had been established in 1922 and dominated the 20th century as a major belligerent in the Cold War.

December 9

1775: The Battle of Great Bridge. A colonial militia defeated British troops in the battle, inflicting about 100 casualties while suffering only one—a minor injury to one of the rebel's thumbs. As a result of the battle, the British withdrew from the colony of Virginia for the remainder of the American Revolutionary War.

1793: Noah Webster publishes the *American Minerva,* the first daily newspaper in New York City. He edited the paper for four years. Webster is best-known for his *Compendious Dictionary of the English Language*. Many of his word spellings caught on, and as a result he is credited with heavily influencing American English.

1835: The Texian Army captures San Antonio, Texas, during the Texas Revolution. Three hundred Texans defeated the Mexican army forces defending the city after a siege that lasted for four days.

1872: P.B.S. Pinchback becomes the first African-American governor of Louisiana. Pinchback, who had served in the Union Army during the American Civil War, became active in Reconstruction politics after the war. He served as president pro tempore of the Louisiana State Senate in 1868, and acting Lieutenant Governor in 1871. He later fought against the segregation of public transit in the state. The case led to the Supreme Court decision Plessy v. Ferguson in 1896.

1897: Marguerite Durand publishes the first issue of *La Fronde* (The Slingshot), a feminist newspaper, in Paris. It was the first newspaper in France to be written and published entirely by women. It ran daily until 1903, and then became a monthly newspaper until 1905.

1935: Gangland assassins murder newspaper editor Walter Liggett. The journalist had been publishing a series of investigative stories about an organized crime family in Minnesota run by Isadore Blumenfeld, aka "Kid Cann." The gangster attempted to bribe Liggett and had him severely beaten to try to make him stop publishing exposes about the crime syndicate. Liggett was not swayed by the attack or repeated threats on his life. Blumenfeld then used his contacts in the Minnesota government to have Liggett arrested for kidnapping and sodomy. Liggett was brought up on charges and tried, but a mistrial was declared when it became clear that the alleged victims were lying on the witness stand. While Liggett was exiting his apartment, Blumenfeld attacked him with a Thompson submachine gun, firing from a passing automobile. Although Liggett's widow and three other witnesses identified Blumenfeld as the shooter, he was acquitted.

1935: Jay Berwanger, a halfback on the University of Chicago football team, wins the first-ever Downtown Athletic Club Trophy. The award was later renamed the Heisman Trophy. It is one of the highest honors in American collegiate athletics.

1965: A fireball streaks across the sky from Michigan to Pennsylvania. The event, known as the Kecksberg UFO Incident, was witnessed by hundreds of people across several states. Astronomers later determined it was probably a meteor bolide (or a meteor that explodes upon impact with the atmosphere).

1965: A *Charlie Brown Christmas* airs for the first time on CBS.

December 10

1684: Isaac Newton sends Edmond Halley a manuscript.
Titled "On the motion of bodies in an orbit," the early paper on
cosmology provided Halley with Newton's ideas about the relations between Kepler's laws of planetary motion. Before he wrote
the paper, Kepler's notions on force and inertia were not yet fully
understood as natural laws. Newton laid out his ideas on cen-
tripetal force (a term that first appeared in the manuscript), cen-
tripetal force, and resistance. Halley was considerably impressed
by Newton's findings and he encouraged the young thinker to
develop his ideas further. Newton followed Halley's advice. His
work on the subject became the *Philosophiae Naturalis Principia
Mathematica* (or the *Principia*), which he published the following
year. It is one of the most important scientific works in human
history. The work laid out the foundations of classical physics, or
the study of how bodies in motion such as planets interact with
one another.

**1884: Mark Twain publishes *Adventures of Huckleberry Finn*
in the United Kingdom.** The novel was published in the United
States two months later. The book is a sequel to *The Adventures
of Tom Sawyer*. It is one of the first novels that was written in
colloquial American English. It helped establish the canon of
American literature.

**1896: The play *Ubu Roi* ("King Turd") opens, and closes,
in Paris.** The comedy was bizarre for its time, and the audience responded to the performance by rioting. It was written by
Alfred Jarry and attacked the abuse of authority by the upper
classes. *Ubu Roi* used slang, puns, and weird patterns of speech
to shock the audience. The play is considered by many critics as
having been the first play in the genre of Absurdist theater.

1901: The first Nobel Prizes are awarded. Wilhelm Röntgen
won the prize for Physics for his discovery of X-rays. Jacobus
van't Hoff was awarded the prize for Chemistry for his work

in chemical thermodynamics. The poet Sully Prudhomme was granted the prize for Literature. The prize for Medicine was given to Emil von Behring for his work in treating diphtheria. The first Nobel Peace Prize was jointly awarded to Jean Henri Dunant for helping found the International Red Cross, and to Frederic Passy for founding the Peace League.

1902: The Aswan Low Dam is completed. At the time, the dam, which spanned the Nile River's first cataract, was the largest of its kind in the world. The dam is 36 meters tall and 1,950 meters in length. In 1946, when it was discovered the Nile was dangerously close to spilling over the top of the Low Dam, the Aswan High Dam was constructed six kilometers upstream from it.

1936: King Edward VIII abdicates the English throne. Edward had decided to marry Wallis Simpson, an American socialite. Simpson was once divorced and in the process of divorcing her second husband when Edward decided to marry her. As she was a commoner, the King was required to relinquish his royal title and throne to do so. Edward's brother George VI succeeded him, and Edward was made the Duke of Windsor. He and Simpson remained married until his death in 1972.

1968: Thieves make off with 300 million yen (about $800,000) from a transport car of the Nihon Shintaku Ginko bank in Tokyo. The thieves posed as police officers and stopped a car carrying the money, telling the driver a bomb was inside. They then escaped with the money. The case remains unsolved.

December 11

1792: The trial of King Louis XVI of France begins. During the French Revolution, the National Convention put the King on trial for treason. Louis was accused of criminally establishing a tyranny in France and destroying the freedom of the Nation. In his defense, he argued that there were no laws that prevented him from doing so. He was further accused of colluding with foreign armies that had invaded France, and he responded by blaming his foreign minister. He attempted to mount a vigorous defense, bringing some of France's greatest legal minds together to help him. But the evidence against him was overwhelming. He had, in fact, invited foreign armies into France in a bid to crush the French Revolution. The Convention found him guilty by a unanimous vote and sentenced him to death. He was executed by guillotine on January 21 in the Place de la Revolution.

1905: The Shuliavka Uprising begins. Workers in Kiev, Ukraine—at the time part of the Russian Empire—rebelled and established the so-called Shuliavka Workers' Republic. All businesses and factories in the city shut down, and the works gathered in the Shuliavka district of the city. Four days later, Russian Imperial troops attacked the rebels and recaptured the city.

1920: British troops burn Cork, Ireland. During the Irish War of Independence, an Irish Republican Army "flying column" ambushed a patrol of British Auxiliary soldiers in the city. The Auxiliaries—known popularly as the "Black and Tans" after the color of their uniforms—were known for brutal tactics and were especially reviled by the Irish public. In retaliation for the attack, the British rampaged through the city, beating civilians and setting buildings on fire. They also attacked firefighters who tried to put out the fires.

1934: Bill Wilson takes his last drink of alcohol. Wilson went on to found Alcoholics Anonymous, an organization for persons suffering from alcoholism to support one another in abstaining from drinking. The organization grew from a meeting in Wilson's living room to include some 20 million members and 10,000 groups around the world. It popularized the Twelve Step Program of remaining sober.

1946: The United Nations establishes UNICEF. The United Nations International Children's Emergency Fund originally provided food and healthcare to children in countries that had been ravaged by World War II. It later shifted to providing humanitarian assistance to mothers and children in war-torn and developing countries.

1968: The Rolling Stones Rock and Roll Circus is filmed. The concert featured rock 'n' roll stars such as Jethro Tull, The Who, The Rolling Stones, and John Lennon and Yoko Ono. The film was initially not aired, in part because the Rolling Stones were recovering from a drug binge and performed poorly. In 1996, it was finally released commercially.

1978: Robbers steal $5.9 million from JFK Airport in New York. The robbers, who were associates of the Lucchese mafia family, made off with $5 million in cash and $875,000 in jewelry from the airport. Mobster Jimmy Burke was suspected but never charged with masterminding the crime. The heist was later a central plot point in the movie *Goodfellas*.

2005: Bernie Madoff goes to prison. Madoff, a stockbroker and investment advisor, was convicted of operating the largest Ponzi scheme in history. He bilked investors out of $64.8 billion before he was caught. Madoff was sentenced to 150 years in prison and had to return about $17 billion gained from the fraud.

December 12

1781: French and British naval forces fight the Second Battle of Ushant. A French convoy was sailing to the Americas to support the American Revolution when it was attacked by a squadron of British ships. Although the British fleet was outnumbered nearly two-to-one, it was able to harass the French ships until a storm forced the French to turn back. Of the forty ships that set out from France, only two made it to the Caribbean. Fifteen transport ships were captured, and the rest returned to France.

1862: A Confederate mine sinks the USS *Cairo*. The *Cairo*, a Union ironclad warship, captured the Confederate Fort Pillow on the Mississippi River. The fort defended Memphis, and its capture allowed the Union to capture the city. While clearing submerged mines in preparation for an attack on Haines Bluff, Confederate soldiers remotely detonated a mine in the river. It was the first ship to be sunk in this manner.

1901: Italian inventor Guglielmo Marconi sends the first radio signal across the Atlantic Ocean. Marconi transmitted the letter "S" in Morse Code (three dots) from a radio station at Signal Hill in St. John's, Newfoundland. It was received by a second station in Cornwall, England.

1946: John D. Rockefeller Jr. donates six blocks of Manhattan property to the United Nations. The site became the location of the UN Headquarters, which was completed in 1952.

1951: Paula Ackerman performs rabbinical functions in Meridian, Mississippi, becoming the first woman in America to do so. Ackerman led the Beth Israel congregation in Meridian until 1953 and later led the Beth-El congregation in Pensacola, Florida.

1963: Jomo Kenyatta becomes the first President of Kenya. As President, he was the first Black person to lead the country's government. Kenyatta, who in the 1930s studied at Moscow's Communist University, University College London, and the London School of Economics, helped mastermind the Mau Mau Uprising against the British colonizers. While he was Prime Minister, Kenyatta helped guide the nation's transition from a colony of the British Empire to an independent republic. His government "Africanized" the country's economy, prohibiting non-Kenyans from owning major industry. He prompted Kenya to join the Organization of African Unity and helped promote Pan-Africanism. Kenyatta died in 1978.

2000: The US Supreme Court hands down a ruling in *Bush v. Gore*. The ruling stopped a recount in the presidential election of that year and awarded victory to George W. Bush. In the November election, Florida initially went to Bush by a margin of just 1,784 votes, less than 0.5% of the votes cast. The result automatically triggered a recount. Secretary of State Katherine Harris, a Republican, halted the recount after several days, and the Gore

BALLOT BOX

campaign sued to force another recount. The case went all the way to the Supreme Court very quickly, but Gore ultimately lost. Bush was inaugurated in January, while Washington DC was engulfed in some of the largest protests of a presidential election in American history.

December 13

1758: The *Duke William* sinks in the North Atlantic. The

English transport ship was carrying Acadians, the descendants of French colonists in North America. The Acadians were being deported to France from Prince Edward Island after control of the region shifted from the French to the English. Among them was Noel Doiron, a leader of the Acadian colonists who had fought the British during Queen Anne's War in the beginning of the 18th century. Doiron also assisted French soldiers during King George's War in the 1740s. When the *Duke William* sank, Doiron and at least 360 other Acadians were killed. The captain of the ship, William Nichols, survived the disaster. He left the Duke William in a smaller cutter ship, saying he was going for help. But when he was rescued by another English ship, they did not return to help the Acadians. Nichols' journal, which recounted the sinking, was published in the 19th century.

1769: Reverend Eleazar Wheelock founds Dartmouth
College. Dartmouth is a private Ivy League college in New Hampshire. Wheelock initially started the college to teach Native Americans the theology of the Anglican Church. In the 20th century, it became a largely secular institution. Since switching to secularism, Dartmouth has become one of the more prominent research universities in North America.

1928: George Gershwin performs *An American in Paris* for the first time at Carnegie Hall. The orchestral piece, which is influenced by jazz themes, was inspired by Gershwin's visit to the City of Love. The score includes not only the usual instruments of a classical symphony orchestra, but also uses saxophones and Parisian taxi horns.

1939: The first Lincoln Continental comes off the assembly line. The car became very popular in the United States in the 1940s, and is considered one of the most beautiful ever made.

1967: King Constantine of Greece attempts to overthrow the military junta ruling the country. Constantine's attempted coup was a failure, and he was forced to flee the country.

1968: US President Lyndon Johnson and Mexican President Gustavo Diaz Ordaz meet at the border bridge in El Paso, Texas. The meeting was part of a ceremony marking the return of the El Chamizal area to Mexico.

2003: American soldiers capture Saddam Hussein near Tikrit, Iraq. The former dictator of Iraq had been a fugitive since the fall of Baghdad in April. He was tried for crimes against humanity and executed in 2006.

December 14

1812: The last remnants of Napoleon's Grande Armée leave Russia. The retreat marked the end of the French invasion of Russia, which had begun in June. Although the French initially won several battles, when the brutal Russian winter set in, they began to die from starvation and disease. The retreat and loss of the war diminished Napoleon's stature in France and later triggered the War of the Sixth Coalition, which resulted in his exile on Elba.

1814: A British fleet defeats a squad of American warships in the Battle of Lake Borgne. Forty-two British gunboats attacked seven US Navy ships in preparation for an attack on New Orleans. The smaller American force was outgunned and quickly overcome. The naval battle lasted for two hours. After the British captured one of the five American gunboats blocking the channel at the mouth of the lake, they turned its guns on the others. Hand-to-hand fighting between British and American marines lasted about five minutes before the Americans were forced to surrender. Although their resistance was short-lived, the Americans inflicted serious damage on the British. Andrew Jackson, the commander of the American troops defending New Orleans, used the time to fortify their positions. Ten days later, the British attacked the city. The battle for New Orleans was the last one fought in the War of 1812. It was fought after the Treaty of Ghent ended the war.

1900: German physicist Max Planck presents his theory of black-body radiation. The theory describes how radiation is emitted by a glowing object. Planck asserted for the first time in human history that light was made up of particles, or quantized packets of energy. The discovery started a scientific revolution and ushered in Quantum Theory.

1901: The Transpacific Telegraph Cable is completed. Ships operated by the Commercial Pacific Cable Company laid

6,912 miles of cable from San Francisco to Manila. The cable operated until October 1951, when the need for extensive repairs made it impossible to keep up.

1907: The *Thomas W. Lawson,* a seven-masted, steel-hulled schooner, sinks in the Atlantic. The *Lawson* was the largest sailing ship without a heat engine ever built. It was caught in a storm off the southeastern coast of England and broke apart. All but two of the eighteen-man crew were lost in the wreck.

1948: Thomas T. Goldsmith and Estle Ray Mann patent a "cathode-ray tube amusement device." It was the first known interactive electronic game. While the game had a video display, it is not considered the first true "video game" because the display was not run by a computer. Players attempted to "hit" targets on the screen with a cathode-ray tube beam.

1962: *Mariner 2* becomes the first space probe sent from Earth to reach Venus. It flew by Venus at about 21,600 miles and sent measurements of the planet's atmosphere back to Earth. The probe lost contact with Earth on January 3, 1963.

2008: Iraqi journalist Muntadhar al-Zaidi throws both of his shoes at US President George W. Bush. Bush, who was speaking at a joint press conference with Iraqi Prime Minister Nouri al-Maliki, was able to dodge both of the shoes. Al-Zaidi was dragged out of the room and beaten by guards. He was sentenced to three years in prison, but released after nine months. Zaidi said he was tortured while he was in prison, and later traveled to Switzerland for injuries sustained during his detention.

December 15

1791: The Virginia General Assembly ratifies the US Bill of Rights. Virginia was the last state to ratify the Bill. After it passed the Assembly, the set of ten amendments to the Constitution then became law.

1864: Union and Confederate troops clash in the Battle of Nashville. The city was defended by the Confederate Army of Tennessee, a force of 30,000 soldiers commanded by John Bell Hood. The Army of the Cumberland, commanded by General George H. Thomas and 55,000 men strong, marched on the city after a fierce naval battle on the Cumberland River. In preparation for the attack on the city, the ironclad USS *Carondelet* and the river monitor *Neosho* attacked a Confederate blockade of the river, but could not drive the Confederate ships away despite a heavy bombardment. On the morning of the 15th, a Union cavalry attacked the Confederates defending the river. Thomas then attacked the Confederate lines on its flanks, but could not turn them. After the first day, the Confederates reinforced their positions. Thomas ordered an all-out attack along the Confederate line, which was routed. Hood retreated south, and Thomas captured Nashville.

1890: Police shoot and kill Sitting Bull, leader of the Lakota Sioux. The police were attempting to arrest Sitting Bull because they were afraid he would join the Ghost Dance movement that was sweeping the Great Plains. The religious movement was uniting Native American tribes against colonization. When Sitting Bull refused to go with the police, they opened fire, killing him. His death preceded a massacre of Ghost Dance followers at Wounded Knee two weeks later.

1905: Pushkin House is founded in St. Petersburg. It is named after Russian novelist and poet Alexander Pushkin, and has served since its inception as the main center of the study of Russian Literature.

1945: Douglas MacArthur issues the Shinto Directive. The order abolished Shintoism as the official religion of the government of Japan. The Allies believed that as the state religion, Shinto had contributed to the nationalism and militarism that gave rise to World War II in the Pacific.

1960: Richard Paul Pavlick is arrested for attempting to assassinate president-elect John F. Kennedy. Pavlick was a former postmaster in Boston who was angry with the government and hated Catholics. Kennedy was the first Catholic elected President of the United States. Pavlick had brought dynamite to Palm Beach in a plan to kill Kennedy in a suicide bombing, but stopped when he saw the president-elect was with his family. He was committed to a mental hospital, and released in 1966.

1961: A war crimes tribunal sentences Adolf Eichmann to death. The Nazi was one of the chief architects of the Holocaust during World War II. Eichmann had fled to Argentina after the war, but was tracked down by the Israeli Mossad in 1960. He was executed in June 1962.

1993: The Downing Street Declaration affirms the right of the Irish people to self-determination. The Declaration stated that Northern Ireland could become part of the Irish Republic if a referendum of Northern Irish people passed such a move. It was a major step in the peace process that ended the Troubles.

December 16

1653: Oliver Cromwell becomes Lord Protector of the Commonwealth of England, Scotland, and Ireland. The period is variously known as the Protectorate or, to monarchists, the Interregnum (period between two royal reigns). Cromwell played a major role in overthrowing King Charles I and abolishing the monarchy in England, and Parliament offered to give him the crown. He ultimately decided against becoming a King, and instead chose the title Lord Protector. Cromwell ruled until his death in 1658, and was succeeded by his son Richard. Nine months after succeeding his father, Richard gave up the position, and two years later, the monarchy was restored by Charles II.

1773: The Boston Tea Party. During tensions leading up to the American Revolution, a group calling themselves the Sons of Liberty boarded ships in Boston Harbor and destroyed chests of tea by throwing them overboard. The action was a protest against the Tea Act of 1773. The law permitted the British East India company to import tea without paying taxes, thus undercutting local merchants who had to pay the tax. The Sons of Liberty were organized by patriots including Samuel Adams, Paul Revere, and John Hancock to fight British taxes imposed on the colonists. Their slogan, "No taxation without representation," became associated with the main causes of the American Revolution. In Boston, they first resisted the Tea Act by preventing British ships from unloading tea for several weeks. Following a mass-meeting about the standoff, about 100 men disguised themselves as Mohawk Indians and marched to the harbor. They threw 342 chests of tea on three ships into the harbor. In response, the British closed the port of Boston and passed a set of punitive laws known as The Intolerable Acts in 1774. The Revolutionary War broke out less than a year later.

1843: John T. Graves discovers octonions. These hypercomplex numbers are an important foundation of mathematical structures. In the 20th century they came to be applied in fields including string theory, special relativity, and quantum logic.

1826: At Nacogdoches, which was then in Mexico, Benjamin W. Edwards declares himself the ruler of the Republic of Fredonia. With his brother Haden, Edwards led a group of white settlers into Texas and attempted to secede from Mexico. The Republic of Fredonia was short-lived. On January 31, a detachment of Mexican army and militiamen captured Nacogdoches. The Haden brothers fled, and did not return until after the Texas Revolution had begun.

1905: The first issue of *Variety* magazine is published. The entertainment trade weekly's first issue had 16 pages and sold for five cents.

1907: The "Great White Fleet" sets out from Hampton Roads, Virginia, to begin its circumnavigation of the globe. The fleet of US Navy ships sailed around the world on the orders of

President Theodore Roosevelt to demonstrate American naval power to its allies and enemies. The fleet completed its voyage on February 22, 1909.

1968: The Second Vatican Council formally revokes the Alhambra Decree. The Decree was issued by Queen Isabella I and King Ferdinand II of Spain in 1492, and ordered all Jews in Spain to be expelled. It had resulted in the conversion or persecution of tens of thousands of Jews.

December 17

1538: Pope Paul III excommunicates Henry VIII of England from the Catholic Church. Henry had appointed himself the Supreme Head of the Church of England in order to be able to divorce his wife Anne Boleyn.

1790: While renovating the Mexico City Cathedral, workers discover the Aztec calendar stone. The stone, which is more than ten feet in diameter and nearly three feet thick, was carved sometime in the late 15th century. Its carved surface depicts events in the history of the universe as described by the Mexica Indians. The calendar stone was mounted on the outside wall of the cathedral.

1835: The Second Great Fire of New York destroys hundreds of buildings across 17 city blocks. It began in a warehouse and was driven by high winds toward the East River. The river was frozen over and firefighters were forced to drill holes in the ice to get water.

1862: General Ulysses S. Grant orders all persons of Jewish origin to be expelled from his military district, an area that included parts of Tennessee, Mississippi, and Kentucky. Grant blamed Jews for illegally selling Southern cotton during the American Civil War. After a public outcry against the order, President Abraham Lincoln revoked it a few weeks later. The order became an issue during Grant's campaign for presidency in 1868. While in office, Grant became the first sitting president to attend a synagogue service, and also appointed more Jews to office than any previous American president.

1892: *Vogue,* a fashion and lifestyle magazine, goes on sale for the first time. Originally a weekly newspaper, the publication shifted to fashion and beauty when it was acquired by Condé Nast in 1905.

1903: The Wright brothers fly at Kitty Hawk. Orville and Wilbur Wright had begun experimenting with heavier-than-air flight about three years earlier. They developed a series of gliders to practice steering a controlled flying machine. They tested these gliders at Kitty Hawk, North Carolina. The brothers selected this site because of the strong breezes that blew in off the Atlantic Ocean and the sandy beaches that made for soft landing surfaces. They tested piloted and unpiloted gliders alike, developing different wing configurations to increase loft and reduce aerodynamic drag during flight. After improving the prototypes so that they could be steered after takeoff, the brothers began working on a biplane that was powered by a propeller. The plane, which was simply named the *Wright Flyer,* took off with Orville at the controls. He flew the *Flyer* for twelve seconds and a distance of 120 feet.

1938: Otto Hahn discovers the nuclear fission of uranium. The discovery was the basis of the development of nuclear energy.

1989: *The Simpsons* premieres on television. The episode, "Simpsons Roasting on an Open Fire," first introduced the family's pet greyhound, Santa's Little Helper. It was the eighth episode of the show, which had previously aired as a segment on *The Tracy Ullman Show.* The first standalone episode was nominated for two Emmy Awards. *The Simpsons* went on to become the longest-running American sitcom and the longest-running scripted primetime television series.

2010: Mohamed Bouazizi, a Tunisian street vendor, sets himself on fire to protest harassment by police. The act became a catalyst for the Tunisian Revolution and the international Arab Spring against autocratic regimes. During the Arab Spring, non-violent demonstrations, civil wars, and foreign intervention brought down governments across North Africa and the Middle East.

December 18

1603: Steven van der Haghen departs the Netherlands with a fleet of ships heading to South Asia. The voyage was the first naval enterprise in support of the Dutch East India Company, which had been established in response to the British East India Company. Each was a multinational corporation that was supported by its home government. Over the next 60 years, the Dutch East India Company engaged in several wars with local governments in its forced extraction of resources.

1621: The English House of Commons attempts to assert its right to deal with matters of the Crown and State without being threatened with retribution or punishment. In response, King James I tore the statement to shreds and immediately disbanded the Parliament.

1719: Thomas Fleet publishes *Mother Goose's Melodies for Children*. Fleet, a publisher in Boston, Massachusetts, said his mother-in-law Elizabeth Vergoose was the source of the children's stories. Other historians have traced the nursery rhymes to both England and France, and their origin remains unclear.

1865: William Seward, the US Secretary of State, proclaims that the 13th Amendment has been adopted. The Amendment outlawed slavery and involuntary servitude throughout the United States, except as punishment for a crime. It was the first of three post-Civil War Reconstruction Amendments. The 14th Amendment extended equal protection of the law to all persons, and the 15th prohibited discrimination in voting rights to citizens on the basis of race, color, or previous enslavement.

1878: John "Black Jack" Kehoe, the last of the Molly Maguires, is executed in Pennsylvania. The Molly Maguires were a secret society of Irish-American coal miners. The society had originated in Ireland in the 1840s as an organization that retaliated against miserable working conditions and tenant evictions with violence. The miners brought the organization with them when they immigrated to America during the Potato Famine. They became known for tacking so-called "Coffin notices," or death threats on the doors of their enemies. When a miners' strike was brutally suppressed by authorities, the Maguires decided to act. They organized a labor union and secretly backed it with intimidation and violence. The president of the Reading Railroad hired the Pinkerton Detective Agency to infiltrate the Maguires. A Pinkerton detective named James McParlan did so, and based on his testimony 20 Maguires were found guilty of various crimes and sentenced to hang. Kehoe was granted a full pardon more than 100 years later by the governor of Pennsylvania.

1936: The first giant panda to come to the United States from China arrives in San Francisco. Explorer Ruth Harkness brought the panda cub, named Su-Lin, to the US.

1958: Project SCORE is launched from Cape Canaveral. It was the world's first experimental communications satellite. SCORE orbited Earth for 13 days before it ran out of power. While it was still in orbit, it broadcast a recording of President Dwight D. Eisenhower wishing holiday greetings to listeners back on Earth.

1961: For the second year in a row, the Associated Press names Wilma Rudolph, an Olympic gold medalist, the Female Athlete of the Year. Rudolph was a track and field star who won 100-and 200-meter dashes in the 1956 and 1960 Olympics.

December 19

1606: Three ships depart England, carrying colonists to Virginia. The *Susan Constant*, *Godspeed*, and *Discovery* belonged to the English Virginia Company, which was chartered by King James I to establish settlements in North America. The colonists founded Jamestown, Virginia, the first of what would become thirteen English colonies and later the United States of America.

1776: Thomas Paine publishes a pamphlet entitled *The American Crisis* in a Pennsylvania newspaper. The essay was an argument for revolution against the British. It included the now-famous line, "These are the times that try men's souls."

1777: The Continental Army settles in at Valley Forge. During the Revolutionary War, the British Army captured Philadelphia, then the US capital. George Washington's attempt at recapturing the city had failed, and he led his 12,000-strong army to Valley Forge, about 18 miles northwest of Philadelphia, to spend the winter. Disease quickly began affecting the troops. Filthy conditions led to outbreaks of typhoid, influenza, smallpox, and dysentery. Food was scarce. Between 1,700 and 2,000 soldiers died from disease and malnutrition at Valley Forge. While encamped there, Washington hired a Prussian drill instructor, Baron Friedrich von Steuben, to train the Continental Army, which was made up of civilian farmers and frontiersmen, in the art of war. Under his direction, the soldiers learned the techniques of 18th century warfare, including volley fire, marching, and drill. After leaving Valley Forge in June, the Continental Army's first engagement at the Battle of Monmouth was a tactical draw but effectively a victory for the Americans' morale.

1793: The French army, under the command of General Dugommier, recapture Toulon, France, from the English. Dugommier used a plan of attack that was devised by a young artillery commander named Napoleon Bonaparte. As a result, Napoleon was promoted to brigadier general at just 24 years old.

1887: Boxers Jem Smith of England and Jake Kilrain of the United States engage in a bare-knuckle boxing match in France that lasts 106 rounds. At the end of two hours, the fight was ruled a draw.

1924: The last Rolls Royce Silver Ghost built in England is sold. The Silver Ghost was a custom touring car, and was first introduced in 1906. At the time it was considered the best automobile ever produced.

1930: Amelia Earhart, a famous American aviator, becomes the first pilot to carry a passenger in an autogyro. The machine was a precursor of the modern helicopter.

1974: The Altair 8800 goes on sale. The machine, which was marketed as a do-it-yourself computer kit, used flashing lights as a display and accepted input as combinations of switches. The Altair is considered the first computer in the personal-computer revolution. The computer language it used, Altair BASIC, was the first product Microsoft ever made.

December 20

1688: William of Orange marches into London with 15,000 soldiers. William's uncle, the English King James II, was quickly overthrown in what became known as "The Glorious Revolution" when William ascended the throne of England with his wife, Queen Mary II. James was later allowed to escape to France.

1699: Peter the Great, the Tsar of Russia, orders the date of the New Year changed from September 1 to January 1. Russians had historically celebrated the New Year in September, but Peter changed the Russian calendar—in which it was the year 7208—to the Julian calendar, joining the rest of Europe. In 1918, the Russian calendar was again changed, when the Bolsheviks moved it from the Julian to the Gregorian calendar.

1803: At a ceremony in New Orleans, the finishing touches are put on the Louisiana Purchase. For just $15 million, the United States acquired territory from France that would eventually become the states of Arkansas, North and South Dakota, Iowa, Missouri, Minnesota, Montana, Colorado, Wyoming, and Louisiana.

1808: A fire destroys the Covent Garden Theatre in London. The theater was rebuilt, but hundreds of scripts, costumes, and stage scenery were lost in the fire as well.

1860: South Carolina becomes the first state to secede from the Union. The South Carolina legislature ratified an article of secession after Abraham Lincoln won the 1860 election for President. South Carolina was soon followed by six other states, who together formed the Confederate States of America. In April, South Carolinian secessionist forces opened fire on Fort Sumter, starting the American Civil War. The state was readmitted to the Union in 1868.

1917: The Soviet Union's secret police, called the Cheka, is founded. The Cheka initially fought against people deemed "counter-revolutionaries" and later carried out the Red Terror, a period of political repression and mass killing at the beginning of the Russian Civil War. The organization was dissolved in 1922 after the end of the Civil War.

1907: Albert Michelson becomes the first American scientist to win a Nobel Prize. Michelson, a physicist, helped contribute to the development of Einstein's theory of relativity.

1938: Vladimir Kosma Zworykin patents the kinescope and the iconoscope. The inventions revolutionized television. The kinescope was a cathode-ray receiver that was used in television and computer monitor screens until the invention of flat screens. The iconoscope was used in TV cameras for fifty years.

1971: Bernard Kouchner and a group of journalists in Paris found *Médecins Sans Frontières,* or Doctors Without Borders. The international organization sends doctors, nurses, and other medical professionals to areas that lack medical infrastructure and to war-torn regions.

1989: The United States invades Panama. In an attempt to overthrow and arrest Panamanian dictator General Manuel Noriega, the United States sent 27,000 troops to the tiny Central American country. Noriega was a former asset of the Central Intelligence Agency, who was paid six figures by the US government for sabotaging left-wing revolutionaries in Nicaragua and El Salvador. He also worked with the Drug Enforcement Agency, but was also laundering money for drug cartels. He had been indicted in an American court for drug trafficking. He was also accused of endangering US nationals living in Panama when American soldiers were stopped at a roadblock and shot. Noriega was captured within a month and flown to the United States.

December 21

1620: William Bradford and a group of colonists known as the Pilgrims land on what is now called Plymouth Rock in Plymouth, Massachusetts.

1844: The Co-operative Movement is started. In Rochdale, England, the Rochdale Society of Equitable Pioneers opened the first consumer co-operative. The business paid dividends to shareholders and established the Rochdale Principles. The Principles, which include voluntary membership and democratic control, are the foundation of the principles on which all co-operatives operate to this day.

1861: President Abraham Lincoln signs a resolution establishing the Navy Medal of Valor. The award, initially presented "for gallantry," eventually became the Medal of Honor. The Medal of Honor is the most prestigious and highest personal military decoration awarded to military personnel in the US.

1872: The HMS *Challenger* sets sail from Portsmouth, England. The ship's expedition, which covered nearly 70,000 nautical miles over four years, made numerous discoveries and established the science of oceanography. The crew measured the ocean's depth and bottom temperature at 360 locations. They also collected biological samples from the sea floor. The *Challenger* expedition was the first to discover an area of the Mariana trench now called the Challenger Deep. At 4,475 fathoms (about five miles) deep, it is one of the deepest places on the planet. Prior to the expedition, the actual depths of far-flung areas of the world's oceans were not known, and not believed to be nearly as deep.

1879: Henrik Ibsen's play *A Doll's House* premieres. The play was highly controversial at the time because it was about the life of a married woman who decides to leave her husband. The play openly questioned the traditional roles of men and women in society. Challenging marriage was scandalous to 19th-century Europeans. It was later recognized as one of the greatest plays ever written.

1913: The crossword puzzle is invented. Arthur Wynne, a British-born journalist who grew up in Pittsburgh, Pennsylvania, was working for the *New York World* when he conceived of the puzzle. He called it a "Word-Cross Puzzle." It became known as a "Cross-Word" the following week as the result of a typesetting error. The puzzles have been called "crosswords" ever since.

1937: *Snow White and the Seven Dwarves* debuts in movie theatres. It was the first feature-length animated movie. Created by the Walt Disney animation company, the film quickly became a classic. It has been re-released numerous times, and it was selected for preservation by the Library of Congress as a "culturally, historically, or aesthetically significant" work in 1989.

1988: Pan Am Flight 103 explodes over Lockerbie, Scotland, with 259 passengers and crew aboard. A bomb was planted on the American passenger plane by Libyan operatives as revenge for a series of military engagements between the US Navy and Libyan ships. In 2003, Libya formally admitted responsibility for the bombing in a letter to the UN Security Council.

2012: The world does not end. In the months leading up to this date, concerns spread that the world might come to an end. The date was the last in a 5,126-year-long cycle in the Mesoamerican Long Count Calendar.

December 22

1708: French forces supported by Native American warriors capture the English settlement at St. John's, Newfoundland, Canada. The capture of the town gave the French control of the Eastern shoreline of Canada.

1807: President Thomas Jefferson signs the Embargo Act into law. During the Napoleonic Wars between Britain and France, the newly-formed United States of America was not yet a military power. Great Britain took advantage of this fact and seized American cargo ships, forcing their crews to serve on British warships in a practice called impressment. In response, the Embargo Act suspended all trade with both Great Britain and France. The embargo was a failure, both in economic and diplomatic terms. Without shipments to and from Europe, the American economy was devastated. Cargo ships rotted in their berths and American farmers could not sell their crops internationally. The public widely began to support smuggling. Ultimately it was repealed, but tensions with Great Britain continued to grow, eventually leading to the War of 1812.

1885: Itō Hirobumi becomes the first Prime Minister of Japan. Hirobumi, a samurai, helped draft the Meiji Constitution, which established the Empire of Japan. He was reelected four times, becoming one of the longest-serving Prime Ministers in Japanese history. In 1909, Korean nationalist An Jung-geun assassinated Hirobumi in retaliation for the annexation of Korea by Japan.

1894: The Dreyfus Affair begins. Captain Alfred Dreyfus was convicted of treason and sentenced to life imprisonment for allegedly giving state secrets to the German Embassy. He was imprisoned in French Guiana for nearly five years before he was exonerated and freed. The Dreyfus Affair is considered a major example of miscarriages of justice and of anti-Semitism.

1900: The Daimler Automobile Company in Germany completes a new 35-horsepower car. Emil Jellinek, the car's designer, named it after his daughter, Mercedes.

1944: Nuts! During the Battle of the Bulge, fought in the final years of World War II, an American force in the town of Bastogne, France, was surrounded and cut off from reinforcements by a German Army. The German commander, Heinrich Freiherr von Lüttwitz, sent word to the American commander, Brigadier General Anthony McAuliffe, requesting his surrender. McAuliffe responded simply saying "Nuts!" The response was typed up and sent to von Lüttwitz as the official response. The Americans held out until January 25, when the weather cleared and allowed Allied air power to resupply them and attack the Germans.

1990: The Brandenburg Gate reopens in Berlin. When the city was divided with the building of the Berlin Wall, the Gate was closed on August 14, 1961. It reopened after the Revolutions of 1989 and demolition of the wall by the people of Berlin.

December 23

1783: George Washington resigns as commander-in-chief of the Continental Army. The act marked the end of Washington's military career and return to civilian life. In resigning, Washington established the precedent of civilian control of the military. This was unprecedented; victorious generals rarely if ever gave up their power after winning a war that upended the political order as the American Revolution had done. The day after resigning, Washington left for Mount Vernon.

1788: The Maryland State Legislature votes to give 100 square miles to the United States government. Two-thirds of that territory became the District of Columbia.

1815: Jane Austen publishes *Emma*. The novel, whose youthful protagonist falls in love with different suitors, was highly regarded and came to be a classic of English literature. In it, Austen explored themes of class, gender, and the prevailing social order.

1888: Vincent Van Gogh cuts his left ear off with a razor. The Dutch painter and impressionist was suffering from severe depression at the time. He later documented the event in his painting *Self-Portrait with Bandaged Ear*.

1912: *Hoffmeyer's Release* debuts in theaters. The silent film was the first installment of the "Keystone Cops" series, which featured bumbling policemen falling over one another in elaborate slapstick routines.

1947: The transistor is demonstrated at Bell Laboratories for the first time. The device, a semiconductor that can amplify or switch electronic signals, led to the development of nearly all electronic devices invented in the latter half of the 20th century. The American physicists John Bardeen, Walter Brattain, and William Shockley shared the 1956 Nobel Prize in Physics for inventing the transistor.

1950: Dr. Richard Lawler performs the first successful kidney transplant at Little Company of Mary Hospital in a suburb of Chicago, Illinois. The transplant was successful, but the patient's body rejected it after ten months. The rejection led to the development of immunosuppressive drugs to prevent this from occurring.

1972: The survivors of the Andes flight disaster are rescued after surviving for 73 days on an alpine glacier. During their ordeal, they were forced to turn to cannibalism to survive, eating parts of the bodies of passengers who died in the plane crash.

1986: After nine days aloft, the *Voyager*, piloted by Dick Rutan and Jeana Yeager, lands at Edwards Air Force Base in California. The Voyager was the first aircraft to fly non-stop around the world without being refueled. When it took off, it was carrying more than three times its weight in fuel.

December 24

1777: English Captain James Cook visits the island of Kiritimati. Cook named it Christmas Island in his journal. It was uninhabited when he arrived. During World War II, the Allies occupied the island, building an airstrip and a radio station.

1800: In what became known as the Plot of the rue Saint-Nicaise, a group of royalists attempt to assassinate Napoleon Bonaparte. The assassins built an explosive device they called the *Machine Infernale* and planted it near a street where Napoleon's carriage would pass. As Napoleon's carriage approached, one of the would-be assassins lit the fuse on the device and fled. A section of the street was blown apart, but Napoleon escaped unscathed.

1818: "Silent Night" is performed for the first time in a parish church in Oberndorf, Austria. Joseph Mohr, the parish priest, wrote the lyrics of the song in 1816. He shared it with the organist, Franz Xaver Gruber. The original copy was lost for many years, and many scholars assumed a famous composer such as Bach or Mozart had written the piece, but Mohr and Gruber were eventually given credit. In 1859, John Young, a priest at Trinity Church in New York City, translated the piece into English. It has since become one of the most recognizable Christmas carols, and has been translated into at least 120 languages.

1851: A huge fire destroys about two-thirds of the 55,000 volumes in the Library of Congress in Washington, DC. Most of Thomas Jefferson's personal library, which he had donated to the Library, was lost in the fire.

1865: A group of former Confederate soldiers found the Ku Klux Klan in Pulaski, Tennessee. The white terrorist organization attacked African Americans and fought to undo the gains made during Reconstruction after the American Civil War. The group's membership swelled to include thousands of white Americans in the early part of the 20th century. Members murdered Civil Rights workers and bombed churches and homes during the 1950s and 1960s.

1914: During World War I, an unofficial ceasefire breaks out along the Western Front. The event became known as the "Christmas truce." It occurred early in the war. French, British and German soldiers faced one another along a line of trenches hundreds of miles long. Following the First Battle of Ypres, which had concluded in November, there was a lull in the fighting as commanders on both sides considered their options. In the week leading up to Christmas, soldiers in close positions left their trenches to occasionally swap cigarettes and chat. During the Christmas truce, the exchanged prisoners, exchanged food and sang Christmas carols together. After their respective commands got word of the Christmas truce, they were forbidden to do so in the following years.

1980: People in the area of Rendlesham Forest, Suffolk, England, report a series of unexplained lights in the sky. The source of the lights was never publicly determined. The event is sometimes called "Britain's Roswell."

December 25

1758: Johann Georg Palitzch sights Halley's Comet. The Comet was predicted by astronomer Edmund Halley. Using the laws discovered by Isaac Newton in his *Principia*, Halley determined that two comets seen in 1531 and 1607 were in fact the same comet returning every 76 years. Based on his calculations, he correctly predicted the comet would return in 1758. Halley died in 1742, and was thus unable to see his prediction be borne out. The comet was named for him, and it will next return on July 28, 2061.

1741: Anders Celsius creates a temperature scale, in which he set the freezing point of water at zero and the boiling point at 100 degrees. The scale is known as the Celsius, or centigrade, scale, and is now part of the International System of Units, or metric system.

1776: George Washington crosses the Delaware River. Washington and his Continental Army crossed the river during the American Revolutionary War to mount a surprise attack on a Hessian garrison at Trenton, New Jersey. The secret crossing was extremely dangerous and challenging. The freezing conditions had filled the river with ice floes that the Americans had to steer heavily-laden rafts with men and equipment around. On Christmas evening, Washington's men assembled and were told they were departing on a secret mission. They silently marched to the river, where they waited for darkness to fall. They loaded into rafts and rowed across the river as a steady snowstorm fell. The last artillery pieces were brought across the river at 3 am. The following morning, Washington's men attacked the Hessian garrison, who were caught entirely by surprise. Three Americans and 22 Hessians were killed in the battle. The Continental soldiers captured 1,000 prisoners, ammunition, and artillery. The victory gave the Army and the Continental Congress a considerable morale boost, and Washington's position as leader of the Revolution was secured.

1826: The Eggnog Riot at West Point is put down. Students had smuggled whiskey into the military academy to make spiked eggnog for a Christmas party. Dozens of students got very drunk. When school administrators attempted to break up the party, the students fought back and began damaging property. Among the rioting cadets were John Archibald Campbell, who would later be a Justice on the US Supreme Court, and Jefferson Davis, later the president of the Confederacy.

1868: US President Andrew Johnson issues an unconditional pardon and general amnesty to all former members of the Confederacy. The former rebels had previously been required to sign an oath of allegiance to the United States. Johnson extended a blanket pardon to everyone without requiring the oath.

1896: American composer John Phillip Sousa completes the piece "Stars and Stripes Forever." The rousing military march is widely considered Sousa's magnum opus. In 1987, the US Congress passed an act that made it the official National March of the United States.

1991: Mikhail Gorbachev resigns as President of the Soviet Union. His reforms, including *perestroika* ("restructuring") and *glasnost* ("transparency") extended freedoms to Soviet citizens that they did not have during the country's authoritarian past. He is credited as being a major figure in the end of the Cold War.

December 26

1854: Hugh Burgess and Charles Watt patent the first process for making paper from wood fiber at a commercial scale. Before their invention, paper was primarily made from rag pulp, which was becoming increasingly expensive. Burgess and Watt's invention made the process much cheaper, making paper more accessible to printers and newspaper publishers. As a result, books and newspapers became more widely available to the public.

1862: In the largest mass execution in American history, 38 Dakota Sioux men are hanged in Mankato, Minnesota. The execution was a reprisal for an uprising by Dakota warriors against White settlements that had begun in August, and was known as the Dakota War of 1862.

1871: Gilbert and Sullivan collaborate for the first time. The pair would go on to enjoy great success in their English-language operas, which included *The HMS Pinafore* and the *Mikado*. Their first collaboration, *Thespis*, did well at the box office, but was considered the pair's "lost" opera because it is not widely remembered.

1906: *The Story of the Kelly Gang* premiers at the Athenaeum Hall in Melbourne, Australia. The silent film is an hour long and was the first continuous film of significant length. It may be considered the first true feature film.

1908: Maire and Pierre Curie announce that they have isolated radium. The husband-and-wife team isolated the radium from radium chloride using electrolysis. Marie Curie was later awarded the Nobel Prize in Chemistry for her achievements in studying radioactive materials.

1919: The Curse of the Bambino. Harry Frazee, the owner of the Boston Red Sox, traded Babe Ruth to the New York Yankees. Ruth went on to become one of the greatest hitters in the history of baseball during his tenure with the Yankees, and was such a great player that Yankee Stadium was known as "The House that Ruth Built." He set multiple records for hits, home runs, and runs batted in. Meanwhile, the Boston Red Sox would not win a World Series for the next 86 years. During the ball club's dry spell, many fans began blaming the lack of a championship on a curse associated with trading away one of the greatest players that ever lived. They called their misfortune "The Curse of the Bambino."

1966: Maulana Karenga first celebrates Kwanzaa. The weeklong celebration honors the culture and heritage of the African diaspora and the Americas.

1991: The Union of Soviet Socialist Republics dissolves. The Supreme Soviet, the country's highest legislative body, passed a resolution officially dissolving the Soviet Union. With it, the Cold War, which had lasted since the end of World War II, drew to a close.

December 27

537: The Hagia Sophia is completed in Constantinople.
The "Place of Wisdom" took five years to finish. Over the centuries, it would variously serve as the main cathedral of Greek Orthodox Christianity, an imperial mosque of the Ottoman Empire, and a museum. At the time it was constructed, it was the largest building ever built. It is considered the height of Byzantine architecture.

1831: Charles Darwin sets sail on the HMS *Beagle*. The voyage lasted for five years and took the young naturalist to South America and the Galapagos Islands. While on the voyage, Darwin read *Principles of Geology* by Charles Lyell, which established the ancient nature of the Earth and asserted the manner in which land formations take shape. The book had a profound effect on Darwin. When he saw fossilized remains of giant ground sloths and other extinct creatures in South America, he began to consider the evolution of species in geological time, over hundreds of centuries. He later observed a diversity of finches in the Galapagos, where every species was very precisely attuned to its particular environmental niche. From his observations, Darwin went on to conceive of the Theory of Evolution by Natural Selection.

1845: Newspaper columnist John O'Sullivan declares that it is the United States' "Manifest Destiny" to annex Texas and the Oregon Country. The term came to embody the widely held belief that the US had a right to expand across all of North America in the following decades.

1900: Carrie Nation, an American prohibitionist and teetotaler, smashes up the bar at the Carey Hotel in Wichita, Kansas. The anti-alcohol crusader took to carrying an axe with her, and would regularly lead crowds to destroy saloons and bars in cities where she was giving speeches in support of Prohibition.

1927: *Show Boat* **opens at the Ziegfeld Theatre on Broadway in New York City.** It was the first true American musical play, and was immediately recognized by critics as having created a new genre of theater.

1935: Regina Jonas is the first woman to be ordained as a rabbi. Jonas was ordained in Offenbach am Main, Germany. When the Nazis came to power, she was arrested and sent to the Theresienstadt concentration camp, where she continued to give rabbinical lectures. In 1944 Jonas was sent to Auschwitz, where she was almost immediately murdered.

2007: Assassins kill Benazir Bhutto, former Prime Minister of Pakistan, at a political rally. Bhutto had already survived another attempt on her life. At the rally, she was campaigning in an upcoming general election when gunmen opened fire and detonated a bomb. At least 24 others were killed as well.

December 28

1793: Thomas Paine is arrested for treason by Robespierre during the French Revolution. Paine, a hero of the American Revolution, had become a fierce supporter of the French Revolution as well. In 1791, Paine wrote *The Rights of Man*, a defense of the French Revolution against its detractors. He then published a series of attacks on conservative writer Edmund Burke, for which he was found guilty of seditious libel in England. By then he had already fled the country for France, where he was elected to the revolutionary French National Convention. Because he was an ally of the Girondist faction of the Revolution, Robespierre, the leader of the rival Montagnard faction, considered Paine a traitor. He was thrown in prison in Paris, and while there, he continued to write, working on *The Age of Reason* from his cell. James Monroe, later President of the United States, managed to secure Paine's release in November 1794. Paine then returned to the United States, but he was ostracized because of his pamphlets that criticized Christianity.

1835: Osceola, leader of the Seminole, begins the Second Seminole War against the US Army. The Seminole took advantage of their extensive knowledge of the Florida swamps to launch guerilla attacks on the Army. The war lasted for nearly seven years and was one of the costliest wars fought by the Army against Native Americans; the Army suffered some 1,600 casualties before it was over.

1893: Brothers Auguste and Louis Lumière screen their first film in the United States. The French inventors and filmmakers were the first to show films in a format that could be viewed by multiple audience members at once, in contrast to Thomas Edison's peephole-style kinetoscopes.

1895: Wilhelm Röntgen publishes his discovery of X-rays. The paper, titled "On A New Kind of Rays," detailed the new type of radiation that the German physicist had detected. He was awarded the 1901 Nobel Prize in Physics for his discovery. The 111th element on the Periodic Table, roentgenium, is named in his honor.

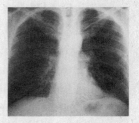

1913: Filming begins on _The Straw Man_. It was the first feature-length film produced in Hollywood, California. It told the story of an Englishman to moved to America and married a Native American woman. The film launched the careers of Samuel Goldfish, who co-founded the Goldwin movie studio, and Cecil B. DeMille.

1958: The Baltimore Colts defeat the New York Giants in the National Football League Championship Game. It was the first NFL playoff game to be decided in sudden death overtime, and is known as "The Greatest Game Ever Played." The Colts took the lead in the first half, 14–3, before the Giants stormed back in the second half, scoring two touchdowns to the Colt's field goal, tying the game at 17. In overtime, the Colts scored a touchdown on a run by Alan Ameche, winning the game.

December 29

1845: The United States annexes the Republic of Texas, following the manifest destiny doctrine. Texas had been the independent Republic of Texas since the Texas Revolution of 1836. It was admitted as the 28th state in the Union.

1851: The first YMCA in America opens in Boston, Massachusetts. The Young Men's Christian Association is an international organization founded in London by Sir George Williams. Its goal was to promote healthy activities for young men in cities. It engaged them in sports, and as a result the Y's mission came to be called "muscular Christianity."

1890: The US 7th Cavalry massacres 300 Lakota Sioux at Wounded Knee Creek, South Dakota. The Wounded Knee Massacre, as it became known, was part of the US Army's effort to suppress the Ghost Dance movement, which was uniting Plains Indians against American expansion into their ancestral lands. Ghost Dancers believed that a Messiah was coming who would make the white man disappear from their lands, bring back the buffalo in great abundance, and return the ghosts of their ancestors to Earth. They wore "ghost shirts" while dancing that they believed would make them bulletproof. A patrol of Cavalry stopped a group of Lakota Sioux who were traveling to a gathering of Ghost Dancers and attempted to disarm them. When one of the Lakota, deaf man named Black Coyote, did not give up his rifle, the soldiers opened fire on the entire group. Women and children were killed along with Lakota warriors, most of whom never had a chance to fire back. In 1973, activists with the American Indian Movement seized the site of the massacre in a standoff that lasted for four months.

1916: Irish author James Joyce publishes *The Portrait of the Artist as a Young Man*. The book uses stream-of-consciousness and other literary devices that would establish Joyce as one of the greatest novelists in the history of English literature.

1940: The German Luftwaffe drops nearly 100,000 incendiary bombs on London. The bombing triggered the Second Great Fire of London and was one of the worst air raids during the Blitz. Hundreds of buildings were destroyed in the fire and 120 civilians, including twelve firefighters, were killed.

1941: US Navy Rear Admiral Ben Morell creates the Seabees. The special construction units were developed in order to build airfields and roads under battlefield conditions during World War II. Their name came from the first letters of their official designation, the Construction Battalions.

1998: Former leaders of the Khmer Rouge, the communist party that ruled Cambodia from 1975 to 1979, formerly apologize for the genocide they directed against their own people. Over one million Cambodians were killed in the genocide. The regime had caused the deaths in part because of agricultural social engineering policies that caused widespread famine. Purges of civilians deemed to be subversive also resulted in thousands being murdered. In 1979 Vietnam invaded Cambodia to remove the regime from power. The resulting war lasted another ten years.

2003: The Akkala Samai language becomes extinct when its last known speaker dies. The language was spoken by the Sami people in the Kola Peninsula of Russia.

December 30

1813: British troops burn Buffalo, New York, during the War of 1812. Earlier in the month, an American army detachment under General George McClure ordered the town of Newark, Canada to be burned as they were evacuating his garrison. Some weeks later, a force of British regulars backed by Canadian militia and Native Americans attacked Buffalo after crossing the Niagara River at midnight. In retaliation for the burning of Newark, the British set fire to Buffalo after capturing the city. All but four buildings were razed.

1853: James Gadsden, the American ambassador to Mexico, and Antonio López de Santa Anna, the President of Mexico, sign the treaty known as the Gadsden Purchase. Mexico ceded 29,670 square miles in parts of present-day Arizona and New Mexico to the United States for $10 million. The Purchase was the final large acquisition of territory in the contiguous United States.

1903: The Iroquois Theatre in Chicago burns down. At least 600 people died in the disaster, which was the deadliest single-building fire in the history of the United States. The theatre had only one entrance, and various levels shared a single stairway. When the fire broke out, the crowd was unable to exit the theatre. As a result of the tragedy, fire codes across the country were revised. Safety curtains and smoke doors were required, theater exits had to be clearly marked, and doors redesigned to be easily opened from the inside.

1922: The Union of Soviet Socialist Republics is formally established. The Soviet Union was dominated by the Russian Communist Party and played a major role in 20th century history. It dominated the world stage until its dissolution in 1991.

1942: Frank Sinatra opens at the Paramount Theatre in New York. The crowd, composed of hundreds of teenaged girls, nearly rioted, and about 400 policemen had to be brought in to prevent them from rushing the stage.

1965: Ferdinand Marcos is inaugurated President of the Philippines. During his 20-year reign, his government became increasingly autocratic, and his family engaged in unprecedented levels of corruption. After he was driven out of power, he was indicted by the United States for embezzling billions of dollars from the Philippine treasury.

2006: Saddam Hussein is executed for war crimes. Hussein had risen to power in the Iraqi Ba'ath Party, taking part in the 1968 coup that brought the party to power. In 1979, after serving as the vice president Iraq for ten years, Hussein seized power. He ruled with an autocratic grip until the US invasion of 2003 toppled him.

December 31

1759: Arthur Guinness signs a nine-thousand-year lease on a brewery in Dublin. The annual rent of the brewery was 45 pounds sterling. Guinness had previously brewed ale at a brewery in County Kildare, Ireland, before moving to Dublin to start his new enterprise. In the 1760s he began experimenting with brewing porter, a darker, heavier beer. After expanding the brewery in the 1790s, Guinness switched to brewing porter exclusively. It wasn't until the 1840s that the company began brewing the stout beer that it is famous for. Although "stout" originally referred to the beer's strong alcohol content, it came to describe the beer's body and dark color. The beer, which is known simply as Guinness, is one of the most successful brands of beer in the world. In 1955, the managing director of the Guinness Breweries, Sir Hugh Beaver, had an argument about what the fastest European game bird is. When he could not find an answer in reference books, he decided to supply a book full of records to settle pub arguments. *The Guinness Book of World Records* has been published since.

1862: Abraham Lincoln signs an act that divides West Virginia from Virginia. During the American Civil War, Virginia seceded from the Union. Delegates at the Virginia Secession Convention who represented the region voted overwhelmingly against secession. They then left the convention and began organizing to form their own state. That state formally joined the Union as West Virginia on June 20, 1863.

1878: In Mannheim, Germany, Karl Benz files a patent for his two-stroke engine. It was the first design of an engine that was reliable. The patent provided Benz with a revenue stream that he used to fund his growing automobile business.

1879: At Menlo Park, Thomas Edison demonstrates the incandescent light bulb to the public for the first time. The bulb was the first to use a filament that is heated until it glows. Its invention caused a revolution in lighting, as gas lamps were phased out and replaced by Edison's bulbs and a power grid to support their use was established across the country.

1942: After five months of brutal fighting, the Japanese troops at Guadalcanal are finally allowed to retreat. American troops finished clearing the island of Japanese soldiers in February. It was the first major defeat of the Japanese by the Allies during World War II.

1951: The Marshall Plan is concluded. Begun in 1948, the Marshall Plan provided more than $13.3 billion to Western Europe following the devastation of World War II. It was named for General of the Army George C. Marshall, and intended to be an economic bulwark against Soviet expansion during the early stages of the Cold War.

1983: The US Justice Department breaks up the Bell System of telephone companies. Often called "Ma Bell," the multistate conglomerate was determined to be a monopoly following an antitrust lawsuit. It was divided into multiple smaller regional telephone companies.

1983: Benjamin Ward becomes the first African-American Police Commissioner of New York City. Ward had started in the force as a patrolman in 1951, and was the first black officer in Brooklyn's 80th Precinct. He retired from the Police Department in 1989.

1999: Boris Yeltsin resigns as the first President of the Russian Federation. He oversaw the country's transition to a market economy following the breakup of the Soviet Union. Yeltsin also proposed a new Russian constitution in 1993. He was succeeded by Vladimir Putin.